12th Russia/CIS/Baltic/Japan Symposium on Ferroelectricity and 9th International Conference on Functional Materials and Nanotechnologies

AA002836

(RCBJSF-2014-FM&NT)

IOP Conference Series: Materials Science and Engineering
Volume 77

Riga, Latvia
29 September - 2 October 2014

ISBN: 978-1-5108-0083-0
ISSN: 1757-8981

Printed from e-media with permission by:

Curran Associates, Inc.
57 Morehouse Lane
Red Hook, NY 12571

proceedings.com

Printed by Curran Associates, Inc. (2015)

For permission requests, please contact the Institute of Physics
at the address below.

Institute of Physics
Dirac House, Temple Back
Bristol BS1 6BE UK

Phone: 44 1 17 929 7481
Fax: 44 1 17 920 0979

techtracking@iop.org

Additional copies of this publication are available from:

Curran Associates, Inc.
57 Morehouse Lane
Red Hook, NY 12571 USA
Phone: 845-758-0400
Fax: 845-758-2634
Email: curran@proceedings.com
Web: www.proceedings.com

Table of contents

Volume 77

12th Russia/CIS/Baltic/Japan Symposium on Ferroelectricity and 9th International Conference on Functional Materials and Nanotechnologies (RCBJSF–2014–FM&NT)
29 September to 2 October 2014, Riga, Latvia

Accepted papers received: 23 February 2015
Published online: 9 March 2015

Preface

Papers

IOP Conf. Series: Materials Science and Engineering **77** (2015) 011001 doi:10.1088/1757-899X/77/1/011001

PREFACE

The joint International Symposium **RCBJSF–2014–FM&NT** successfully has united two international events – 12th Russia/CIS/Baltic/Japan Symposium on Ferroelectricity (**RCBJSF-12**) and 9th International Conference Functional Materials and Nanotechnologies (**FM&NT-2014**).

The **RCBJSF** symposium is a continuation of series of meetings on ferroelectricity, the first of which took place in Novosibirsk (USSR) in 1976. **FM&NT** conferences started in 2006 and have been organized by Institute of Solid State Physics, University of Latvia in Riga. In 2012 the International program committee decided to transform this conference into a traveling Baltic State conference and the FM&NT–2013 was organized by the Institute of Physics, University of Tartu, Estonia.

In 2014 the joint international symposium **RCBJSF–2014–FM&NT** was organized by the Institute of Solid State Physics, University of Latvia and was part of **Riga - 2014, the European Capital of Culture** event.

The purpose of the joint Symposium was to bring together scientists, students and high-level experts in solid state physics, materials science, engineering and related disciplines. The number of the registered participants from 26 countries was over 350. During the Symposium 128 high quality scientific talks (5 plenary, 42 invited, 81 oral) and over 215 posters were presented.

All presentations were divided into 4 parallel sessions according to 4 main topics of the Symposium:

Ferroelectricity, including ferroelectrics and multiferroics, pyroelectrics, piezoelectrics and actuators, integrated ferroelectrics, relaxors, phase transitions and critical phenomena

Multifunctional Materials, including theory, multiscale and multiphenomenal material modeling and simulation, advanced inorganic, organic and hybrid materials

Nanotechnologies, including progressive methods, technologies and design for production, investigation of nano- particles, composites, structures, thin films and coatings

Energy, including perspective materials and technologies for renewable and hydrogen energy, fuel cells, photovoltaics, LEDs, OLEDs.

Based on these reports, 48 papers are included in this volume of *IOP Conference Series: Materials Science and Engineering*.

Additional information about **RCBJSF–2014–FM&NT** is available at the homepage http://www.fmnt.lu.lv.

The Organizing Committee would like to thank all the speakers, contributors, session chairs, referees and other involved staff for their efforts in making the **RCBJSF–2014–FM&NT** successful.

Sincerely, organizers of the event
Andris Sternberg
Liga Grinberga
Anatolijs Sarakovskis
Martins Rutkis

IOP Conf. Series: Materials Science and Engineering **77** (2015) 011001 doi:10.1088/1757-899X/77/1/011001

The Symposium photo

RCBJSF-12 Organizing Committee

Honorary chairmen
J. Kobayashi, Y. Ishibashi (Japan)
B.A. Strukov (Russia)
J. Grigas (Lithuania)

Chairmen
Y. Akishige (Japan)
A. Sternberg (Latvia)

Vice-chairmen
S. Kojima (Japan)
A.S. Sigov (Russia)
J. Banys (Lithuania)

Members
N. Ikeda, M. Itoh, M. Iwata, S. Mori, M. Shimizu, H. Suzuki, S. Wada (Japan)
E. Birks (Latvia)
V. Samulionis (Lithuania)

V.M. Fridkin, S.G. Lushnikov, V.Ya. Shur, A.S. Sidorkin, O.G. Vendik (Russia)
Yu.M. Vysochanskii (Ukraine)

FM&NT-2014 Organizing Committee
Chairman
A. Sternberg (Latvia)
Vice chairman
M. Rutkis (Latvia)
Members
J. Banys (Lithuania)
G. Borstel (Germany)
N. E. Christensen (Denmark)
R. A. Evarestov (Russia)
C. G. Granqvist (Sweden)
D. Hovik (Norway)
M. Kirm, E. Nommiste (Estonia)
J. Kulda (France)
W. Lojkowski (Poland)
I. Thorbjornsson (Iceland)

RCBJSF-12 Program Committee
Chairmen
N. Ikeda (Japan)
E. Kotomin (Latvia)
Members
J. Kano, H. Mashiyama, S. Mori, K. Ohwada, M. Takesada, H. Taniguchi, S. Tsukada, R. Wang, H. Yokota (Japan)
E. Klotins (Latvia)
R. Grigalaitis, E. Tornau (Lithuania)
I.N. Flerov, S.A. Gridnev, V.P. Sakhnenko, S. Vakhrushev, T.R. Volk, V.I. Zinenko (Russia)
M.D. Glinchuk, R. Vlokh (Ukraine)

FM&NT-2014 Program Committee
Chairman
A. Sarakovskis (Latvia)
Members
L. Berzina-Cimdina, J. Grabis, L. Grinberga, E. Kotomin, L. Skuja, M. Springis, J. Zicans (Latvia)

Local Organizing Committee
Chairlady
L. Grinberga
Members
A. Sarakovskis, J. Grube, R. Siatkovskis, A. Sivars, P. Lesnicenoks, A. Knoks, A. Antuzevičs, K. Klismeta, M. Kundzins, L. Klebaha

IOP Conf. Series: Materials Science and Engineering **77** (2015) 011002 doi:10.1088/1757-899X/77/1/011002

Peer review statement

All papers published in this volume of *IOP Conference Series: Materials Science and Engineering* have been peer reviewed through processes administered by the proceedings Editors. Reviews were conducted by expert referees to the professional and scientific standards expected of a proceedings journal published by IOP Publishing.

IOP Conf. Series: Materials Science and Engineering **77** (2015) 012001 doi:10.1088/1757-899X/77/1/012001

Ab initio simulations on Frenkel pairs of radiation defects in corundum

A Platonenko, S Piskunov, Yu F Zhukovskii and E A Kotomin

Institute of Solid State Physics, University of Latvia, 8 Kengaraga Str, Riga, Latvia
E-mail: alexander.platonenko@gmail.com

Abstract. Large scale first principles periodic calculations based on the density functional theory within the localized atomic orbital approach (DFT-LCAO) using the hybrid exchange-correlation potential B3PW have been performed in order to study the structural and electronic properties of radiation-induced Frenkel pairs O_i+V_O in corundum crystal. As an initial approach, we have used conventional 2×2×1 supercell for defective α-Al_2O_3 lattice containing 120 atoms. After relaxation of the ideal supercell structure, the optimized $d_{O_i\text{-}V_O}$ distance has been found to be ~4.5 Å while the formation energy of Frenkel pair has achieved 11.7 eV. The interstitial O_i atom, both single and a component of O_i+V_O pair, spontaneously forms a dumbbell with the adjacent atom in the regular oxygen sublattice ($d_{O_i\text{-}O}$ = 1.404 Å) with the induced charge -1.1 *e*. On the whole, possibilities of supercell model for proper description of Frenkel pairs with changing inter-defect distance and space orientation inside corundum crystal are rather limited. This is why the alternative cluster model must be developed for this aim.

1. Introduction

Corundum (α-Al_2O_3) is important radiation-resistant material with potential applications for components of diagnostics, breeder blanket and in future fusion reactors as coating to avoid the light gases permeation, as well as corrosion produced by lithium-based alloys [1]. Induced changes in structural and optical properties of radiation-exposed α-Al_2O_3 crystalline materials are mainly associated with oxygen vacancies V_O and complementary Frenkel pairs of defects (O_i+V_O) [2]. Despite technological importance of corundum crystalline structures, their fusion applications and prospects as a laser and dosimeter material, the point, *e.g.*, electronic, defects have not yet comprehensively studied theoretically [2]. The main reasons for this are the following: the complicated atomic structure of α-Al_2O_3 (Fig. 1) as well as the semi-covalent and semi-ionic chemical bonding as was experimentally observed using X-ray scattering [3]. Moreover, after formation of defect pair in perfect corundum crystal, a dramatic lowering of its initial symmetry $R\bar{3}c$ (local space group D_{3d}^{6}) occurs, thus, calculations of Frenkel pairs require significant computational time and resources.

Limited number of theoretical studies on point defects and Frenkel pairs in corundum is available, where empirical, semi-empirical and first-principle methods were used [2, 4-7]. While there is information about some mechanisms of defect formation, recombination and clustering, the reported formation energies of oxygen vacancy in corundum vary in a wide range from 4.9 to 13.3 eV [8-10]. Interpretation of experimental data also is still discussable theme [11].

In this study, we have consequently performed first principles calculations on (*i*) perfect corundum structure, (*ii*) that containing single point defects, such as neutral oxygen vacancy V_O and interstitial oxygen atom O_i, (*iii*) that containing Frenkel pair O_i+V_O , as well as (*iv*) that containing dumbbell O_i+O_{reg} pair in the case of α-Al_2O_3 structures containing either single O_i impurity (as described earlier for defective MgO bulk [12]) or Frenkel pair. Using optimization procedure, we have determined equilibrium locations and configurations of V_O, O_i, and O_i+V_O inside corundum lattice. We have tested basis sets of Al and O atoms for adequate description of α-Al_2O_3 properties: lattice and elastic constants as well as cohesion energies and band gaps, in order to choose the most appropriate model and computational details, *e.g.*, exchange-correlation functional [13]. We also discuss limits for application of periodic supercell model for adequate description of radiation defects in corundum. The paper is organized as follows: details of calculations, *e.g.*, atomistic models of defective α-Al_2O_3 are discussed in Section 2, main results obtained for optimized Frenkel pairs in corundum are analyzed in Section 3, while main conclusions drawn in this study are summarized in Section 4.

2. Computational details

Large-scale *ab initio* calculations have been performed combining the linear combination of atomic orbitals (LCAO) and hybrid exchange-correlation Becke-Perdew-Wang (B3PW) functional as implemented in *CRYSTAL'14* computer code [13]. Such hybrid functional allows us to perform very accurate calculations of the band gap and other properties as compared to those obtained using other hybrid functionals (Table 1). The basis sets of Al and O have been selected and optimized in order to achieve reasonable compromise between the calculation accuracy and computational time. We have selected the corresponding basis sets with smaller number of Gaussian-type functions: (*i*) for aluminum, we have used the effective core pseudopotential (ECP) with $3s^23p^1$ external shell as developed by Causá [14], (*ii*) while for oxygen, we have applied the full basis set in the form of 6*s*-2111*sp*-1*d* developed by Baima *et al* [15]. For calculation on the formation energy of free oxygen molecule, this basis set has been extended by adding *sp*-function with diffuse exponents.

Table 1. Lattice parameters *a* and *c*, band gap and formation enthalpy as well elastic constant for perfect corundum calculated using different hybrid and pure DFT exchange-correlation potentials.

Method (functional)	*a*, Å	Δ*a*,* %	*c*, Å	Δ*c*,* %	Band gap, ε_g, eV	$\Delta\varepsilon_g$,* %	Formation entalphy, E_{form}** eV	ΔE_{form}* %	Elastic constants, GPa [19] C_{11}	Δ,* %	C_{44}	Δ,* %
Experimental	4.76 [16]		12.99 [16]		8.8 [17]		-17.36 [18]		496.5		146.15	
B3PW	4.775	0.32	12.979	-0.02	8.39	-4.7	-19.45	12.0	494.2	-0.5	146.4	0.2
B3LYP	4.792	0.67	13.036	0.41	8.55	-2.8	-20.19	16.3	488.4	-1.6	139.5	-4.5
HSE-sol	4.744	-0.35	12.884	-0.75	8.25	-6.3	-19.70	13.5	516.8	4.1	150.2	2.8
PWGGA-HF25	4.761	0.02	12.944	-0.30	8.86	0.7	-19.41	11.8	456.6	-8.0	148.2	-6.9
PBE (DFT)	4.788	0.59	12.987	0.04	6.01	-31	-18.69	7.7	469.5	-5.4	142.3	-2.6
PWGGA (DFT)	4.816	1.18	13.072	0.70	6.31	-28.3	-18.69	7.7	497.6	0.2	148.0	1.2

*as compared to experimentally found values written in previous columns;

$$** \; E_{form} = E_{tot}^{Al_2O_3(perfect)} - 4E_{tot}^{Al} - 6E_{tot}^{O} \quad (1)$$

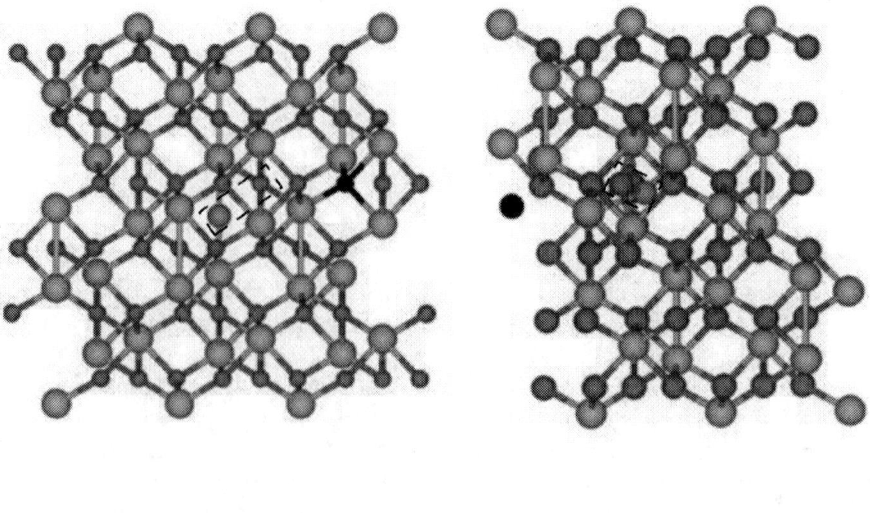

IOP Conf. Series: Materials Science and Engineering **77** (2015) 012001 doi:10.1088/1757-899X/77/1/012001

Figure 1. Images of Frenkel pair in initial and optimized corundum structures (left and right panels, respectively). V_O vacancy is shown as a black ball while interstitial O_i atom is shown as a blue (dark grey) one. Configurations of *H* center (or dumbbell) are shown as a dotted rectangle.

Formation energies for various defects in α-Al_2O_3 have been calculated using following equations:

$$E_{form}^{O_i} = E_{tot}^{O_i} - E_{tot}^{Al_2O_3(perfect)} - E_{tot}^{\frac{1}{2}O_2}, \quad (2)$$

$$E_{form}^{V_O} = E_{tot}^{V_O} - E_{tot}^{Al_2O_3(perfect)} + E_{tot}^{\frac{1}{2}O_2}, \quad (3)$$

$$E_{form}^{pair} = E_{tot}^{pair} - E_{tot}^{Al_2O_3(perfect)}. \quad (4)$$

Within the SCF procedure, the accuracies (tolerances) 10^{-7}, 10^{-7}, 10^{-7}, 10^{-7}, 10^{-14} have been chosen for calculations of Coulomb and exchange integrals [13]. Effective charges on atoms have been estimated using Mulliken population analysis [20]. The integration over Brillouin zone in the reciprocal space has been performed with a 4×4×4 Pack-Monkhorst grid. When modeling the Frenkel pairs in corundum bulk, the 2×2×1 supercell model containing 120 atoms has been chosen (Fig. 1). All the calculations were performed with the total optimization of geometry. Obviously, the band gaps of perfect corundum, as presented in the upper part of Table 1 obtained using various hybrid DFT+HF methods, are much closer to the experimental value than those obtained within pure DFT method.

3. Results and analysis

Firstly, we have performed first principles calculations on single oxygen neutral point defects. Interstitial O_i atom initially introduced to the octahedral center between the two nearest oxygen triangles across α-Al_2O_3(0001) plane (left panel of Fig. 1) has been displaced towards one of the apexes shown by dashed rectangle after energy optimization, with energy gain of 3.84 eV (qualitatively analogously to displacement of O_i atom from the cube-centered position in MgO unit cell estimated earlier [12]). Initial configuration of vacancy V_O has been chosen by such a way, to provide its location between the central oxygen planes of corundum supercell. Vacancy was created by fully removing O atom from crystal structure. As a result of geometry optimization, V_O point defect has been displaced from its initial position too. Influence of both defects on the nearest neighbor atoms is shown in Table 2 (note that the two of four Al neighbors are located closer to the oxygen vacancy than the other two, the same is true for oxygen neighbors).

Table 2. Structural relaxation of the lattice area around arbitrary neutral defect in the Frenkel pair (Fig. 1). For arbitrary defect, its distances (in Å) to 1*NN*-4*NN* atoms in the perfect crystal (located at the nearest and next-nearest neighboring atomic spheres) are shown first, followed by the changed distances (Δd) and changes of induced charges (Δq_{Al} or Δq_O) after the geometry optimization.

Type of site or defect	1*NN*	Δd%	Δq, *e*	2*NN*	Δd%	Δq, *e*	3*NN*	Δd%	Δq, *e*	4*NN*	Δ%	Δq, *e*
Interstitial site*	1.92	(2 Al)	–	1.98	(6 O)	–	2.76	(6 Al)	–	3.29	(6 O)	–
Interstitial O_i atom**	1.87	-2,6	0.0	2.15	8.59	+0.07	2.82	2.17	-0.02	3.3	1.02	0.00
Regular O site*	1.86, 1.97	(2+2 Al)	–	2.52 2.62	(2+2 O)	–	2.72, 2.87	(4+4 O)	–	3.22, 3.43	(3+3 Al)	–
V_O vacancy	1.77, 1.99	-4.84, 1.02	-0.37, -0.23	2.50, 2.60	-0.79, -0.76	+0.02	2.72, 2.86	0.0, -0.67	+0.02	3.23, 3.42	1.0, 1.0	0.00

*brackets in rows contain numbers and chemical symbols of the nearest atoms as follows from Fig. 1a;
**results for O_i impurity atom are obtained for its location at the center of oxygen octahedron

Formation energies of single point defects are 4.21 eV for interstitial oxygen atom defined by Eq. (2), possessing induced charge -0.5 *e*, as well as 7.60 eV for oxygen vacancy defined by Eq.(3)

respectively. Since there is noticeable change in charges of Al atoms (-1.2 *e* in total for 4 Al atoms) of 1*NN* sphere of the vacancy we can assume that it is positively charged. Sum of both defect formation energies is 11.81 eV, what is reasonable for formation of well separated (>15 Å) Frenkel pair.

In perfect corundum crystal, the effective atomic charges have been found to be -1.0 *e* for O and 1.5 *e* for Al atoms, respectively, while considerable contribution of the covalency to the chemical bonding have been found to be too: 0.386 *e* and 0.262 *e* (for smallest values of d_{Al-O} equal to 1.858 Å and 1.968 Å, respectively). The Frenkel pair has been modeled inside the conventional 2×2×1 supercell with 4.42 Å distance between V_O vacancy and O_i atom. Interstitial O_i atom also forms a dumbbell with adjacent regular O atom after energy optimization ($d_{Oi\text{-}O}$ = 1.404 Å) with total charge −1.1 *e*. Mulliken population analysis shows that the bond population between Al atom and dumbbell oxygen atoms is almost twice smaller than that in regular Al–O bond (0.198 *e vs.* 0.386 *e*, respectively). After geometry relaxation O_i–V_O distance has been found to be ~4.4 Å. Al atoms inside the first coordination sphere around V_O vacancy possess smaller charges (1.12–1.27 *e*) as compared to that in perfect corundum (1.5 *e*), so it seems that vacancy tends to be slightly positively charged. In the next coordination spheres, induced charges towards V_O vacancy has been found to be almost neglecting. The formation energy of Frenkel pair in α-Al_2O_3 has been obtained to be 11.7 eV. It is close to sum of formation energies for single O_i and V_O defects obtained separately as mentioned above 11.81 eV (4.21 eV + 7.60 eV). Values of E_{form}^{pair} found earlier were reported in a range 4.87–8.27 eV [10], whereas in other theoretical studies, this value achieved 13.3 eV [8,9]. Still, our result may be overestimated because of defect-defect interactions in supercell model.

For further calculations, we are going to use conventional supercell of 3×3×1 extension, where initial length of the Frenkel pair can be chosen as a more realistic value (7.0–7.2 Å), as also to compare results and evaluate defect-defect interaction in periodic model. As to further increase of the conventional supercell for simulation of Frenkel pairs, $d_{O_i-V_O}$, distance in which can achieve 15–17 Å (as observed in experiments), it cannot be realized in principle. This is why, in order to provide further possibility to simulate a series of Frenkel pairs in corundum with consequently increasing $d_{O_i-V_O}$ length, we plan to perform large-scale cluster model calculations using the *NWChem* code [21]. Moreover, the first-principles formalism realized in this code foresees the possibility to consider not only neutral defects as in periodic calculations but also charged and excited defects.

4. Conclusions

First-principle periodic calculations have been performed in order to study the structure and formation energies of point and pair-wise defects in corundum. It has been found that the interstitial oxygen atom in both absence and presence of V_O in α-Al_2O_3 can form dumbbell *H*-centers like in MgO bulk [12]. Behavior of oxygen vacancy is similar in both cases: it tends to be positively charged.

In order to simulate well-separated Frenkel pairs of varying length as well as charged and excited defects, a cluster model will be used in the next studies, which also excludes artifacts caused by the periodic network of defect-defect interactions.

Acknowledgements

This work has been carried out within the framework of the EUROfusion Consortium and has received funding from the European Union's Horizon 2020 research and innovation programme under grant agreement number 633053. The authors thank also Latvian LZP grant No.237/2012 for financial support. The views and opinions expressed herein do not necessarily reflect those of the European Commission. Authors thank R Vila and A I Popov for fruitful discussions.

References

[1] F.Mota, C.J.Ortiz, R. Vila, N. Casal, A. Garcia, A. Ibarra. 2013 *J Nucl Mater* **442** 5699
[2] Jacobs P W M and Kotomin E A 1994 *J Amer Ceram Soc* **77** 2505

IOP Conf. Series: Materials Science and Engineering **77** (2015) 012001 doi:10.1088/1757-899X/77/1/012001

[3] Lewis J, Schwarzenbach D, and Flack H D 1982 *Acta Crystallogr A* **38** 733

[4] Zhukovskii Yu F, Sokol A A, Kotomin E A, Catlow C R A, and Nieminen R M 1997 *J Phys Cond Matt* **9** 3559

[5] Kotomin E A, Popov A I, and Stashans A 1998 *Phys Status Solidi B* **207** 69

[6] Catlow C R A, James R, Mackrodt W C, and Stewart R F 1982 *Phys Rev B* **25**, 1006

[7] Grimes R W, 1994 *J Amer Ceram Soc* **77** 378

[8] Hine N D M, Frensch K, Foulkes W M C, and Finnis M W 2009 *Phys Rev B* **79** 24122

[9] Janetzko F, Evarestov R A, Bredow T, and Jug K 2004 *Phys Status Solidi B* **241** 1032

[10] Matsunaga K, Tanaka T, Yamamoto T, and Ikuhara Y 2003 *Phys Rev B* **68** 085110

[11] Evans B D 1995 *J Nucl Mater* **219** 202

[12] Brudevoll T, Kotomin E A, and Christensen N E 1996 *Phys Rev B* **53** 7731

[13] Dovesi R, Saunders V R, Roetti C, Orlando R, Zicovich-Wilson C M, Pascale F, Civalleri B, Doll K, Harrison N M, Bush I J, D'Arco Ph, Llunell M, Causá M, and Noël Y 2014 *CRYSTAL14 User's Manual* (University of Torino)

[14] Causá M, Dovesi R, and Roetti C 1991 *Phys Rev B* **43** 11937

[15] Baima J, Erba A, Rerat M, Orlando R, and Dovesi R 2013 *J Phys Chem C* **117** 12864

[16] Lutterotti L and Scardi P 1990 *J. Appl. Crystallogr* **23** 246

[17] French R H 1990 *J. Am. Ceram. Soc.* **13** 471

[18] Zumdahl S S 2009 *Chemical Principles* (6th Edition, Houghton Mifflin Co, Boston, New York)

[19] Gladden J R, Maynard J D, So J H, Saxe P, and Le Page Y 2004 *Appl Phys Lett* **85** 392

[20] Mulliken R S 1955 *J Chem Phys* **23** 1833

[21] Valiev M, Bylaska E J, Govind N, Kowalsi K, Straatsma T P, Van Dam H J J, Wang D, Nieplocha J, Apra E, Windus T L, and De Jong W A 2010 *Comput Phys Commun* **181** 1477

IOP Conf. Series: Materials Science and Engineering **77** (2015) 012002 doi:10.1088/1757-899X/77/1/012002

The Peculiarities of Halogens Adsorption on A^3B^5(001) Surface

A Bakulin[1,2], S Kulkova[1,2], O.E. Tereshchenko[3], A Shaposhnikov[2] and I Smolin[1]

[1] Institute of Strength Physics and Materials Science of Siberian Branch Russian Academy of Science, pr. Akademichesky 2/4, Tomsk 634021, Russia
[2] National Research Tomsk State University, pr. Lenina 36, Tomsk 634050, Russia
[3] Institute of Semiconductor Physics, pr. Lavrentieva 13, Novosibirsk 630090, Russia

E-mail: kulkova@ispms.tsc.ru

Abstract. Theoretical study of the F, Cl, Br, I adsorption on GaAs(001) surface is presented. The most stable configurations of halogens on Ga-rich ζ-(4×2) reconstruction are determined with increasing of adatoms concentration. The bonds weakening is found more significant for the F and Cl atoms which can induce stationary etching whereas I-induced changes in Ga-As binding energy is not sizable and its adsorption lead to the surface passivation.

1. Introduction

The investigation of halogen adsorption on semiconductor surfaces is very important from both scientific and technological point of view [1]. Since halogens and halogen compounds are widely used in the etching technologies it is necessary to understand the mechanism of their interaction with semiconductor surfaces on the microscopic level [2]. On other hand, the halogens are good model adsorbates to study their chemisorption on the covalently bonded materials caused by relative simple nature of their bonds. Semiconductors A^3B^5 are widely used in microelectronics, optoelectronics, spintronics, etc. The (001) surface is of practical interest in this regards. Halogen (F, Cl, I) adsorption on the (001) surface with ζ-(4×2) reconstruction for GaAs and InAs were studied in our papers [3, 4] at low halogen coverage of 0.125 monolayer (ML). The influence of the halogen concentration on the atomic and electronic structure of the (001) surface has not been studied for A^3B^5 semiconductors theoretically. The adsorption of 0.75 and 1.0 ML of F, Cl, Br and I was considered on Ge(001) surface in [5, 6]. It was found that the stable atomic structures are almost the same for considered halogens. The only difference was found in the position of adatoms above the substrate [5]. However, the F-covered surface exhibits slightly different features because fluorine local density of states near the Fermi level was significantly smaller than those of Ge, Cl and I atoms.

In the present paper the influence of halogen adsorption on the atomic structure of GaAs(001) surface and surface atoms bonding as a function of halogen coverage have been studied.

2. Computational details

Atomic and electronic structures of A^3B^5(001) surface were calculated using projector augmented wave (PAW-DFT) method [7, 8]. Generalized gradient approximation (GGA-PBE) for the exchange-correlation functional was employed. The surface was simulated by periodic slabs with a supercell containing eight atomic layers of semiconductor and vacuum gap of ~10 Å. The bottom anion layer was passivated by pseudo-hydrogen atoms with fractional charge Z equal to 0.75*e* that allows us to

IOP Conf. Series: Materials Science and Engineering **77** (2015) 012002 doi:10.1088/1757-899X/77/1/012002

make this layer bulk-like. Two bottom semiconductor layers and pseudo-hydrogen layer were fixed during the structure optimization, whereas the atomic positions in other layers were relaxed using damping Newton method up to the forces at atoms ~0.01 eV/Å. The binding energy (E_b) of halogen adatom with the semiconductor surface was calculated using equation:

$$E_b = -(E_{halogen/A3B5} - E_{A3B5} - E_{halohen}), \quad (1)$$

where $E_{halogen/A3B5}$ and E_{A3B5} are the total energies of both halogen-adsorbed system and adsorbate-free semiconductor surface, respectively, and $E_{halogen}$ is the total energy of halogen atom.

3. Results and discussion

The ζ-(4×2) reconstruction of GaAs(001) surface is known the most stable in the Ga-rich limit [9]. Our calculations also confirm its stability among considered structures with (4×2) geometry [10, 11]. The calculations of F, Cl and I binding energies at low coverage [3, 4] reveal that halogen adatom prefers to be bonded in the M_1-site above Ga dimerized atoms (figure 1). Table 1 summarizes binding energies for several positions above Ga and As surface atoms in which halogens behave differently. The mentioned above trend is valid for Br as well. Note that it was not studied in our papers [3, 4]. Binding energy of Br in the M_1-site is lower than that of Cl but it is higher than that of I, in agreement with assumption of electronegativity concept. It should be noted that halogen adsorption influences on the bond lengths between Ga surface atoms and neighboring As atoms. The change of bond lengths depends considerably on the adsorbate sites [4]. For instance, the bond length in Ga dimer increases by 0.08-0.10 Å when I, Br, Cl and F are adsorbed in the M_1-site. One can expect that the increase in the dimer bond lengths contributes to the halogen-induced decrease in the binding energies of surface atoms and as a consequence, leads to monohalide desorption. To confirm this assumption we estimated the influence of halogen on the binding energy of Ga or As surface atoms with the substrate.

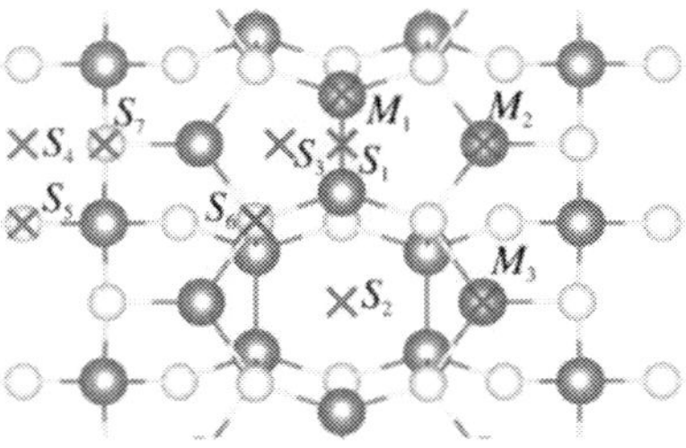

Figure 1. Adsorption sites (croses) of halogens on the ζ-GaAs(001)-(4×2) surface. Ga and As atoms are shown by dark and white balls.

Table 1. Binding energies (E_b) of halogens and corresponding gallium or arsenic monohalides (in eV) with ζ-GaAs(001)-(4×2).

	Halogen adsorption sites	M_1	M_2	M_3	S_6	S_7
F	E_b(halogen-surface)	**4.74**[a]	**4.71**	**4.52**	3.51	4.07
	E_b(monohalide-surface)	2.22	2.70	3.06	4.14	4.98
Cl	E_b(halogen-surface)	**2.86**	**2.82**	2.56	1.87	2.38
	E_b(monohalide-surface)	2.25	2.75	3.15	4.16	4.85
Br	E_b(halogen-surface)	**2.51**	2.49	2.29	1.58	2.09
	E_b(monohalide-surface)	2.25	2.75	3.14	4.25	4.82
I	E_b(halogen-surface)	1.86	1.78	1.59	1.07	1.54
	E_b(monohalide-surface)	2.30	2.82	3.23	4.44	4.87

[a] Cases when E_b of halogen with surface is greater than that of Ga monohalide with surface are shown by bold numbers.

In order to demonstrate the "back bonds" weakening, we removed the Ga or As atoms from the clean ζ-GaAs(001)-(4×2) surface and the Ga- or As-monohalides from the halogen adsorbed system as well. Table 1 shows that in the M_1-site the bonding of GaF, GaCl and GaBr molecules with the substrate is considerably weaker than the bonding energy of halogen atom (F, Cl, Br) with surface. On the contrary, the binding energy of GaI with substrate is of 0.44 eV higher than that of I with Ga dimerized atom. In the latter case, one can expect that I adsorption should lead to the surface passivation with subsequent desorption of GaI induced by temperature activation. This conclusion is in agreement with the experimental results [1, 12]. The same behavior occurs in the M_2-site where only minor difference in the binding energies of halogen atoms compared to M_1-site was calculated. At the same time when halogen atoms are adsorbed in S_6 or S_7 sites their binding energies decrease whereas the binding energies of As monohalides increase considerably. It means that the removing of the corresponding As monohalides from the semiconductor substrate is not energetically favorable with respect to halogen desorption. The calculated decrease in the binding energy (ΔE_b) of Ga atom with the nearest surface atoms averaged over the three bonds is 0.86-0.88 eV for M_1-site. Almost the same change of ΔE_b is found if halogens are adsorbed above Ga nondimerized atoms. It should be noted that initially the binding of Ga dimerized atom with substrate is weaker than that of Ga nondimerized atoms by 0.19 eV and 0.31 eV in M_2 and M_3 sites, respectively. For all halogens adsorbed in the S_7-site ΔE_b is in the range of 0.59-0.62 eV. The binding of As atom with substrate in the S_7-site is stronger than that of Ga one in the M_1-site by 0.62 eV therefore the decrease of ΔE_b is smaller in case of halogen adsorption above As-sites. Thus, the formation of bond between halogen and Ga surface atoms on GaAs(001) surface results in decrease of Ga binding energy with substrate. Almost the same values of ΔE_b (0.78–0.81 eV) were calculated for I, Cl and F adsorption on ζ-InAs(001)-(4×2) in the M_1-site [4]. In general, our results reveal that in case of the Ga-terminated GaAs(001) surface Ga atoms should be removed from the surface layer by halogens at the first step of etching. This result was confirmed by the experimental observation for the I/GaAs(001) interface: desorption of Ga-halides observed at I concentration below 0.5 ML, while above this value, only iodine molecules were desorbed [12]. It is seen from Table 1 that in case of F the transition from the chemisorption state to etching is possible whereas an additional energy is needed in case of less electronegative halogens.

Figure 2 presents the atomic structures of several Cl covered systems. Two Cl adatoms can be bonded with both Ga dimer atoms [13]. In this case the position of Ga surface dimer shifts toward vacuum by 2.00 Å (2.03 Å [13]). The average binding energy of Cl adatoms (3.06 eV/atom) is higher than that in case of single adsorbed chlorine in the M_1-site. It is necessary to point out that halogen-induced decrease of Ga dimer binding energy with the nearest surface atoms averaged over the four bonds, is 0.89 eV. The adsorption of two Cl atoms above Ga dimerized atoms reveals the decrease of the Ga–As bond energy by 0.07 eV, however this decrease is more pronounced (0.16 eV) in case of F adsorption on GaAs(001) and it is 0.19 eV on InAs(001). The adsorption of two Cl atoms in S_1 and S_2 sites (figure 1) is found to be more preferable than in M_1-M_1 sites (figure 2a). In this case the breaking of Ga dimer is occurred with shift of Ga atoms toward vacuum by 1.76 Å. The increase of the binding energy of Cl adatoms in S_1-S_2 with respect to adsorption in M_1-M_1 is 0.39 eV.

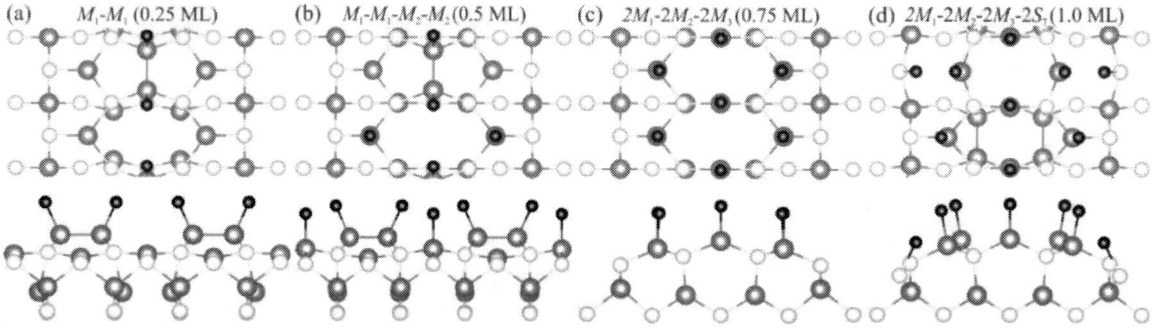

Figure 2. Atomic structures of Cl covered ζ-GaAs(001)-(4×2) surface up to 1 ML (top and side view).

Further we considered the configuration with two additional Cl adatoms adsorbed in M_2-sites (figure 2b). It is seen that chlorine bonding in both M_2-sites alongside with M_1-sites does not lead to breaking of Ga surface dimer bond as in case of Cl adsorption in M_1-M_1 sites. All positions above Ga surface atoms are occupied at the Cl coverage of 0.75 ML and as a consequence the transition to the (4×1) structure is observed (figure 2c). Finally, several structures with eight Cl adatoms which form the bond with dimerized, nondimerized Ga atoms as well as As atoms were considered also. One of these structures is shown in figure 2d. In case of simultaneous adsorption of two Cl atoms on one Ga atom in the M_2 or M_3 sites one of Cl adatom relaxes toward position in the missing dimer row (S_7 in figure 1). The formation of Ga dichloride can be obtained when two Cl adatoms are bonded with Ga in the M_1-site, however this structure is less stable than that shown in figure 2d. The average binding energies between substrate and Cl atoms with increase of their concentration are listed in table 2. In general, the binding energy of Cl adatom weakly depends on halogen coverage in agreement with Ref.

[5]. Table 2 presents the change in the binding energy of Ga atoms with substrate relative to its value on the clean surface (4.87 eV). In order to estimate ΔE_b we removed only one GaCl molecule from M_1-site with increase of Cl coverage. It is seen that maximal ΔE_b is observed for Cl coverage of 0.75 ML that corresponds to the adsorption in the most preferential sites, while decrease of ΔE_b above 0.75 ML reflects the saturation of unfavorable sites (As-sites) and possibly influence of adsorbates lateral interaction. The space limitation for this paper does not allow to discuss the change of the electronic structure with increase of halogen coverage that will be performed in our forthcoming paper. Note, that the "back bonds" weakening can be explained by the adsorbate–induced charge redistribution in top Ga-As atoms similar to those observed for electropositive alkali atoms on GaAs(001) surface [14].

Table 2. Binding energy of Cl adatom, Cl-Ga bond length, the binding energy change of Ga surface atoms as a function of the Cl coverage.

Coverage	0.125 ML	0.25 ML	0.50 ML	0.75 ML	1.00 ML
E_b (eV/atom)	2.86, 2.87 [13]	3.06	3.12	3.05	3.05
$d_{Cl\text{-}Ga}$ (Å)	2.22, 2.23 [13]	2.20	2.20	2.18	2.17
ΔE_b (eV)	2.62	2.69	2.80	2.89	2.63

4. Conclusion

The comparative analysis of halogen atom adsorbsion on ζ-GaAs(001)-(4×2) surface were performed by ab-initio PAW-DFT method. Our results reveal that halogen adsorption leads to the weakening of Ga-As surface bonds and this effect is most pronounced with increase of electronegativity of halogens. It is shown that the transition from the chemisorption to etching state is occurred in case of F whereas adsorption of I leads to passivation of Ga-rich GaAs(001) surface. The consideration of Cl adsorption on GaAs(001) surface shows that the increase of halogen coverage above 0.75 ML weakly changes the binding energy of surface Ga atoms.

Acknowledgment

The work was supported by the Russian Foundation for Basic Research under Grant 13-02-98017_r_a and Tomsk State University Competitiveness Improvement Program. Numerical calculations were performed on the SKIF-Cyberia supercomputer at the National Research Tomsk State University.

References

[1] Simpson W C and Yarmoff J A 1996 *Annu. Rev. Phys. Chem.* **47** 527
[2] Wang W K, Simpson W C and Yarmoff J A 2000 *Phys. Rev.* B **61** 2164
[3] Bakulin A, Eremeev S, Tereshchenko O, Chulkov E and Kulkova S 2011 *IOP Conf. Ser.: Mater. Sci. Eng.* **23** 012015
[4] Bakulin A V, Kulkova S E, Eremeev S V and Tereshchenko O E 2014 *J. Phys. Chem. C* **118** 20097
[5] Mikolajczyk P and Stankiewicz B 2010 *Appl. Surf. Sci.* **256** 4822
[6] Mikolajczyk P and Stankiewicz B 2009 *Cent. Eur. J. Phys.* **7** 279
[7] Blöchl P E 1994 *Phys. Rev. B* **50** 17953
[8] Kresse G and Joubert J 1999 *Phys. Rev. B* **59** 1758
[9] Lee S H, Moritz W and Scheffler M 2000 *Phys. Rev. Lett.* **85** 3890
[10] Bakulin A V, Kulkova S E, Eremeev S V and Tereshchenko O E 2013 *Surf. Sci.* **615** 97
[11] Tereshchenko O E, Bakulin A V, Kulkova S E and Eremeev S V 2013 *J. Phys. Chem. C* **117** 9723
[12] Tereshchenko O E, Paget D, Toropetsky K V, Alperovich V L, Eremeev S V, Bakulin A V, Kulkova S E, Doyle B P and Nannarone S 2012 *J. Phys. Chem. C* **116** 8535
[13] Lee S M, Lee S H and Scheffler M 2004 *Phys. Rev. B* **69** 125317
[14] Tereshchenko O E, Alperovich V L and Terekhov A S 2004 *JETP Letters* **79** 131

IOP Conf. Series: Materials Science and Engineering **77** (2015) 012003 doi:10.1088/1757-899X/77/1/012003

On time-resolved approach for phonon assisted interband transitions

E Klotins

Institute of Solid State Physics, University of Latvia, 8 Kengaraga str., Riga, LV-1063, Latvia

E-mail: klotins@cfi.lu.lv

Abstract. Photoexcited dynamics of electrons and holes in two-band dielectric, with special emphasis on back reaction of phonons are developed by combining the quantum electrodynamics and Baker-Campbell-Hausdorff (BCH) canonical transformation. These methods create an explicit time-domain representation of photoinduced processes and contribute in unifying phonon-assisted description of distribution functions of electron and hole quasiparticles for the description of observable effects of photoinduced processes in dielectrics.

1. Introduction

Determining the properties of electron-hole (e-h) pairs excited at interaction of electromagnetic radiation with matter is of major interest to many branches of solid state physics. Currently, there are no general efficient methods to deal with this challenging problem that involves joint kinetics of e-h pairs, arbitrary dispersion of the constituting electrons and holes, and their interaction with phonons. Early approaches to this problem categorized as time-convolutionless projection operator techniques were introduced in [1] and provide a systematic way to go beyond the Markovian approximation for quantum two-level systems in a bosonic reservoir [2]. However, these approaches are technicaly complex and inconvenient for systems with band structure [3].

Complementary developments are based on quantum field theory (QFT) [4] addressed to meta-stable vacuum, quasiparticle representation of electron and positron, and the associated mathematical technique for the dynamic Schwinger effect. Its extension to lattice systems has lead to a fundamentally new QFT approach to the solid state physics. Similarity between the kinetics of quantum field systems with meta-stable vacuum and the e-h kinetics in solid state is that in both cases the vacuum state is not uniquely defined and the quasiparticle representation is introduced by physical reasons. With application to solids, the vacuum state must obey the concepts of band theory, the conventional quantum mechanics, as well as requirements to the Lorentz invariance and special relativity.

To this end, the Lagrangian density, conventionally derived for a free particle, is addressed to the electron–hole pair and the associated field functions are harmonized with individual dispersion relations of these non-interacting quasiparticles [5]. The electromagnetic radiation is considered as spatially uniform within the atomic scale and is presumably switched on at infinite past. The resulting interband transition obeys the principle of least action for e–h birth and annihilation operators constituting a background for their distribution functions and for the kinetics of observables [6]. Obvious benefits of the QFT approach are reduction of conventional restrictions for the parameters of electromagnetic radiation and non-Markovian nature of the short time e-h kinetics.

A topic, still unresolved and essential for more realistic models, is the Coulomb interaction between photoexcited electrons and holes. Formally this interaction is categorized as back-reaction of phonons which must be distinguished from the impact of thermal (boson) bath at finite temperatures. Available option is to implement additional terms to the QFT Hamiltonian as attempted on the basis

IOP Conf. Series: Materials Science and Engineering **77** (2015) 012003 doi:10.1088/1757-899X/77/1/012003

of the Baker-Campbell-Hausdorff (BCH) canonical transformation[7,8]. However, to justify this *ad hoc* approximation one needs in a connection between conventional and QFT approaches as a disadvantage.

In the present work we continue our previous attempts as based on a self-consistent description of electrons, holes and the associated phonons. Formally, the back reaction of phonons is accounted for in Lagrangian density on equal grounds with electromagnetic radiation and bridges the conventional band theory with QFT. We are limited within the nondissipative approximation and neglect the interaction with thermal bath.

We start from the free field case to sketch the notations and derivation of the associated Lagrange-Hamilton approach (Section 2). Systematic derivation of field functions is reconsidered for arbitrary dispersion of electrons and holes as a background for further extension toward the back-reaction of phonons (Section 3). A survey of BCH in context of Lagrange – Hamilton approach is given in (Section 4). Conclusions are in (Section 5). We also include Appendix A where the harmonization with dispersion relations and the transition to field operator representation is given. Natural units, $\hbar = c = 1$, are used.

2. Model description: the free field

Derivation of the interband transitions kinetics in the context of e-h pair is thoroughly discussed in the literature [5,6] focussed on quadratic dispersion. However, in order to introduce the notation and to emphasize details specific for the back reaction of phonons we reconsider the survey following the free field, electromagnetic radiation, and phonons stages. The description keeps oscillator-like equation, the Lagrange – Hamilton, and principle of least action representations maintained.

The model two-band system is defined by e-h pair fixed to hypersurfaces in energy - quasi-momentum space with electron and hole states in the conduction and valence bands [5] as

$$E_c = \frac{\Delta}{2} + \varepsilon_c(\mathbf{p}), \quad E_v = -\frac{\Delta}{2} - \varepsilon_v(\mathbf{p}), \tag{1}$$

where the static dispersion laws of the kinetic energy $\varepsilon_c(\mathbf{p})$ and $\varepsilon_v(\mathbf{p})$ are either assumed on the phenomenological basis or calculated by methods of nonrelativistic quantum mechanics. The band gap Δ and the dispersion laws assigned to conduction and valence states $\varepsilon_{c,v}(\mathbf{p})$ remain presumably invariable quantities, and two Schrödinger equations associated with equations (1) read as [6]

$$\left(\begin{vmatrix}\hat{E}\\ \hat{E}\end{vmatrix} - \begin{vmatrix}\Delta/2 + \varepsilon_e(\hat{\mathbf{p}}) & 0\\ 0 & -\Delta/2 - \varepsilon_c(\hat{\mathbf{p}})\end{vmatrix}\right)\Psi(\mathbf{x},t) = 0 \ . \tag{2}$$

The electron and hole states are correlated by charge and momentum balance and allow for a joint description in terms of the general dispersion law [5]

$$(E - E_e)(E - E_v) = \left[E - \frac{\Delta}{2} - \varepsilon_c(\mathbf{p})\right]\left[E + \left(\frac{\Delta}{2}\right) + \varepsilon_v(-\mathbf{p})\right] = 0 \, . \tag{3}$$

For quadratic dispersion (3) reads as

$$(E - E_e)(E - E_v) = \left[E - \frac{\Delta}{2} - \frac{\mathbf{p}^2}{2m_e}\right]\left[E + \left(\frac{\Delta}{2}\right) + \frac{\mathbf{p}^2}{2m_h}\right] = 0 \, . \tag{4}$$

The physical picture under (4) is that at very low electron density the Coulomb interaction between the electrons and ions and the phonons and ions is accounted for by effective mass of the conduction and valence states in the quadratic dispersion law $\varepsilon_{c,v} = \mathbf{p}^2/(2m^*_{c,v})$. This effective mass approximation has substantial calculation advances as based on simple relations between the quadratic dispersion law and its quantized form.

In this section we follow [5] and emphasize specific features emerging beyond the effective mass and in the presence of the back reaction of phonons. The first quantized free-field equation of motion is given by relation

$$\left[\hat{E}-\frac{\Delta}{2}-\varepsilon_c(\hat{\mathbf{p}})\right]\left[\hat{E}+\left(\frac{\Delta}{2}\right)+\varepsilon_v(\hat{\mathbf{p}})\right]\Psi(\mathbf{x},t)=0 \quad (5)$$

where $\Psi(\mathbf{x},t)$ is the two body wave function, $\hat{E}=i\,\partial/\partial t$,and $,\hat{\mathbf{p}}=-i\,\nabla(\mathbf{x})$.
The first quantized free-field equation of motion is given by relation

$$\left[\hat{E}-\frac{\Delta}{2}-\varepsilon_c(\hat{\mathbf{p}})\right]\left[\hat{E}+\left(\frac{\Delta}{2}\right)+\varepsilon_v(-\hat{\mathbf{p}})\right]\Psi(\mathbf{x},t)=0, \quad (6)$$

where $\Psi(\mathbf{x},t)$ is obeing the kinetic equation

$$\left(\left(-\frac{\partial^2}{\partial t^2}\right)+\left(i\frac{\partial}{\partial t}\right)\varepsilon_v(-\hat{\mathbf{p}})-\left(\frac{\Delta}{2}\right)^2-\frac{\Delta}{2}\varepsilon_v(-\hat{\mathbf{p}})-\varepsilon_c(\hat{\mathbf{p}})\left(i\frac{\partial}{\partial t}\right)-\varepsilon_c(\hat{\mathbf{p}})\left(\frac{\Delta}{2}\right)-\varepsilon_c(\hat{\mathbf{p}})\varepsilon_v(-\hat{\mathbf{p}})\right)\Psi(\mathbf{x},t)=0 \quad (7)$$

Path to the relativistic Lagrange-Hamilton description starts with transformation to the oscillator-like equation having uniform differential form obtained by the phase transformation toward the auxiliary function $\Phi(\mathbf{x},t)$

$$\Psi(\mathbf{x},t)=\exp\left\{\frac{i}{2}[A(\hat{\mathbf{p}})]t\right\}\Phi(\mathbf{x},t)\ . \quad (8)$$

The oscillator-like equation reads as

$$\partial_0\Phi^*(\mathbf{x},t)\partial^0\Phi(\mathbf{x},t)-\Omega^2(\hat{\mathbf{p}})\Phi^*(\mathbf{x},t)\Phi(\mathbf{x},t)=0 \quad (9)$$

and the effective frequency operator

$$\Omega(\hat{\mathbf{p}},t)=\frac{1}{2}\left[\varepsilon_v(-\hat{\mathbf{p}})+\varepsilon_e(\hat{\mathbf{p}})+\Delta\right]. \quad (10)$$

With complex Klein-Gordon field in mind the Lagrangian density for the auxiliary function $\Phi(\mathbf{x},t)$is given by

$$\mathcal{L}(\mathbf{x},t)=\frac{1}{\Delta}\left[\frac{\partial}{\partial t}\Phi(\mathbf{x},t)\frac{\partial}{\partial t}\Phi^*(\mathbf{x},t)-\Omega^2(\hat{\mathbf{p}})\Phi(\mathbf{x},t)\Phi^*(\mathbf{x},t)\right] \quad (11)$$

The factor $1/\Delta$ proceed from the specific correspodence principle: at $\Delta\to\infty$ the derived relations turn into the ordinary quantum mechanical analogies for independent bands. The total field function $\Psi(\mathbf{x},t)$ is restored by inverse phase transformation

$$\Phi(\mathbf{x},t)=\exp\left[-\frac{it(\varepsilon_v(\hat{\mathbf{p}})-\varepsilon_c(\hat{\mathbf{p}}))}{2}\right]\Psi(\mathbf{x},t)$$
$$\Phi^*(\mathbf{x},t)=\exp\left[\frac{it(\varepsilon_v(\hat{\mathbf{p}})-\varepsilon_c(\hat{\mathbf{p}}))}{2}\right]\Psi^*(\mathbf{x},t) \quad (12)$$

As a result the Lagrangian density for the field function $\Psi(\mathbf{x},t)$ reads as

$$\mathcal{L}(\mathbf{x},t)=\frac{1}{\Delta}\left(\left|\frac{i(\varepsilon_v(\hat{\mathbf{p}})-\varepsilon_c(\hat{\mathbf{p}}))}{2}\Psi^*(\mathbf{x},t)+\dot{\Psi}^*(\mathbf{x},t)\right|^2+\Omega^2(\hat{\mathbf{p}})|\Psi|^2\right)\ . \quad (13)$$

The derived entities are canonical moments

$$\pi = \frac{\partial \mathcal{L}}{\partial \dot{\Psi}} = -\frac{i(\varepsilon_c(\hat{\mathbf{p}}) - \varepsilon_v(\hat{\mathbf{p}}))}{2\hbar}\Psi^*(\mathbf{x},t) + \dot{\Psi}^*(\mathbf{x},t) \,, \tag{14}$$

$$\pi^* = \frac{\partial \mathcal{L}}{\partial \dot{\Psi}^*} = +\frac{i(\varepsilon_c(\hat{\mathbf{p}}) - \varepsilon_v(\hat{\mathbf{p}}))}{2\hbar}\Psi(\mathbf{x},t) + \dot{\Psi}(\mathbf{x},t) \,, \tag{15}$$

and the free Hamiltonian density

$$\mathcal{H} = \pi\dot{\Psi} + \pi^*\dot{\Psi}^* - \mathcal{L}. \tag{16}$$

Expansion of the Lagrangian density (13)

$$\mathcal{L}(\Psi,\Psi^*;\dot{\Psi},\dot{\Phi}^*) = \frac{1}{\Delta}\left\{\left[\frac{(\varepsilon_c(\hat{\mathbf{p}}) - \varepsilon_v(\hat{\mathbf{p}}))^2}{4}\Psi\Psi^* + \frac{i(\varepsilon_c(\hat{\mathbf{p}}) - \varepsilon_v(\hat{\mathbf{p}}))}{2}\Psi\dot{\Psi}^* - \frac{i(\varepsilon_c(\hat{\mathbf{p}}) - \varepsilon_v(\hat{\mathbf{p}}))}{2}\Psi^*\dot{\Psi} + \dot{\Psi}\dot{\Psi}^*\right] - \Omega^2(\hat{\mathbf{p}})\Psi^*\Psi\right\} \tag{17}$$

exhibits spatial derivative terms higher than second order and creates a self-interacting potential supplementary to the canonical Klein-Gordon form. The corresponding condition for the self-interacting potential being small is $\varepsilon_{c,v} << \Delta$ with presumably holds for the energy of radiation of the order of band gap.

3. Decomposition of field functions for e-h pairs

Decomposition is the first step toward the derivation of field functions. Unlike the free Klein-Gordon particle constrained only by relativistic requirements, the particle in lattice environment is constrained by e-h dispersion relations. Into plane wave basis [5]

$$\Psi(\mathbf{x},t) = \frac{1}{\sqrt{V}} \sum_{\mathbf{p}} \int dE\, \tilde{\Psi}(E,\mathbf{p}) e^{-iEt+i\mathbf{p}\mathbf{x}} \tag{18}$$

the decomposition is defined on the hypersurfaces specified by the roots of (3) , $\mathbf{p}$ is the crystal quasi-momentum, $p_n = (2\pi / L)n_k$, $n_k = 0,1,2,...$, and $Et - \mathbf{p}\cdot\mathbf{x} = px$ is 4-vector scalar product. Within this vector field approach both direct and complex conjugate field functions applay as

$$\Psi(\mathbf{x},t) = \frac{1}{\sqrt{V}} \sum_{\mathbf{p}} \left(\int dE \tilde{\Psi}(E,\mathbf{p}) e^{-iEt}\right) e^{i\mathbf{p}\mathbf{x}} \,, \tag{19}$$

$$\Psi^*(\mathbf{x},t) = \frac{1}{\sqrt{V}} \sum_{\mathbf{p}} \left(\int dE \tilde{\Psi}^*(E,\mathbf{p}) e^{iEt}\right) e^{-i\mathbf{p}\mathbf{x}} \,. \tag{20}$$

While the general dispersion law (3) is defined on grid, we not introduce new notations and write the Fourier transformation as

$$\tilde{\Psi}(E,\mathbf{p}) = \delta\left[E - \frac{\Delta}{2} - \varepsilon_c(\mathbf{p})\right]\left[E + \frac{\Delta}{2} + \varepsilon_v(-\mathbf{p})\right]\psi(E,\mathbf{p}) = \delta(g(E,\mathbf{p}))\psi(E,\mathbf{p}) \,, \tag{21}$$

$$\tilde{\Psi}^*(E,\mathbf{p}) = \delta\left[E - \frac{\Delta}{2} - \varepsilon_c(\mathbf{p})\right]\left[E + \frac{\Delta}{2} + \varepsilon_v(-\mathbf{p})\right]\psi^*(E,\mathbf{p}) = \delta(g(E,\mathbf{p}))\psi^*(E,\mathbf{p}) \,. \tag{22}$$

Inserting relations (21) , (22) into the (19) , (20) the field functions read as

$$\Psi(\mathbf{x},t) = \frac{1}{\sqrt{V}} \sum_{\mathbf{p}} \left(\int_{-\infty}^{\infty} dE \delta(g(E,\mathbf{p})) f(E,\mathbf{p})\right) e^{+i\mathbf{p}\mathbf{x}} \,, \tag{23}$$

$$\Psi^*(\mathbf{x},t) = \frac{1}{\sqrt{V}} \sum_{\mathbf{p}} \left(\int_{-\infty}^{\infty} dE \delta(g(E,\mathbf{p})) f^*(E,\mathbf{p})\right) e^{-i\mathbf{p}\mathbf{x}} \,. \tag{24}$$

IOP Conf. Series: Materials Science and Engineering **77** (2015) 012003 doi:10.1088/1757-899X/77/1/012003

where $f(E,\mathbf{p}) = \psi(E,\mathbf{p})e^{-iEt}$ and $f^*(E,\mathbf{p}) = \psi^*(E,\mathbf{p})e^{iEt}$. The delta function in the integrands in (23) , (24) , formally is defined as a distribution for the function, and the integral in (19) coincides on the integral form of generalized scaling property. Harmonization of field functions with general dispersion (3) and the transition to field functions in operator representation is given in Appendix A (A27), (A28):

$$\hat{\Psi}(\mathbf{x},t) = \frac{(2\pi)^{3/2}}{V} \sum_{\mathbf{p}} \left(\sqrt{\frac{\Delta}{2\,\Omega(\mathbf{p})}} \left(\hat{a}_e(\mathbf{p},t) + \hat{a}_h(-\mathbf{p},t) \right) \right) e^{i\mathbf{px}} \quad , \tag{25}$$

$$\hat{\Psi}^+(\mathbf{x},t) = \frac{(2\pi)^{3/2}}{V} \sum_{\mathbf{p}} \left(\sqrt{\frac{\Delta}{2\,\Omega(\mathbf{p})}} \left(\hat{a}_e^+(\mathbf{p},t) + \hat{a}_h^+(-\mathbf{p},t) \right) \right) e^{-i\mathbf{px}} \quad . \tag{26}$$

In these terms the time derivatives of field functions reads in operator representation as:

$$\dot{\hat{\Psi}}(\mathbf{x},t) = \frac{(2\pi)^{3/2}}{V} \sum_{\mathbf{p}} \left(\sqrt{\frac{\Delta}{2\,\Omega(\mathbf{p})}} \left(-i(\Delta/2 + \varepsilon_c(\mathbf{p}))a_e(\mathbf{p},t) + i(\Delta/2 + \varepsilon_v(\mathbf{p}))a_h(-\mathbf{p},t) \right) \right) e^{i\mathbf{px}} , \tag{27}$$

$$\dot{\hat{\Psi}}^*(\mathbf{x},t) = \frac{(2\pi)^{3/2}}{V} \sum_{\mathbf{p}} \left(\sqrt{\frac{\Delta}{2\,\Omega(\mathbf{p})}} \left(i(\Delta/2 + \varepsilon_c(\mathbf{p}))a_e^*(\mathbf{p},t) - i(\Delta/2 + \varepsilon_v(\mathbf{p}))a_h^*(-\mathbf{p},t) \right) \right) e^{-i\mathbf{px}} \quad . \tag{28}$$

Calculations, associated with the principle of least action starts with the free Hamiltonian

$$\mathcal{H}_{free} = \pi\dot{\Psi} + \pi^*\dot{\Psi}^* - \mathcal{L} \, . \tag{29}$$

The total Hamiltonian is obtained from free Hamiltonian inserting dynamic dispersion in field function and canonical momentum terms by replacement $\mathbf{p} \to \mathbf{P} = \mathbf{p} - q\mathbf{A}$. The 4-potential $\mathbf{A} = \{0,0,0,A_3\}$ corresponds the electromagnetic radiation approximed by electric field $\varepsilon = -\dot{\mathbf{A}}$. As a result the Lagrangian density reads as

$$\mathcal{L} = \pi\dot{\Psi} + \pi^*\dot{\Psi}^* - \mathcal{H}_{tot} \quad . \tag{30}$$

In quasiparticle representation the Lagrangian density (30) comprises products of field functions which are complex conjugate in the space coordinate and that, in total, become functions on field operators. Their values are found by variational derivatives of the action $\delta S = 0$ [6]. The corresponding differential equations for field operators (A23)-(A26) are Heisenberg-like, first order in time derivative, and comprise time dependent coefficients so turning the problem to conventional algebra [8 (Eqs.(32-36))].

Going back to the electron-hole problem in presence of electric field it is convenient to go to the dynamic dispersion. Replacement the dynamic dispersion contributes into the dispersion laws $\varepsilon_c(\mathbf{p}) \to \varepsilon_c(\mathbf{P}_c(t))$ and $\varepsilon_v(\mathbf{p}) \to \varepsilon_v(\mathbf{P}_v(t))$, and it means that the hypersurfaces (1) becomes time dependent. The effect of dynamic dispersion on field operators (A23)-(A26) appears through transition amplitudes proportional to the ratio $\dot{\Omega}/\Omega$ and turns to zero at $\mathbf{A} = 0$.

4. Electron-phonon and hole-phonon interaction

The two band model, reconsidered in sections 2 and 3 is based on the concept of lattice potential as a source of the band structure and the e-h pairs emerging only under the electromagnetic radiation. Interaction of electrons and holes with the ionic lattice vary their kinetics and is categorized as the back – reaction of phonons. Phonons excited due the thermal bath at finite temperatures are omitted in this manuscript.

The electronic part of this back reaction is derived in [7,8] as sums over creation and annihilation operators derived by conventional quantum mechanics [9]. Propagation of the back reaction of phonons starts at the Lagrange – Hamilton stage as supplementary to the electromagnetic radiation. As a consequence, the back reaction contributes in the least action and, subsequently, in kinetic equations for the field operators.

In terms of conventional quantum mechanics (in this section the reduced Planck constant $\hbar$ is restored) the system is distinguished by Hamiltonian:

$$H = \sum_{\mathbf{p}} E_c(\mathbf{p}) a_e^+(\mathbf{p}) a_e(\mathbf{p}) + \sum_{\mathbf{p}} E_v(\mathbf{p}) a_h^+(\mathbf{p}) a_h(\mathbf{p}) + \sum_{\mathbf{q}} \hbar\omega_{\mathbf{q}} c_{\mathbf{q}}^+ c_{\mathbf{q}} + \sum_{\mathbf{pq}} V_{(e)\mathbf{q}} \left(a_{e,\mathbf{p}+\mathbf{q}}^+ a_{e,\mathbf{q}} c_{\mathbf{q}} + c_{\mathbf{q}}^+ a_{e,\mathbf{p}-\mathbf{q}}^+ a_{\mathbf{q}} \right) + \sum_{\mathbf{pq}} V_{(h)\mathbf{q}} \left(a_{h,\mathbf{p}+\mathbf{q}}^+ a_{h,\mathbf{q}} c_{\mathbf{q}} + c_{\mathbf{q}}^+ a_{h,\mathbf{p}-\mathbf{q}}^+ a_{\mathbf{q}} \right), \tag{31}$$

where $E_{c(v)}(\mathbf{p})$ is dispersion of electrons (holes) and $V_{(e,h)\mathbf{q}}$ are matrix elements weighting the combinations of creation and annihilation operators. [9]. In the context of the back reaction, only the last two terms with mixed electron and hole operators are relevant. The mathematical technique transforming Eq.(14) into a pure electronic problem is Baker-Campbell-Hausdorff canonical transformation (BCH) [7] that gives following constituents of electron-phonon and hole-phonon interaction due to exchange of virtual phonons [8]:

$$\boldsymbol{e} = \sum_{\mathbf{p},\mathbf{p}',\mathbf{q}} \frac{V_{e,\mathbf{q}}^2}{2} \left(\frac{1}{E_{el}(\mathbf{p}'+\mathbf{q}) - E_{el}(\mathbf{p}) - \hbar\omega_{\mathbf{q}}} + \frac{1}{E_{el}(\mathbf{p}) - E_{el}(\mathbf{p}-\mathbf{q}) - \hbar\omega_{\mathbf{q}}} \right) a_{e,\mathbf{p}'+\mathbf{q}}^+ a_{e,\mathbf{p}'} a_{e,\mathbf{p}-\mathbf{q}}^+ a_{e,\mathbf{p}}, \tag{32}$$

$$\boldsymbol{eh} = \sum_{\mathbf{p},\mathbf{p}',\mathbf{q}} \frac{V_{e,\mathbf{q}} V_{h,\mathbf{q}}}{2} \left(\frac{1}{E_{el}(\mathbf{p}+\mathbf{q}) - E_{el}(\mathbf{p}) - \hbar\omega_{\mathbf{q}}} + \frac{1}{-E_h(\mathbf{p}-\mathbf{q}) + E_h(\mathbf{p}) - \hbar\omega_{\mathbf{q}}} \right) a_{e,\mathbf{p}+\mathbf{q}}^+ a_{e,\mathbf{p}} a_{h,\mathbf{p}-\mathbf{q}}^+ a_{h,\mathbf{p}}, \tag{33}$$

$$\boldsymbol{h} = \sum_{\mathbf{p},\mathbf{p}',\mathbf{q}} \frac{(V_{h,\mathbf{q}})^2}{2} \left(\frac{1}{E_h(\mathbf{p}'+\mathbf{q}) - E_h(\mathbf{p}') - \hbar\omega_{\mathbf{q}}} + \frac{1}{-E_h(\mathbf{p}-\mathbf{q}) + E_h(\mathbf{p}) - \hbar\omega_{\mathbf{q}}} \right) a_{h,\mathbf{p}'+\mathbf{q}}^+ a_{h,\mathbf{p}'} a_{h,\mathbf{p}-\mathbf{q}}^+ a_{h,\mathbf{p}}, \tag{34}$$

$$\boldsymbol{he} = \sum_{\mathbf{p},\mathbf{p}',\mathbf{q}} \frac{V_{h,\mathbf{q}} V_{e,\mathbf{q}}}{2} \left(\frac{1}{E_h(\mathbf{p}'+\mathbf{q}) - E_h(\mathbf{p}') - \hbar\omega_{\mathbf{q}}} + \frac{1}{-E_{el}(\mathbf{p}-\mathbf{q}) + E_{el}(\mathbf{p}) - \hbar\omega_{\mathbf{q}}} \right) a_{h,\mathbf{p}+\mathbf{q}}^+ a_{h,\mathbf{p}} a_{e,\mathbf{p}-\mathbf{q}}^+ a_{e,\mathbf{p}}. \tag{35}$$

The underlaying physics is that back reaction changes the kinetic energy of electron-hole system on equal grounds with the electromagnetic radiation. Formally it means that dynamic dispersion laws $\varepsilon_c(\mathbf{p}) \to \varepsilon_c(\mathbf{P}_c(t))$ and $\varepsilon_v(\mathbf{p}) \to \varepsilon_v(\mathbf{P}_v(t))$ turns to

$$\varepsilon_c(\mathbf{p}) \to \varepsilon_c(\mathbf{P}_c(t)) + \boldsymbol{P}_{e,int} \tag{36}$$

and

$$\varepsilon_v(\mathbf{p}) \to \varepsilon_v(\mathbf{P}_v(t)) + \boldsymbol{P}_{h,int} \tag{37}$$

where $\boldsymbol{P}_{e,int} = \boldsymbol{e} + \boldsymbol{eh}, \boldsymbol{P}_{h,int} = \boldsymbol{h} + \boldsymbol{he}$ are inputs calculated by nonrelativistic quantum mechanics[9].

5. Summary

Self-consistent description of electrons, holes and the associated phonons is addressed to the short time kinetics of interacting of electromagnetic radiation with matter and obeying requirements of quantum mechanics, quantum field theory and special relativity. Special attention is paid to the allowed time dependence of radiation which, unlike the existing models, is arbitrary. Another property crucial for high speed practical needs is the time response condensing full history of the material.

The state of art relies on well-known theoretical approaches to the free particles. With application to solids distinguished by the lattice of surrounding atoms these approaches become more complex and are not completely understood.

What is new in the presented research is extension of the model of noninteracting electrons and holes to the back reaction of phonons as based on disconnection of the electron – phonon and the electron hole – phonon interaction in purely electronic and hole parts, and putting extra terms in Lagrange – Hamilton approach. Formally these extra terms contribute on equal grounds with electromagnetic radiation with nonperturbative mathematical structure of the solution maintained.

IOP Conf. Series: Materials Science and Engineering **77** (2015) 012003 doi:10.1088/1757-899X/77/1/012003

The necessary inputs are available within nonrelativistic quantum mechanics and commercial codes while the key output is distribution function for electron – electron hole pair determining the observables of interest.

Funding statement
This work was supported in part by National Research Program of Latvia Nr. 2014. 10-4/VPP-2/ and Grant Nr.237/2012.

Appendix
The delta function in the integrands in (23), (24), formally is defined as a distribution for the function, and coincides with the integral form of generalized scaling property written as

$$\int_{-\infty}^{\infty} f(E,\mathbf{p})\delta(g(E,\mathbf{p}))dE = \sum_i \frac{f(E_i,\mathbf{p})}{|g'(E_i)|} = \frac{f(E_1,\mathbf{p})}{|g'(E_1)|} + \frac{f(E_2,\mathbf{p})}{|g'(E_2)|} \tag{A1}$$

and

$$\int_{-\infty}^{\infty} f^*(E,\mathbf{p})\delta(g(E,\mathbf{p}))dE = \sum_i \frac{f^*(E_i,\mathbf{p})}{|g'(E_i)|} = \frac{f^*(E_1,\mathbf{p})}{|g'(E_1)|} + \frac{f^*(E_2,\mathbf{p})}{|g'(E_2)|} \tag{A2}$$

where

$$g(E,\mathbf{p}) = \left[E - \frac{\Delta}{2} - \varepsilon_c(\mathbf{p})\right]\left[E + \frac{\Delta}{2} + \varepsilon_v(\mathbf{p})\right] \tag{A3}$$

The roots of $g(E,\mathbf{p})$ read as

$$E_1 = \Delta/2 + \varepsilon_c(\mathbf{p}) \tag{A4}$$

$$E_2 = (-\Delta/2 - \varepsilon_v(\mathbf{p})). \tag{A5}$$

Derivative of $g(E,\mathbf{p})$ is equal to the derivative of dispersion relation (3)

$$\frac{\partial g(\mathbf{p},E)}{\partial E} = 2E - \varepsilon_c(\mathbf{p}) + \varepsilon_v(\mathbf{p}) \tag{A6}$$

Substituting (A4) and (A5) into (A6) gives

$$g'(E_1) = (\Delta + \varepsilon_c(\mathbf{p}) + \varepsilon_v(\mathbf{p})) \tag{A7}$$

$$g'(E_2) = (-\Delta - \varepsilon_c(\mathbf{p}) - \varepsilon_v(\mathbf{p})) \tag{A8}$$

and the common denominator

$$|g'(E_1)| = |g'(E_2)| = (\Delta + \varepsilon_c(\mathbf{p}) + \varepsilon_v(\mathbf{p})) = 2\Omega(\mathbf{p}) \tag{A9}$$

Relation (A9) is the on shell condition for 4-field.

The numerators in (A1), (A2) start with $f(E_i,\mathbf{p})$ $i = 1,2$ function. For direct field functions

$$f(E_1,\mathbf{p}) = \psi\left(\frac{\Delta}{2} + \varepsilon_c(\mathbf{p})\right)e^{-i\left(\frac{\Delta}{2}+\varepsilon_c(\mathbf{p})\right)t} = a_e(\mathbf{p})e^{-i\left(\frac{\Delta}{2}+\varepsilon_c(\mathbf{p})\right)t} \tag{A10}$$

$$f(E_2,\mathbf{p}) = \psi\left(-\frac{\Delta}{2} - \varepsilon_v(\mathbf{p})\right)e^{-i\left(-\frac{\Delta}{2}-\varepsilon_v(\mathbf{p})\right)t} = a_h(-\mathbf{p})e^{-i\left(-\frac{\Delta}{2}-\varepsilon_v(\mathbf{p})\right)t} \tag{A11}$$

For complex conjugate field functions

$$f^*(E_1,\mathbf{p}) = \psi^*\left(\frac{\Delta}{2} + \varepsilon_c(\mathbf{p})\right)e^{+i\left(\frac{\Delta}{2}+\varepsilon_c(\mathbf{p})\right)t} = a_e^* e^{+i\left(\frac{\Delta}{2}+\varepsilon_c(\mathbf{p})\right)t} \tag{A12}$$

$$f^*(E_2,\mathbf{p})=\psi^*\left(-\frac{\Delta}{2}-\varepsilon_v(\mathbf{p})\right)e^{+i\left(-\frac{\Delta}{2}-\varepsilon_v(\mathbf{p})\right)t}=a_h^*(-\mathbf{p})e^{+i\left(-\frac{\Delta}{2}-\varepsilon_v(\mathbf{p})\right)t} \tag{A13}$$

Going back to (A1) and (A2) and inserting (A9) – (A13) gives

$$\Psi(\mathbf{x},t)=\frac{(2\pi)^{3/2}}{V}\sum_{\mathbf{p}}\left(\frac{1}{|\Delta+\varepsilon_c(\mathbf{p})+\varepsilon_v(\mathbf{p})|}\left(\psi((\Delta/2+\varepsilon_c(\mathbf{p}))\right)e^{-i(\Delta/2+\varepsilon_c(\mathbf{p}))t}+\psi((-\Delta/2-\varepsilon_v(\mathbf{p}))\right)e^{i(\Delta/2+\varepsilon_v(\mathbf{p}))t}\Big)\right)e^{i\mathbf{px}} \tag{A14}$$

$$\Psi^*(\mathbf{x},t)=\frac{(2\pi)^{3/2}}{V}\sum_{\mathbf{p}}\left(\frac{1}{|\Delta+\varepsilon_c(\mathbf{p})+\varepsilon_v(\mathbf{p})|}\left(\psi^*((\Delta/2+\varepsilon_c(\mathbf{p}))\right)e^{i(\Delta/2+\varepsilon_c(\mathbf{p}))t}+\psi^*((-\Delta/2-\varepsilon_v(\mathbf{p}))\right)e^{-i(\Delta/2+\varepsilon_v(\mathbf{p}))t}\Big)\right)e^{-i\mathbf{px}} \tag{A15}$$

As it follows, the correspondence between effective frequency operator $\Omega(\hat{\mathbf{p}})=\dfrac{\Delta+\varepsilon_v(\hat{\mathbf{p}})+\varepsilon_c(\hat{\mathbf{p}})}{2}$ and the effective frequency $\Omega(\mathbf{p})=\dfrac{|\Delta+\varepsilon_c(\mathbf{p})+\varepsilon_v(\mathbf{p})|}{2}$ valid for quadratic dispersion is maintained in case of arbitrary dispersion. As to the field amplitudes (A14), (A15), they are renormalized as

$$\psi((\Delta/2+\varepsilon_c(\mathbf{p})))\Rightarrow a_e(\mathbf{p}) \tag{A16}$$

and similarly for other field operators. As a result, the field functions reads as

$$\Psi(\mathbf{x},t)=\frac{(2\pi)^{3/2}}{V}\sum_{\mathbf{p}}\left(\frac{1}{2\,\Omega(\mathbf{p})}\left(a_e(\mathbf{p})e^{-i(\Delta/2+\varepsilon_c(\mathbf{p}))}+a_h(-\mathbf{p})e^{i(\Delta/2+\varepsilon_v(\mathbf{p}))t}\right)\right)e^{i\mathbf{px}}\quad , \tag{A17}$$

$$\Psi^*(\mathbf{x},t)=\frac{(2\pi)^{3/2}}{V}\sum_{\mathbf{p}}\left(\frac{1}{2\hbar\,\Omega(\mathbf{p})}\left(a_e^*(\mathbf{p})e^{\frac{+i(\Delta/2+\varepsilon_c(\mathbf{p}))t}{\hbar}}+a_h^*(-\mathbf{p})e^{\frac{-i(\Delta/2+\varepsilon_v(\mathbf{p}))t}{\hbar}}\right)\right)e^{-i\mathbf{px}}\ . \tag{A18}$$

Formally the field functions are redefined by dividing the factor $1/(2\,\Omega(\mathbf{p}))$in (A17) and (A20) by the factor $1/\sqrt{\Delta\,2\Omega(\mathbf{p})}$ resulting in

$$\Psi(\mathbf{x},t)=\frac{(2\pi)^{3/2}}{V}\sum_{\mathbf{p}}\left(\sqrt{\frac{\Delta}{2\,\Omega(\mathbf{p})}}\left(\tilde{a}_e(\mathbf{p})e^{i(\Delta/2+\varepsilon_c(\mathbf{p}))t}+\tilde{a}_h(-\mathbf{p})e^{i(\Delta/2+\varepsilon_v(\mathbf{p}))t}\right)\right)e^{i\mathbf{px}}\ , \tag{A 19}$$

$$\Psi^*(\mathbf{x},t)=\frac{(2\pi)^{3/2}}{V}\sum_{\mathbf{p}}\left(\sqrt{\frac{\Delta}{2\,\Omega(\mathbf{p})}}\left(\tilde{a}_e^*(\mathbf{p})e^{i(\Delta/2+\varepsilon_c(\mathbf{p}))t}+\tilde{a}_h^*(-\mathbf{p})e^{-i(\Delta/2+\varepsilon_v(\mathbf{p}))t}\right)\right)e^{-i\mathbf{px}}\quad , \tag{A20}$$

Where $\tilde{a}_{e,h}=(2\,\Delta\,\Omega(\mathbf{p}))^{-1/2}a_{e,h}$ and time derivatives from field functions (A19), (A20) read as (the punctuation ~ is omitted in further calculations):

$$\dot{\Psi}(\mathbf{x},t)=\frac{(2\pi)^{3/2}}{V}\sum_{\mathbf{p}}\left(\sqrt{\frac{\Delta}{2\,\Omega(\mathbf{p})}}\left(-i(\Delta/2+\varepsilon_c(\mathbf{p}))a_e(\mathbf{p})e^{-i(\Delta/2+\varepsilon_c(\mathbf{p}))}+i(\Delta/2+\varepsilon_v(\mathbf{p}))a_h(-\mathbf{p})e^{i(\Delta/2+\varepsilon_v(\mathbf{p}))}\right)\right)e^{i\mathbf{px}}\quad , \tag{A21}$$

$$\dot{\Psi}^*(\mathbf{x},t)=\frac{(2\pi)^{3/2}}{V}\sum_{\mathbf{p}}\left(\sqrt{\frac{\Delta}{2\hbar\,\Omega(\mathbf{p})}}\left(i\frac{(\Delta/2+\varepsilon_c(\mathbf{p}))}{\hbar}a_e^*(\mathbf{p})e^{\frac{+i(\Delta/2+\varepsilon_c(\mathbf{p}))t}{\hbar}}-i\frac{(\Delta/2+\varepsilon_v(\mathbf{p}))}{\hbar}a_h^*(-\mathbf{p})e^{\frac{-i(\Delta/2+\varepsilon_v(\mathbf{p}))t}{\hbar}}\right)\right)e^{-i\mathbf{px}}\ . \tag{A22}$$

Next concern is new time-dependent electron and hole operators:

$$a_e(\mathbf{p})e^{-i(\Delta/2+\varepsilon_c(\mathbf{p}))t}\Rightarrow\hat{a}_e(\mathbf{p},t)\quad , \tag{A23}$$

$$a_h(-\mathbf{p})e^{i(\Delta/2+\varepsilon_v(\mathbf{p}))t}\Rightarrow\hat{a}_h(-\mathbf{p},t)\quad , \tag{A24}$$

$$a_e^*(\mathbf{p})e^{i(\Delta/2+\varepsilon_c(\mathbf{p}))t}\Rightarrow\hat{a}_e^+(\mathbf{p},t)\quad , \tag{A25}$$

$$a_h^*(-\mathbf{p})e^{-i(\Delta/2+\varepsilon_v(\mathbf{p}))t}\Rightarrow\hat{a}_h^+(-\mathbf{p},t). \tag{A26}$$

Finally, the field functions in operator representation are derived from (A19), (A20) and reads as

$$\hat{\Psi}(\mathbf{x},t)=\frac{(2\pi)^{3/2}}{V}\sum_{\mathbf{p}}\left(\sqrt{\frac{\Delta}{2\,\Omega(\mathbf{p})}}\left(\hat{a}_e(\mathbf{p},t)+\hat{a}_h(-\mathbf{p},t)\right)\right)e^{i\mathbf{px}}\quad , \tag{A27}$$

$$\hat{\Psi}^+(\mathbf{x},t)=\frac{(2\pi)^{3/2}}{V}\sum_{\mathbf{p}}\left(\sqrt{\frac{\Delta}{2\,\Omega(\mathbf{p})}}\left(\hat{a}_e^+(\mathbf{p},t)+\hat{a}_h^+(-\mathbf{p},t)\right)\right)e^{-i\mathbf{px}}\ . \tag{A28}$$

IOP Conf. Series: Materials Science and Engineering **77** (2015) 012003 doi:10.1088/1757-899X/77/1/012003

References

[1] Nakajama S 1958 *Progr. Theor. Phys.***20** 948
[2] Zwanzig R 1960 *J. Chem. Phys.* **33** 1338
[3] Smirne A and Vacchini B 2010 *Phys. Rev.* A **82**, 022110
[4] Schmidt S, Blashke D, Röpke G K, Smolyansky S A, Prozorkevich A V and Toneev VD 1998 *International Journal of Modern Physics* E, 7 709
[5] Smolyansky S A, Bonitz M, Tarakanov A V *2010 Physics of Particles and Nuclei* **41**, No. 7.1075
[6] Smolyansky S A, Tarakanov A V and Bonitz M 2009 Plasma *Phys.* **49**(7-8) 575–584, DOI: 10.1002/ctpp.200910058
[7] Klotins E and Zvejnieks G 2013 *physica status solidi* (c) Volume **10**, Issue 4, pages 705–708, DOI: 10.1002/pssc.201200872
[8] Klotins E 2014 *Lithuanian Journal of Physics* Vol. 54, No. 4, pp. 217–226
[9] Bruus H and Flensberg K 2002 *Introduction to Many-body Quantum Theory in Condensed Matter Physics*, Copenhagen

IOP Conf. Series: Materials Science and Engineering **77** (2015) 012004 doi:10.1088/1757-899X/77/1/012004

Ab-initio Study of Cation-rich InP(001) and GaP(001) Surface Reconstructions and Iodine Adsorption

A Bakulin[1,2], A Ponomarev[1] and S Kulkova[1,2]

[1] Institute of Strength Physics and Materials Science of the Russian Academy of Sciences, Siberian Branch, pr. Akademichesky 2/4, Tomsk 634021, Russia
[2] National Research Tomsk State University, pr. Lenina 36, Tomsk 634050, Russia

E-mail: bakulin@ispms.tsc.ru

Abstract. Atomic and electronic structures for a number of InP and GaP (001) surface geometries were studied within the density functional theory (DFT) in order to reexamine the energy stability of surface reconstructions in the cation-rich limit. It is shown that in both cases the mixed dimer (2×4) reconstruction is the energetically favored in the cation rich limit. The ζ(4×2) reconstruction has the lowest surface energy among considered (4×2) surface structures. Comparative theoretical study of iodine adsorption on the semiconductor surface with ζ(4×2) and mixed dimer (2×4) structures was performed. It was shown that iodine prefers to be bonded with dimerized cation atoms irrespective on the surface reconstruction.

1. Introduction

It is well-known that A^3B^5 semiconductors are widely used in the production of devices such as light-emitting diodes, light detectors, integrated circuits, solar cells, etc. The cation-rich surface geometries attract a special attention because they are preferable for the deposition of ferromagnetic metals and alloys [1] that may be relevant to the nanotechnology of magnetic materials and spintronics [2]. For many technological applications atomically smoothed surface morphology are needed. In order to obtain such surface it is necessary to develop the "digital etching" technique. Usually halogens and halogen-containing molecules are used for this goal. As was shown in [3] the formation of strong ionic bonding between halogen and cation with the charge transfer via depletion of filled orbitals of the surface anion atoms leads to a weakening of bonds in the surface layer of semiconductor. Namely this condition is important in the initial stage of dry etching process on the semiconductor surface. The comparative study of halogen adsorption on the ζ-(4×2) reconstruction was performed for GaAs and InAs in [3, 4] and Ge [5, 6]. In the latter case the halogen coverage of 0.75 and 1.0 monolayer was studied. Less attention was paid to halogen adsorption on other semiconductors. As was shown in [4, 7] iodine having a lower electronegativity and a larger covalent radius provides a delicate selective action depending on the composition and reconstruction of the semiconductor surface. In this work we present the calculations of the surface reconstructions stability for GaP and InP as well as the comparative study of iodine adsorption on the (001) surface in the cation-rich limit.

2. Method of calculation

Atomic and electronic properties of the InP and GaP (001) surfaces were calculated using the projector augmented-wave (PAW) method [8, 9] and corresponding PAW potentials supplied with the VASP package [10]. The generalized gradient approximation (GGA-PBE) [11] for the exchange-correlation functional was used. The surface was simulated using the periodic slab geometry with a supercell containing eight-nine atomic layers of semiconductor and vacuum gap of ~10 Å. Earlier works [12, 13] performed on oxide and semiconductor systems showed that the thickness of the metal slab becomes important in determining the electronic structure. A surface energy is converged quickly with increasing number of atomic layers. In general, a slab consisting of six-eight layers describes well cation-rich $A^3B^5(001)$ surface. The saturation of dangling bonds of semiconductor compounds was performed using pseudo-hydrogen atoms approach suggested by K. Shiraishi [14]. The bottom P layer was passivated by pseudo-hydrogen atoms (two atoms per P atom) with fractional charge Z equal to $0.75e$ that allows us to make this layer bulk-like. The arrangement of pseudo-hydrogen atoms was determined by the energy minimization. The resulting bond length between P and pseudo-hydrogen atom was 1.469 Å. Two bottom semiconductor layers and pseudo-hydrogen layer were fixed during the structure optimization, whereas the atomic positions in other layers were relaxed using damping Newton method until the forces at atoms ~0.01 eV/Å. A 3×6×1 grid of k points, obtained by the Monkhorst-Pack method [15] was used for Brillouin zone integration for surface reconstruction (4×2), a 6×3×1 grid for (2×4) structures, as well as a 3×3×1 grid for c(4×4) structures. The plane wave cutoff was 500 eV. The calculation for the clean $A^3B^5(001)$ surfaces were carried out using theoretical bulk lattice constants of 5.53 Å for GaP and 6.00 Å for InP. Following linear dimensions of supercells for (2×4) and (4×2) surface geometries $a\sqrt{2}\times 2a\sqrt{2}\times 4a$ and $2a\sqrt{2}\times a\sqrt{2}\times 4a$ were used while for the c(4×4) geometry the size of supercell was ($2a\times 2a\times 4a$). Since slab thickness is $2a$ the vacuum gap of ~$2a$ is enough avoid self-interaction between neighbor slabs. Other computational details can be found in our paper [16].

The binding energy (E_b) of iodine adatom with semiconductor substrate was calculated using equation:

$$E_b = -(E_{I/A^3B^5} - E_{A^3B^5} - E_I), \quad (1)$$

where E_{I/A^3B^5} and $E_{A^3B^5}$ are the total energies of both iodine-adsorbed system and adsorbate-free semiconductor surface, respectively, and E_I is the total energy of isolated halogen atom. The halogen coverage of 0.125 monolayer (ML) is considered only in this work.

The formation energy of semiconductor surfaces was estimated using equation

$$E_f = E_{tot} - \sum_i^2 n_i \mu_i , \quad (2)$$

where n_i is a number of atoms of species i, with i = Ga (In) or P and μ_i represents the chemical potential of cation or anion which is restricted by upper limits $\mu_i \leq \mu_i$(bulk). Other computational details can be found in [16].

3. Results and discussion

To determine the atomic structure of the lowest cation-rich reconstruction, we have reexamined the stability of both InP and GaP (001) surface reconstructions. The calculations of the relative formation energies of about 20 different surface structures in dependence on the chemical potential of cation were performed. It should be noted that the thermodynamic equilibrium condition between surface and substrate atoms supposes that μ_{AB}(bulk) = $\mu_A + \mu_B$. The lower limit for cation chemical potential can be obtained by the addition of the heat of formation of semiconductor (ΔH): $\mu_i + \Delta H$. The reconstructions with the lowest surface energies, which are energetically preferable among the considered ones, are given in figure 1 by solid lines. The chemical potential of cation is changed within the range $-\Delta H < \Delta\mu < 0$. The calculated value of ΔH for InP is 0.45 eV, which is close to the value of 0.43 eV obtained in [17]. It is necessary to emphasize that in [17] for the determination of upper and lower limits of chemical potentials on the surface were used the following bulk values of μ for In, P and InP: -3.270 eV, -6.028 eV, 9.728 eV, respectively, that allowed us to estimate above mentioned value of ΔH for InP. In case of GaP, ΔH is 0.85 eV that agrees satisfactory with the value of 0.91 eV [18]. For determination of ΔH the chemical potential of black phosphorus was used. The calculated ΔH are significant less than experimental ones: 0.73±0.05 eV for InP and 1.07±0.02 eV for GaP [19]. It can be due to ΔH in [19] was determined relative to amorphous phase of red phosphorus which is less stable than the black one. In general, our diagrams are in good agreement with the previously obtained ones [18].

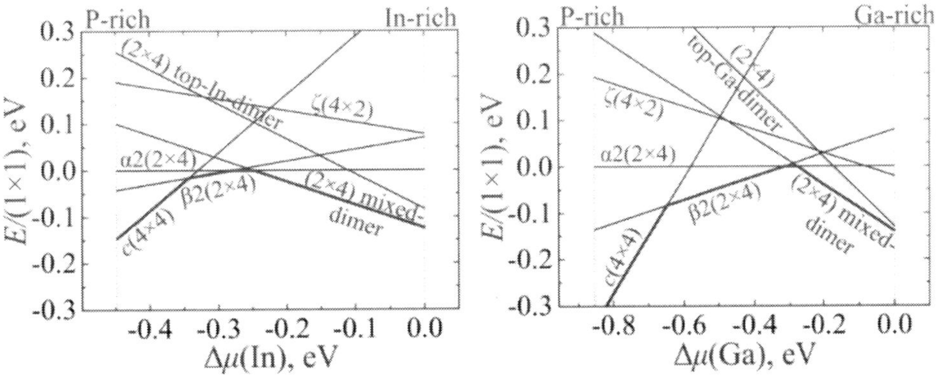

Figure 1. Relative formation energy per (1×1) unit cell for InP (left) and GaP (right) surface reconstructions depending on the In (Ga) chemical potential. Dashed lines denote P anion- and In(Ga) cation-rich limits of the thermodynamically allowed range of $\Delta\mu$.

In the P-rich limit, our calculations confirm that the $c(4\times4)$ structure with three P dimers in the surface layer is stable for both semiconductor surfaces. The same can be said about the β2(2×4) structure with two P dimers in the surface layer and one P dimer in the third layer. Mixed-dimer (2×4) reconstruction is the most stable structure in the cation rich limit for both semiconductors whereas the structure with the top Ga dimer was found to be stable in case of GaP in [18]. The surface energy of the mixed dimer (2×4) structure is only of 0.01 eV/(1×1) and 0.04 eV/(1×1) lower than that with the top-In(Ga)-dimer structure for GaP and InP, respectively. Phase α2(2×4) can be also stable but in a very narrow region of the cation concentration that agrees well with earlier results [18]. It should be stressed that the ζ(4×2) reconstruction is the most stable among structures with (4×2) geometry in the cation-rich limit. The same trend was obtained for GaAs and InAs in [3, 4]. The calculations show a 118 meV/(1×1) gain for the mixed-dimer (2×4) structure with respect to ζ(4×2) in the Ga-rich limit GaP(001) that is in good agreement with the value of ~115 meV/(1×1) [18]. This gain is of 202 meV/(1×1) in case of InP(001) but it was only 130 meV/(1×1) in [18]. In the latter case ΔH was found to be twice more (-0.91 eV) than the value calculated in the present paper. In general, our diagrams (figure 1) support the preference of the mixed dimer (2×4) structure for the (001) surface of GaP and InP semiconductors in the cation-rich limit.

Iodine adsorption was considered on both ζ(4×2) and mixed dimer(2×4) reconstructions in the positions given in figure 2. Let's to discuss firstly I adsorption on ζ(4×2). Iodine binding energies on this reconstruction are listed in table 1. The values obtained in [3] for GaAs and InAs are given also for comparison. It is seen that I prefers to be adsorbed in the M_1-site above dimerized In(Ga) surface atoms on both InP(001) and GaP(001) like in case of GaAs and InAs. It is interesting that the I adsorption is only minor preferable in the M_1-site over that in the S_2-site between adjacent surface dimers on the InP(001), however this difference is higher (0.26 eV) in case of GaP(001). The difference of 0.21-0.23 eV in I adsorption energies in the bridge dimer S_1-site and M_2-site were calculated whereas the adsorption of I in the M_2-site was found to be preferable of 0.10 eV than that in the S_1-site in case of GaAs [4]. Note that almost the same values were obtained for both sites on ζ-InAs(001)-(4×2) surface [3]. It is seen from table 1 that the smallest I binding energy in the M_3-site among considered on top sites above cation atoms (M_1-M_3) is consistent with results obtained for other semiconductors in [3]. Iodine adsorption on top sites above P is found to be less preferable.

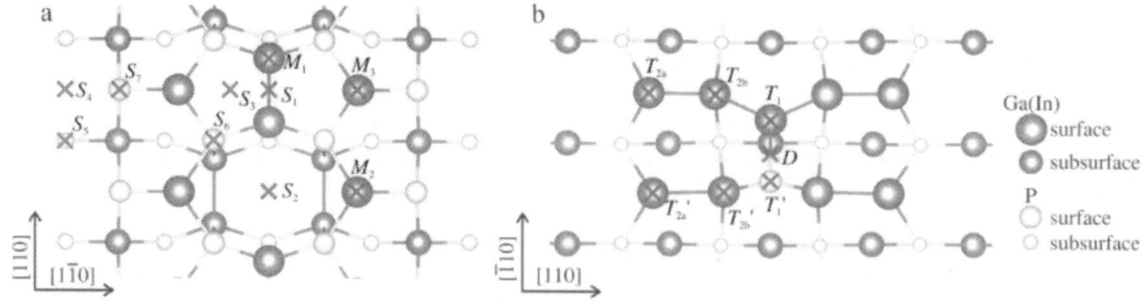

Figure 2. Iodine adsorption sites on ζ(4×2) (a) and mixed-dimer (2×4) (b) reconstructions.

Table 1. Iodine binding energies (eV) on ζ-(001)-(4×2) structure (figure 2a) for A^3B^5 semiconductors.

I sites	M_1	M_2	M_3	S_1	S_2	S_3	S_4	S_5	S_6	S_7
I-InP	2.43	1.99	1.76	2.20	2.41	1.79	1.31	0.81	1.33	1.66
I-GaP	2.19	1.79	1.53	2.02	1.93	1.53	0.74	0.83	1.01	1.52
I-InAs [3]	2.07	1.99	1.68	2.00	1.49	1.42	1.48	1.25	1.42	1.90
I-GaAs [3]	1.86	1.78	1.59	1.68	0.89	0.95	1.13	0.89	1.07	1.54

The interaction of iodine with semiconductor surface results in a large displacement of surface atoms. At all sites, except S_7, halogens pull the atoms involved into interaction from their initial position toward vacuum. A large displacement of the In dimer atom (~1.76 Å) toward vacuum in the M_1-site and opposite displacement (−0.39 Å) toward bulk in case of S_7-site is occurred. The displacements of In nondimerised atoms in the M_2 and M_3 sites are smaller (0.70-0.75 Å) as in case of I adsorption on InAs(001) surface [3]. Indium dimer lies by 0.43 Å higher than the three-fold coordinated P atoms when iodine is adsorbed in the bridge S_1-site. The interaction of iodine with InP(001) in the S_2-site results in a large displacement of surface In dimer as well. It locates by 1.07 Å higher that the corresponding P surface atoms whereas it lies by 0.70 Å lower than P surface atoms on pure InP(001) surface. In general, the atomic displacements of surface atoms in considered sites driven by halogens are slightly larger on ζ-InP(001)-(4×2) with respect to ζ-GaP(001)-(4×2). The same trend was found for InAs and GaAs on this reconstruction.

The adsorbate sites on the mixed dimer (2×4) reconstruction are basically cation ones (figure 2b). It is seen from table 2 that iodine binding energies on top sites above cation atoms are larger than that above P dimer atom. It is necessary to point out that iodine shifts differently during relaxation from T_{2b} and T_{2b}' sites which are nearest to mixed dimer: it relaxes to the bridge site between cation atoms from T_{2b}'-site nearest to P atom whereas I shifts slightly towards cation of the mixed dimer from T_{2b}-site. The bridge D-site (figure 2b) is found to be unstable because iodine relaxes towards T_1-site.

Table 2. Iodine binding energies (eV) on the mixed-dimer (2×4) structure (figure 2b) for InP and GaP semiconductors.

I sites	T_1	T_1'	T_{2a}	T_{2a}'	T_{2b}	T_{2b}'
I-InP	2.01	1.40	2.28	2.12	2.19	2.25
I-GaP	1.89	1.33	2.18	2.15	2.20	2.23

Finally, we considered I adsorption on the P-rich β2-InP(001)-(2×4) reconstruction. It was found also that the most preferential site for I adsorption is above In atom in the subsurface layer. The binding energy of 1.96 eV was calculated in this case. The analysis of the electron energy spectra and local densities of states shows similar behavior of the electron characteristics upon I adsorption as was found earlier [3, 4]. In particular, the charge transfer of 0.40-0.42e was calculated inside Bader volumes for the M_1-site on the $\zeta(4\times2)$ reconstruction and almost the same values was obtained in case of the T_1-site on the mixed dimer (2×4) reconstruction as well.

4. Conclusion

The present total energy calculations using PAW method within DFT confirm the stability of the mixed dimer (2×4) structure on both InP and GaP semiconductors (001) surface in the cation-rich limit. The existence of (2×4) surface structure in the cation-rich limit demonstrates a general tendency in minimization of the surface energy for $A^3B^5(001)$ surfaces. The $\zeta(4\times2)$ reconstruction has the smallest surface energy among considered surface structures with (4×2) geometry. Comparative study of I adsorption on InP(001) and GaP(001) surfaces with $\zeta(4\times2)$ and mixed dimer (2×4) reconstructions was performed at the first time. The on-top cation dimer M_1-site was found energetically most favorable for iodine adsorption on $\zeta(4\times2)$ reconstruction. Our study reveals an ionic mechanism of iodine bonding with the semiconductor substrate irrespective on the surface reconstruction.

Acknowledgment

The work was supported by the Russian Foundation for Basic Research under Grant 13-02-98017_r_a and Tomsk State University Competitiveness Improvement Program. Numerical calculations were performed on the SKIF-Cyberia supercomputer at the NR TSU.

References

[1] Xu Y B, Kernohan E T M, Freeland D J and Ercole A J A C 1998 *Phys. Rev.* B **58** 890
[2] Tereshchenko O E, Lamine D, Lampel G, Lassailly Y, Li X, Paget D and Peretti J 2011 *J. Appl. Phys.* **109** 113708
[3] Bakulin A V, Kulkova S E, Eremeev S V and Tereshchenko O E 2014 *J. Phys. Chem.* C **118** 20097
[4] Bakulin A, Eremeev S, Tereshchenko O, Chulkov E and Kulkova S 2011 *IOP Conf. Ser.: Mater. Sci. Eng.* **23** 012015
[5] Mikolajczyk P and Stankiewicz B 2009 *Cent. Eur. J. Phys.* **7** 279
[6] Mikolajczyk P and Stankiewicz B 2010 *Appl. Surf. Sci.* **256** 4822
[7] Tereshchenko O E, Eremeev S V, Bakulin A V and Kulkova S E 2010 *JETP Letters* **91** 466
[8] Blöchl P E 1994 *Phys. Rev.* B **50** 17953
[9] Kresse G and Joubert J 1999 *Phys. Rev.* B **59** 1758
[10] Kresse G and Furthmüller J 1996 *Phys. Rev.* B **54** 11169
[11] Perdew J P, Burke K and Ernzerhof M 1996 *Phys. Rev. Lett.* **77** 3865
[12] Eglitis R I 2014 *Int. J. Mod. Phys.* B **28** 1430009
[13] Shah G A, Radny M W, Smith P V and Schofield S R 2012 *J. Phys. Chem.* C **116** 6615
[14] Shiraishi K 199 *J. Phys. Soc. Jpn.* **59** 3455
[15] Monkhorst H J, Pak J D 1976 *Phys Rev B* **13** 5188
[16] Bakulin A V, Kulkova S E, Eremeev S V and Tereshchenko O E 2013 *Surf. Sci.* **615** 97
[17] Castleton C W M and Mirbt S 2003 *Phys. Rev.* B **68** 085203
[18] Schmidt W G 2002 *Appl. Phys.* A **75** 89
[19] Yamaguchi K, Takeda Y, Kameda K and Itagaki K 1994 *Mater. Trans.* **35** 596

IOP Conf. Series: Materials Science and Engineering **77** (2015) 012005 doi:10.1088/1757-899X/77/1/012005

The consideration of virial corrections in the diffusion equations

D. Levin and E. Trutnev
Tula State University, department of physics, Tula, Russia

E-mail: levin@physics.tsu.tula.ru, trutnevs@mail.ru.

Abstracts. The diffusion phenomenon is known to be a universal physical process and the basis of many industrial technologies for advanced materials. Abnormalities of diffusionhave been observed which cannot be described by existing theory. The diffusion of atoms of the introduction (boron) possessing great energy of dissolution, results in significant deformation of the crystal lattice of iron in the radius of the first coordination spheres. Mayer's approach to the solution of the problem of getting diffusion equation with viral corrections, indicating the account of interparticleinteraction in thise systems, has been offered in this work.In theongoing project the model of diffusion of boron impurities, strongly interacting with the lattice has beendescribed.Further directions in the solution of this problem have been suggested. To explain the anomalies observed in the diffusion mechanism of implementation a comparison of theoretical calculations has been attempted, in particular, in the system iron-boron.

1. Introduction

Diffusion is one the most general physical processes and phenomena underlying at the basisa lot of transformations and processes in solids. The ordinary diffusion theory does not explain a number of effects during the diffusion of the interstitial atoms with high dissolution energy, causing a considerable deformation of the crystal lattice in the first coordination spheres. Such atoms strongly interact with the nearby atoms of the solvent crystal lattice and with each other, that becomes considerably apparent in a number of special effects [1], particularly, these are anomalies of atom redistribution within the diffused zone, a special character of the influence of lattice defects on the diffusion kinetics, diffusion under tensionand so on. The account of the interatomic interaction influence on the diffusion kinetics is of great significance due to the small sizes of the diffused zones with great magnitudes of concentration gradients, typical of diffusion processes on a micro- and submicron-scale.

At present there are some experimental diffusion data in the systems with limited solubility Fig.1 [1, 2]. In the systems with even little chemical impurity concentration one should not neglect the interaction of the impurity atoms with each other. It is necessary to take into account the interaction of the solvent atoms, forming the crystal lattice in which the diffusion process takes place.

Indeed, in many cases, by the study of diffusion redistribution of components of the solid solution anomalies have been detected whichhaven't been receivedwith help of existing diffusion theory. For example, in the references [1, 2] the results obtained in the system with limited solubility iron-boron have been presented. While on one hand, in [1] it hasn't been revealed two phases. Having been shown in Fig. 1 the sectionof the diffusion zone is a solid solution and 'the jump'on $C(x)$ can't be explained within the theory of diffusion.

The objective of this work is to identify models of impurity diffusion, strongly interacting with the lattice while on one hand, on the other hand the clarification of the connection diffusion equation by the configuration integral and the interatomic interaction. An attempt has been made to compare

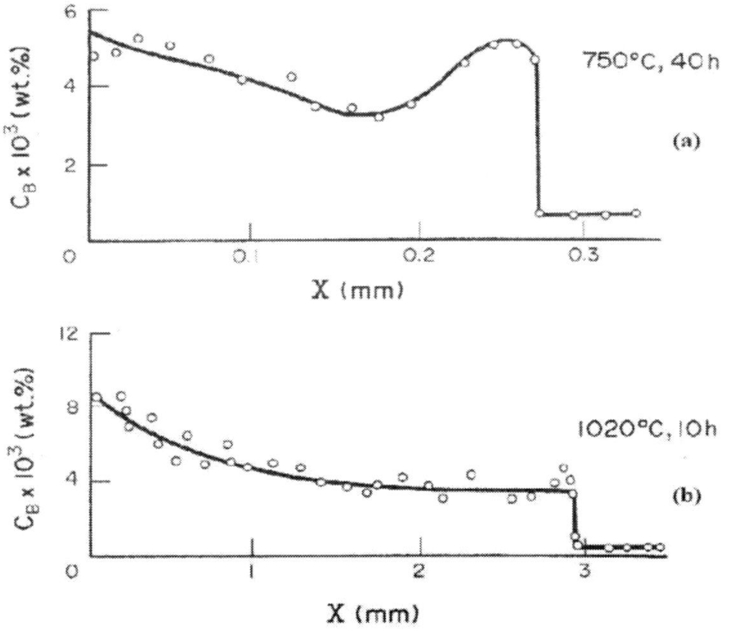

Fig.1. The boron distribution as a function of the diffusion layer depth in iron after annealing in the temperature ranges α-Fe (a) and γ-Fe (b) [1]

theoretical calculations for the explanation of the anomalies observed in the diffusion of interstitial mechanism (the system iron-boron).

2. Theory

To study the diffusion processes in the complex systems, which properties are obviously different from the model of ideal gas, a number of approaches has been developed by present time. For example, in accordance with the theory of Mayerfor concentrated solid solutions [3] the expression for the chemical potential is used

$$\mu(T,p,n_i) = \mu_i^0(T,p) + kT\ln a(n_i), \quad (1)$$

where instead of the concentration of n_i the activity of component $a(n_i)$ has been introduced, depending on the concentration of n_i and other state parameters and the standard chemical potential μ_i^0 is the function that depends on the temperature T and pressure p.

Being based on Mayer approach [3], the expression for the chemical potential of such systems was obtained, which must contain, in addition summand chemical potential of an ideal classical gas, the member, expressing the interatomic interaction. The eq. of state in the form of a virial expansion is known [3, 4] to have the form of eq.

$$\frac{pv}{\theta} = 1 - \sum_{n\geq 1} \frac{n}{n+1}\beta_n \rho^n, \quad (2)$$

where $\rho = \dfrac{N}{V}$ is the density of particles (atoms), β_n are the irreducible group integrals (virial coefficients), p is the pressure, $\theta = kT$, where k, is Boltzmann's constant.

Virial coefficients correspond to the interactions between two, three, etc. molecules sequentially. We have reduced the problem ofvirial equation of state for solid solution consisting of N particles to the number of special problems of interaction between two, three particles, and etc., similarly as the problemswith gases aresolved.

Our calculation has been limited with the first virial coefficient, considering only pairwise interactions. According to [3] the chemical potential in the form of a virial decomposition has the form

$$\mu = \mu_0 - \theta\sum_{n\geq 1} \beta_n \left(\frac{N}{V}\right)^n, \quad (3)$$

where $\mu_0 = \theta \ln C$ is the standard chemical potential, $C = \dfrac{N}{V}$ the concentration of the substance.

Solid implementation solutions with bcc lattice on the basis of α-Fe, metals V and VI groups are characterized with a strong distortion created by the interstitial atoms and significantly features of its mechanical and physical properties [5].

The Morse potential has been chosen alternatively the potential of the Lennard-Jones (LJ). The potential of LJ is two-parameter, so it has very limited possibilities for variations of the macroscopic parameters of the simulated material. In fact, it can satisfy the value of only one macroscopic parameter. It can be, for example, the modulus of elasticity or ultimate strength in statics, the speed of propagation of longitudinal waves etc. On the other hand, this potential describes the properties of some substances (primarily crystalline inert gases) and the interaction strength of the Van der Waals forces very accurately. The potential of LJ is widely used as a classical model potential, especially in the works, which main objective is the study of General physical laws without getting the exact quantitative results.

The Morse potential has three parameters that allows to satisfy the value of two independent macroscopic parameters, such as modulus of elasticity and ultimate strength in statics. The simplicity and clarity of calculating the parameters of the potential makes it very convenient to use. Another advantage compared with the potential of LJ is faster decay at a distance, which is handy if you are modeling it is necessary to consider the interaction of only the nearest particles. The Morse potential is given by formula [5]:

$$\Phi(r) = D\left[e^{-2\aleph(\frac{r}{b}-1)} - 2e^{-\aleph(\frac{r}{b}-1)}\right]. \tag{4}$$

This potential is different from LJ potential by a faster (exponential instead of power) decrease up to the infinity, it hasn't singular point in zero, but what is more important in the considered case is the presence of the dimensionless parameter $\aleph$. The definition of all three parameters of the (4): $\aleph$ defines the width of the potential well, D is the depth of the potential well (the minimum potential energy of interaction between two atoms of boron), b is the minimum distance between a pair of particles, when the force of interaction is zero, i.e., $b = r_{\min}$.

According to [2] the second group of the integral is defined by expression:

$$b_2 = \frac{1}{2} \cdot 4\pi \int_0^{\infty} r^2 f(r) dr,$$

where $f(r) = e^{-\Phi(r)/\theta} - 1$ - Mayer function. $\Phi(r)$ is the Morse potential.

Performing Mayer function expansion, we have transformed the expression, standing under the integral to mind

$$r^2 f(r) = \sum_{n=1}^{\infty} \frac{r^2(-1)^n}{\theta^n} \cdot \frac{\Phi^n(r)}{n!}.$$

Performing some mathematical calculations, we have obtained the expression for the first virial coefficient

$$\beta_1 = 2b_2 = -\frac{4\pi b^3}{\aleph^3} S, \tag{5}$$

where $S = \sum_{(n)} \frac{(-1)^n D^n}{\theta^n} \sum_{k=0}^{n} \frac{(-2)^{k+1}}{k!(n-k)!(2n-k)^3} \cdot e^{\aleph(2n-k)}$ and the symbol $\sum_{(n)}$ denotes summation over nearest environment of the boron atom in the bcc lattice of iron.

According to (3) the chemical potential of our system will be determined

$$\mu = \theta \ln C + \frac{4\pi b^3 \theta}{\aleph^3} S \cdot C, \tag{6}$$

i.e., the expression (6) can be represented in the form

$$\mu = kT \ln C - kT\beta_1 C. \tag{7}$$

Using formulas, which relate the concentration in mass percent concentration in mass relative units, we have found the number of boron atoms in 1 m^3 of iron. The minimum distance between B atoms in the lattice of Fe has been found by the formula: $r_{\min} = \sqrt[3]{\frac{1\,\text{m}^3}{N_B}} = 2.604 \cdot 10^{-9}\,\text{m}$,

where $N_B = 5.66 \cdot 10^{25}$ is the number of boron atoms in the lattice of iron, $r_{\min} > a$ where a is the lattice parameter of iron, $a = 2.866$ Å [6].

As a variant of the consideration of interaction of the boron atom with neighboring atom of iron we have chosen the Morse potential (with the introduction of additional screened multiplier for forces of attraction).

Researching convergence of these series (5) we can see, that d'Alamberts criterion must converge. But in (5) we get divergent series. Which feature is that by involution $\Phi(r)$ in n the (5) contents multiplier $e^{\aleph(2n-k)}$ increases with n ($n \geq k$). To construct a integrand function $r^2 f(r)$, we shall see a big and thin peak by $r = r_{\min}$. Considering its area we get a very big value, that we can't accept because it hasn't got any physical sense. As a rule for Morse's potential cut-off radius is entered, that means that the distance where there is an interaction, it is zero. Using cut-off radius that decreases the number of neighbours, interacting with every atom.

We do screening of Morse's potential (multiply component, which is resposable for attracting forces to screened multiplier $e^{-\lambda r}$). Such interstitial impurity like boron, creates very strong mechanical tensions in crystal lattice of iron [2, 7]. Radius of action of these tensions are several $r_{\min}$, the interaction of the atoms is assumed to depend from the interatomic distances [8]. That's why by numerical calculations β_1 we must do the integration to non-infinitum, but to the order $r = 3r_{\min}$. The calculation from the three spheres is optimal, because the energy contribution over three spheres is very small [8, 9]. In this case the procedure of numerical calculation in Maple allows to our mind, not to miss this peak, because it is very narrow. Taking the screening for (4) we have been got

$$\Phi(r) = D\left[e^{-2\aleph\left(\frac{r}{r_{\min}}-1\right)} - 2e^{-\aleph\left(\frac{r}{r_{\min}}-1\right)} \cdot e^{-\lambda r}\right], \tag{8}$$

Where D is the minimal potential energy of interaction between boron atom and iron atom. In studying of sources, we haven't identified any specific values for the minimum potential energy of interaction between atoms of iron and boron. But on the basis the average interaction energy, we have assumed that $D = 10 \cdot 10^{-19}$ J [10, 11]. Parameter λ is found by inspection. Grüneisen constant is known [5] to be connected with parameter $\aleph$ by the relation $\Gamma = \frac{\aleph}{2}$. Taking for boron Γ = 3, then $\aleph$ has been used by the simulation $\aleph = 2\Gamma = 6$.

The calculation has been performed in Maple. Let us choose $\lambda = 1.95 \cdot 10^7$, $\beta_1 = 212.3916$. The graphics of potential $\Phi(r)$ (8) and the integrand function $r^2 f(r)$ are presented in Fig. 2 and 3. β_1 corresponding to the interaction between the iron atom and boron is $\sim 10^{-14}$, so that we neglect this interaction.

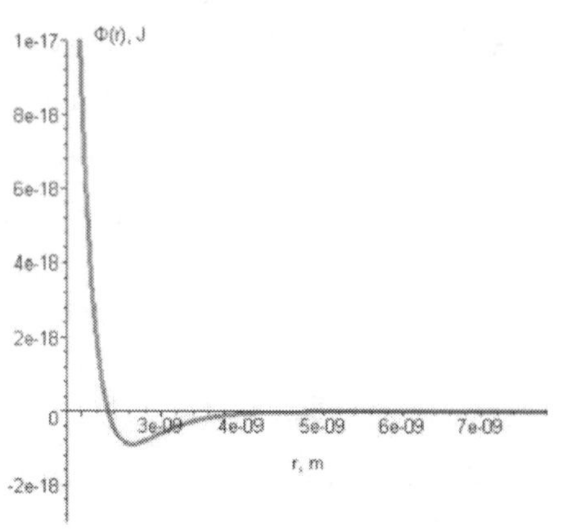

Fig.2 The graphic of Morse potential (8)

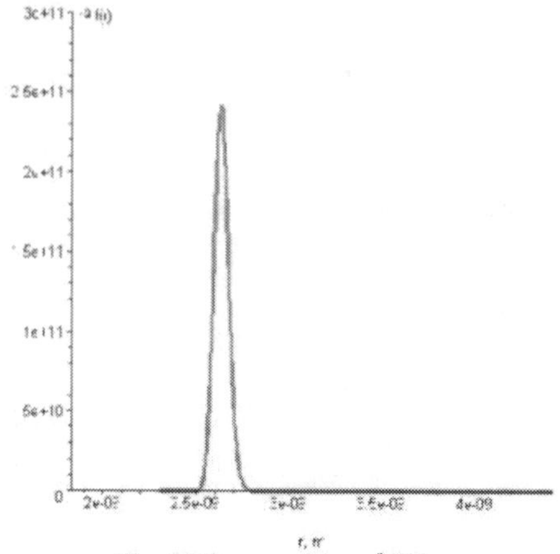

Fig.3 The graphic $r^2f(r)$

3. Results

According to the chosen approach, the diffusion equation has the form

$$\frac{\partial C}{\partial t} = D\frac{\partial}{\partial x}\left((1-\beta_1 C)\frac{\partial C}{\partial x}\right). \qquad (9)$$

The initial and boundary conditions have the form

$$C(x,0) = \begin{cases} C_1, x \le 0 \\ 0, x > 0 \end{cases};\ C(0,t) = C_1,\ C(1,t) = 0.$$

The solution of (9) is presented in Fig. 4. When the initial concentration impurity is $C_1 = 5\cdot 10^{-3}, 5\cdot 10^{-4}$, respectively.

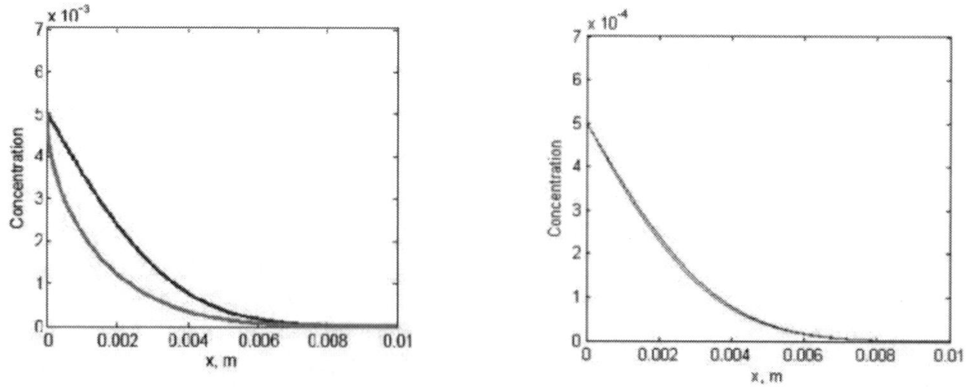

Fig.4 The solution of (9) when $\beta_1 = 0$ (blue) and $\beta_1 = 200$ (red)

Its solution is slightly different from Fick's diffusion equation. In the framework of this model a linear dependence of the diffusion coefficient on the concentration has beenobtained.We have restricted by the pair interaction, i.e. in the virial series is one component of β_1. This result is evident in the approximation of weak interaction of the diffusion of atoms with the atoms of the lattice. For calculating a diffusion coefficient of boron in iron $D = 2.7\cdot 10^{-10}$ m^2/s has been taken [12]. In our calculation $\beta_1 C \ll 1$.

4. Conclusions

The problem of the interaction between the two boron atoms among the atoms of iron must be considerate. These two atoms are not neighbors, i.e. they do not interact directly. This problem is described in the framework of a phenomenological theory of diffusion, we have used the same Mayer approach to calculate the chemical potential of the impurity atom and the virial coefficient of this impurity in a solid. The procedure for obtaining expressions for $\mu = \left(\frac{\partial F}{\partial N}\right)_{\theta V}$ and F is the free energy should be modified for the case of a solid body. Our next task is to obtain expressions virial coefficient in the solid, but on the basis of the statistical sum of a defect-free crystal (for phonon gas), considering that during the interaction of phonons contribute to the interaction of atoms.

This work was supported by the grant of RFBR № 13_08_97545_r_center_a

References

[1] Grinberg E M 1992 *The application of track autoradiography for studying diffusion of boron in iron*, Nucl. Tracks Radiat. Meas., **20**, No. 2 pp 273-276
[2] Grinberg E M 1997 *Metal science of structural boron steel* (Moscow: Misis) pp 58-67 [Гринберг Е.М. Металловедение борсодержащих конструкционных сталей]
[3] Mayer J, Goeppert-Mayer M 1980 *Statistical Mechanics* (Moscow: Mir) pp 259-266, (in Russian)
[4] Kvasnikov I A 2002 *Statistical Physics*, **2** (Moscow: URSS) pp 390-393, (in Russian)
[5] Krivtsov A M 2009 *Theoretical mechanics. Elastic and thermal properties of ideal crystals* (Sanct-Pitersburg: Izdatelstvo Politehnicheskogo universiteta) pp. 143, (in Russian)
[6] Drits M E 1985 *The properties of elements*: reference book (Moscow: Metallurgy) p. 463, (in Russian)
[7] Yeremeev V S 1984 *Diffusion and tension* (Moscow: Energoatomizdat) pp 27-30, (in Russian)
[8] Yashin A V, Sinitsa N V 2008 Studying of areas of superstructure damage under uniaxial dynamic deformation in the alloy Ni_3AL. Proceedings of the V International scientific - technical conference 10 - 13 November 2008, MOSCOW YOUNG SCIENTISTS – 2008 part 3 MIREA p. 160-163, (in Russian)
[9] Tkachev P V, Krivtsov A M 2012 *Using of the Morse potential to describing of the dependence of slabbing strength of metals on the rate of deformation*. Nanosystems: Physics, Chemistry, Mathematics, 3(5), pp 0-5, (in Russian)
[10] Iskakov B M, Bakranova D I 2011 *The parameters of the Morse potential bcc metals* (Moscow: Workshop the XXI International meeting on' Radiation Solid State Physics") pp 701-708, (in Russian)
[11] Zaliznyak V E 2012 *Interatomic interaction in metals having a volume-centered cubic lattice*. Nanosystems: Physics, Chemistry, Mathematics, 3(6), pp 64-69, (in Russian)
[12] Grigorjev I S, Meilichov E Z 1991 Physical quantities: reference book (Moscow: Energoatomizdat) p. 1232, (in Russian)

Silver nanoprisms self-assembly on differently functionalized silica surface

J Pilipavicius, A Chodosovskaja, A Beganskiene and A Kareiva

Vilnius University, Faculty of Chemistry, Department of Inorganic Chemistry Naugardukas str. 24, Vilnius, Lithuania LT-03225

E-mail: Jurgis.pilipavicius@chf.vu.lt

Abstract. In this work colloidal silica/silver nanoprisms (NPRs) composite coatings were made. Firstly colloidal silica sols were synthesized by sol-gel method and produced coatings on glass by dip-coating technique. Next coatings were silanized by (3-Aminopropyl)triethoxysilane (APTES), N-[3-(Trimethoxysilyl)propyl]ethylenediamine (AEAPTMS), (3-Mercaptopropyl)trimethoxysilane (MPTMS. Silver NPRs where synthesized via seed-mediated method and high yield of 94 ±15 nm average edge length silver NPRs were obtained with surface plasmon resonance peak at 921 nm. Silica-Silver NPRs composite coatings obtained by self-assembly on silica coated-functionalized surface. In order to find the most appropriate silanization way for Silver NPRs self-assembly, the composite coatings were characterized by scanning electron microscopy (SEM), dynamic light scattering (DLS), water contact angle (CA) and surface free energy (SFE) methods. Results have showed that surface functionalization is necessary to achieve self-assembled Ag NPRs layer. MPTMS silanized coatings resulted sparse distribution of Ag NPRs. Most homogeneous, even distribution composite coatings obtained on APTES functionalized silica coatings, while AEAPTMS induced strong aggregation of Silver NPRs.

1. Introduction

Noble metal nanoparticles have been investigated and applied in many research areas including chemical catalysis [1], biochemistry [2, 3], optics [4-6], and electronics [7]. Currently, scientists are focused on anisotropic Silver and Gold nanoparticles properties and their application in various fields [8-10]. There are many references about synthesized silver and gold nanocubes [11], nanododecahedron [12], nanodiscs [13] and nanoprisms [9]. Silver nanoprisms (NPRs) are one of the most interesting nanoparticle type, because of wide tuneability and high intensity of surface plasmon resonance (SPR) band [14, 15]. Manipulating particle thickness and edge length, it is possible to obtain SPR peak position in 400 – 1300 nm range, when conventional spherical particles can only exceed up to 600 nm [16]. These properties opens the gateway for application in near infrared working technologies, like telecommunications [16] or surface enhanced Raman spectroscopy (SERS) [17].

Our research is focused on synthesis of Silver nanoprism composite materials. One of the approaches are silica/Ag NPRs coatings. Currently many papers are available on production various types of composites with Ag NPRs. Most of them are about integration those particles to different polymer matrices [18, 19], but still lacking references on use of inorganic matrix. Our selected way is make silica-silver nanoprism coatings by self-assembly of particles on silanized silica surface. *Xue, C.,*

IOP Conf. Series: Materials Science and Engineering **77** (2015) 012006 doi:10.1088/1757-899X/77/1/012006

et al. [20, 21] already did some research on this topic. However, information about how different surface affects Ag NPRs self-assembly are still missing.

In this work we investigated how differently prepared silica surface affects assembly of silver NPRs. Therefore we synthesized a Silver NPRs with average edge lengths of 94 ±15 nm, made silica coatings by sol-gel technique, silanized surfaces using different APTES, AEAPTMS MPTMS, and deposited NPRs on silanized surface. Produced composite coatings were evaluated by scanning electron microscopy (SEM) and UV-VIS spectroscopy.

2. Experimental

2.1. Chemicals and Materials

Tetraethyl orthosilicate (TEOS, 98%), (3-Aminopropyl)triethoxysilane (APTES, 99%), N-[3-(Trimethoxysilyl)propyl]ethylenediamine (AEAPTMS, 98%), (3-Mercaptopropyl)trimethoxysilane (MPTMS 99%), Silver nitrate ($AgNO_3$, 99%), Poly(sodium 4-styrenesulfonate) (PSSS, 1000000 Mw), Sodium Borohydride ($NaBH_4$, 99%), L-Ascorbic Acid ($C_6H_8O_6$, 99%) were purchased from Sigma-Aldrich and used without further purification. Absolute ethanol (99.9 %) prepared by refluxing with preheated CaO, and after distilled.

2.2. Synthesis of silver NPRs.

Silver NPRs were made by seed-mediated method which include two steps – producing silver nanoseeds and nanoprism growing.

Making silver seeds. In this experiment, silver seed are obtained by mixing aqueous trisodium citrate (20 mL, 25 mM), PSSS (1 ml, 0.5 g/L), freshly prepared $NaBH_4$ (1.2 mL, 10 mM) solutions, followed by addition of aqueous $AgNO_3$ (20 mL, 0.5 mM). Addition rate of $AgNO_3$ was 2 mL/min. The rate was maintained using syringe pump.

Growing of NPRs. Silver NPRs are produced by mixing 35 ml distilled water, aqueous ascorbic acid (0.525 mL, 10 mM) solution, 0.560 mL of prepared silver seeds solution, followed by addition of aqueous $AgNO_3$ (21 mL, 0.5 mM) at a rate of 1 mL/min. Obtained Silver seeds and NPRs solutions characterized by UV-VIS spectroscopy (Perkin Elmer Lambda 35), Scanning Electron Microscopy (SEM, Hitachi FE-SEM SU-70).

2.3. Synthesis of colloidal silica and coating procedure

Colloidal silica sols in ethanol were made by base catalyzed TEOS hydrolysis and condensation. For that purpose TEOS, Ammonia, Water, and Ethanol were mixed with molar ratio 1:0.2:2.3:38 respectively. Sols were left at room temperature for 10 days. Particle size was determined by Dynamic Light Scattering (DLS, Malvern Instruments Zetasizer Nano ZS) method.

For coating process microscope soda-lime glass slides were used as substrate. Substrates were cleaned by immersing them in piranha solution (volume ratio 1:4 of H_2O_2 and H_2SO_4 respectively) for 15 min. Then substrates were washed with deionized water few times, and dried by leaving in desiccator for 24 h at room temperature. Coatings were made by Dip-Coating technique (KSV Dip Coater D). For all samples were used constant 40 mm/min withdrawal speed. Coated samples were ready for the next step – surface silanization.

2.4. Surface silanization and deposition of Silver NPRs.

Thermal silanization. Coatings and uncoated substrates were functionalized by alkoxysilanes – APTES, AEAPTMS and MPTMS. Same silanization procedure were used for all silanes. In first step three different 3% alkoxysilane solutions in toluene were prepared. Then substrates were immersed in prepared solution and refluxed for 12 hours under argon atmosphere. After silanization procedure, substrates were washed several times with pure toluene and acetone.

For deposition of Silver NPRs, functionalized substrates were immersed in previously prepared Silver NPRs colloidal solution and kept for 7 days. The composite coatings were gently washed with distilled water and left to dry in room temperature.

3. Results and discussions

3.1. Synthesis of Silver NPRs

For silver nanoprism synthesis was selected the seed-mediated method, which was previously suggested by Aherne, D., et al. [14]. Synthesized particles by this method are sufficiently monodisperse, and comparing with photoconversion method [22] is performed faster and easier.

In synthesis method Sodium citrate and PSSS were used as particle shape determining materials. Silver NPRs were synthesized in high yield, with edge length about 94 ±15 nm. Average particle sizes were calculated form SEM micrographs (fig 1.) As can be seen from fig. 1, obtained particles are plate-like structure and almost all have trigonal or hexagonal shape, but varies in size. Particle shape and size plays significant role in their optical properties. SPR band of high aspect ratio Ag particles (plate-like structures) like nanoprisms, nanohexagons or nanodiscs generally are in longer wavelengths (700-1200 nm). Higher aspect ratio of particle gives red-shifting of SPR band. In case of Silver nanoprisms, significant role plays tip length of the prism – sharper prisms give more red-shifted SPR band, and duller – blue-shifted. Therefore nanohexagons also can be considered as prisms having dull tips.

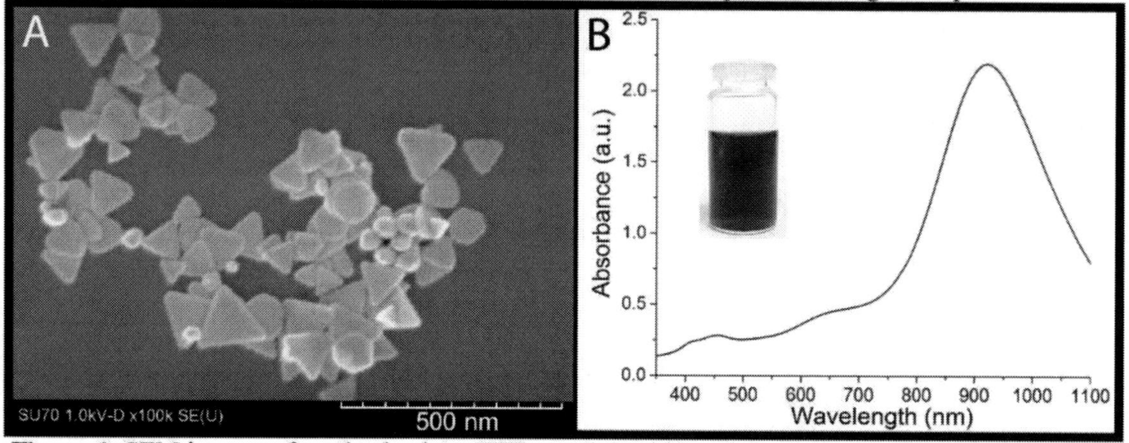

Figure 1. SEM images of synthesized Ag NPRs particles (A) and absorption spectra of synthesized Ag NPRs solution (B)

Obtained colloidal Silver NPRs solution are relatively stable and no significant changes were observed in few months. Single SPR band (fig. 1) shows high intensity absorption at 921 nm, which gives a blue color of NPRs colloidal solution. Red-shifted SPR peak indicates presence of high aspect ratio anisotropic particles like triangular NPRs in solution. SPR peak is slightly broadened, but it is common for all plasmonic particles which has absorption in near infrared [8].

3.2. Synthesis of colloidal silica, fabrication of coatings and surface silanization

As mentioned in experimental section, in this research base catalyzed silica sol were used. During this reaction, usually forms silica spheres which diameter can be varied form nanometer to submicrometer scale. Exact size of the spheres is controlled by initial ratio of starting materials [23]. Accordingly, 3% (w/w) monodisperse colloidal silica in ethanol, with 30 ±11 nm average sphere diameter were made by selecting appropriate ratio of standing materials [24]. After sol synthesis, coatings were formed using dip-coating method and samples were kept for next step – surface functionalization.

In this work we selected thermal silanization method. By this technique, ideally, only one layer of silane is formed on coating surface. Coated substrates functionalized by – APTES, AEAPTMS and MPDMS. Coatings made of 20-30 nm silica particles have higher surface area and more hydroxyl groups comparing with activated bare glass surface which leads to higher amount of attached silane molecules during silanization procedure. This surface silanization method is well-studied, however silanized surface must be characterized to make sure, that surface functionalization were succeeded. In our case, to make sure that silanization process was successful, water contact angles (CA) and surface free energy (SFE) measurements were made (table 1).

Table 1. Measured water contact angle and SFE values of silanized coatings

Sample name	Modifying silane	Contact angle, deg.	Dispersive Surface free energy, mN/m	Polar Surface free energy, mN/m	Total Surface free energy, mN/m
JV011	None	40.20	36.80	24.27	61.07 ±4,45
JVm1	APTES	69.75	31.16	9.82	40.98 ±5,09
JVm2	AEAPTMS	66.75	25.94	13.55	39.48 ±4,60
JVm3	MPTMS	54.84	37.55	14.94	52.49 ±5,31

For SFE measurements we selected Harmonic Mean (Wu) method which gives more accurate results when surface have high SFE values. In order to calculate surface free energy, in contact angle measurements we used four solvents – diiodomethane, water, glycerol and ethylene Glycol.

As can be seen form table 1, all silanized coatings have higher CA and smaller total SFE values, comparing with non-silanized coating. As result of high amount of polar hydroxyl groups on surface, highest wetting properties have non-functionalized silica films. Due to polar amine groups, APTES and AEAPTMS silanized coatings must give better wetting properties, but grafted carbon chain can give higher water contact angle value and smaller surface free energy values. MPTMS modified coatings has slightly better wetting properties and higher surface free energy comparing with APTES and AEAPTMS functionalized coatings.

3.3. Silver NPRs self-assembly on silanized surfaces

In this section we investigated silver NPRs self-assembly on differently silanized silica coatings. We tried to find out which silanized surface is more suitable for assembly of NPRs. Thus, as mentioned in experimental section, in order to obtain self-assembled silver NPRs layer, functionalized coatings were vertically immersed in plastic vials with containing silver NPRs solution and kept for 1 week at room temperature. After this procedure, samples were ready for UV-Vis measurements.

Firstly, we took pictures of samples (fig. 2) to visually evaluate how silver NPRs assembled on silanized silica coatings.

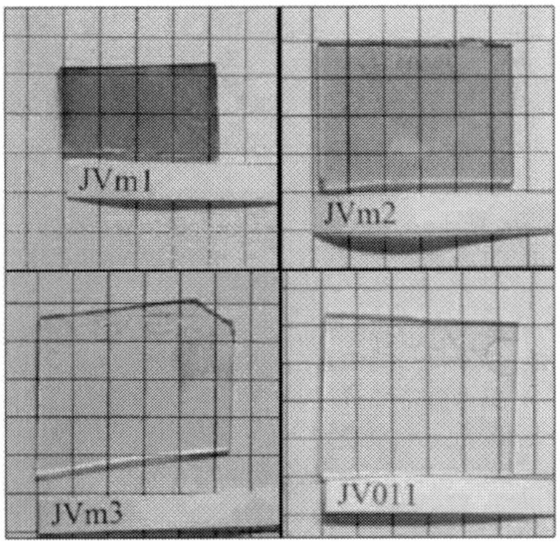

Figure 2. Images of Ag-silica composite coatings

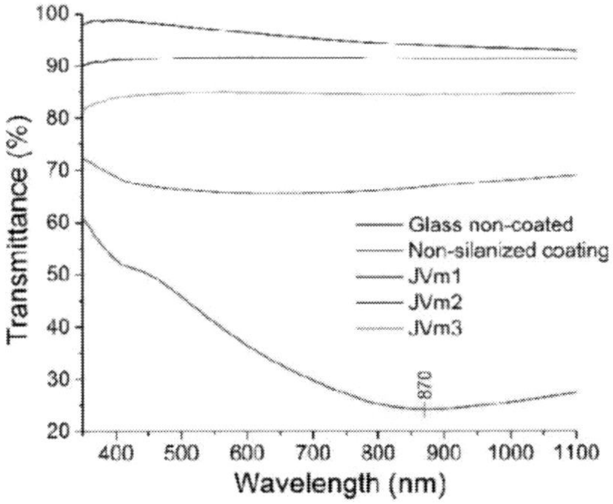

Figure 3. Transmittance spectra of Ag-silica composite coatings

On APTES silanized coating layer of assembled Ag nanoparticles is clearly visible. The latter layer gives intensive blue color. In UV-Vis (fig. 3) transmittance spectra's is visible strong absorbance around 870 nm in case of APTES functionalized coating. Also are noticeable, that this SPR peak is broadened and blue shifted comparing with the Ag NPRs solution. This may be caused by change of surrounding media of the particle – dielectric environment and refractive index of surrounding affects position of SPR [25]. In case of AEAPTMS functionalized coatings, Ag nanoparticles layer become blue-greyish. It can be result of large amount of primary and secondary amine groups, which may induce stronger attraction and later agglomeration of Ag nanoparticles. Thus, no SPR peak, only broad absorbance are visible in UV-Vis spectra of this sample. Ag NPRs layer is barely visible on MPTMS modified and unmodified coatings.

In addition coatings were investigated by SEM. SEM images of differently silanized-Ag NPRs self-assembled coatings are presented in figure 4.

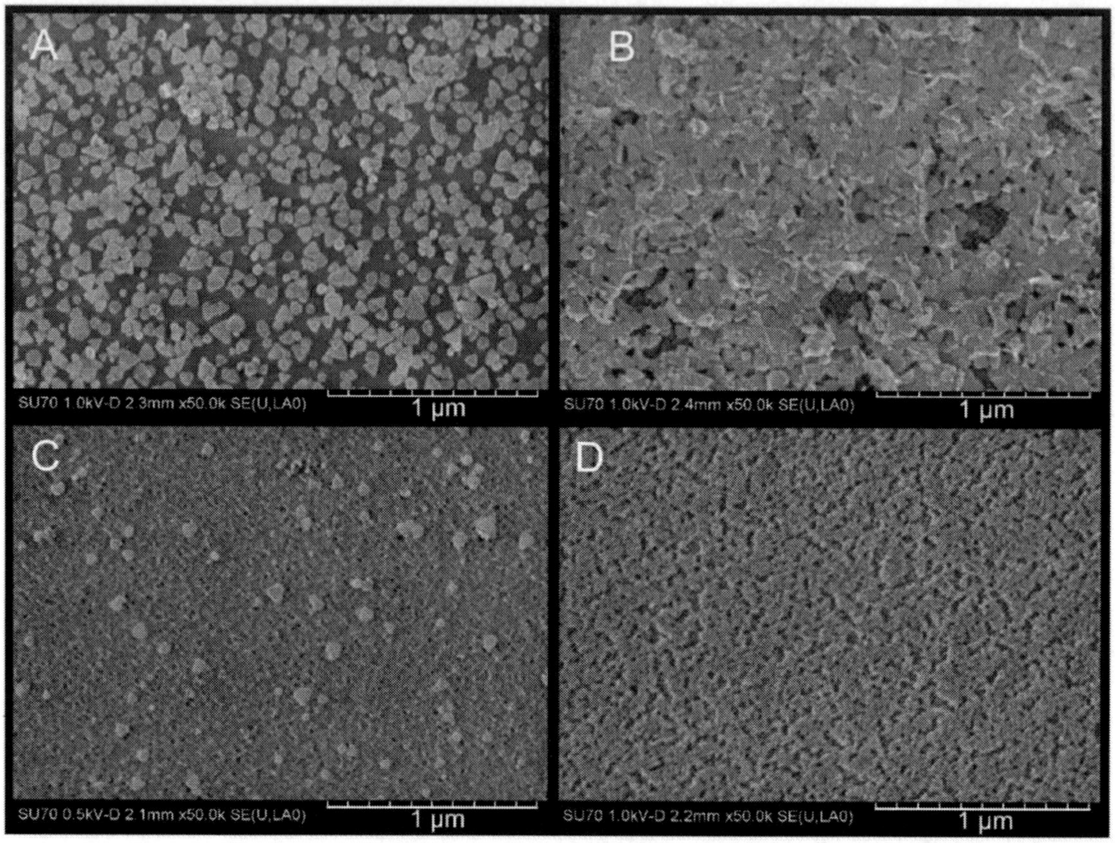

Figure 4. SEM images of Ag-silica composite coatings: silanized by APTES (A), AEAPTMS (B), MPTMS (C) and non-silanized (D)

Most even arrangement of Ag NPRs are visible on APTES functionalized coating – particles are arranged mostly as monolayer and noticeable aggregation is minimal. The image of AEAPTMS functionalized surface shows aggregation of particles and uniform layer no longer forms. Consequently, this layer of Ag NPRs looks grayish and no SPR is observable in UV-Vis spectra. Only separate, wide spread particles are observed on MPTMS functionalized coating. The UV-Vis spectra data show weak absorbance of unmodified coating and no Ag NPRs are observed on SEM image.

4. Conclusions

This research showed that most homogeneous silica-Ag NPRs coatings are obtained when surfaces were silanized with APTES. However, AEAPTMS silanized coatings tend to aggregation of silver NPRs, while MPTMS functionalized coatings have low affinity to Ag NPRs. Nevertheless, we successfully made functionalized silica coatings by sol-gel method and formed Ag NPRs layer on them via self-assembly process. This type of coatings can be applied for further research in SERS, or used for making multilayer coatings to obtain Ag NPR- silica embedded matrixes. On the other hand, the influence of other surface parameters to Ag NPRs self-assembly, like surface roughness, silanization method and etc., should be further investigated.

Acknowledgment

Postdoctoral fellowship of A.C. is being funded by European Union Structural Funds project "Postdoctoral Fellowship Implementation in Lithuania" within the framework of the Measure for Enhancing Mobility of Scholars and Other Researchers and the Promotion of Student Research (No. SF-PD-2012-12-31-0397) of the Program of Human Resources Development Action Plan.

References

[1] Signori, A.M., K.D. Santos, R. Eising, B.L. Albuquerque, F.C. Giacomelli, and J.B. Domingos 2010 *Langmuir* **26** 17772-17779
[2] Abalde-Cela, S., P. Aldeanueva-Potel, C. Mateo-Mateo, L. Rodriguez-Lorenzo, R.A. Alvarez-Puebla, and L.M. Liz-Marzan 2010 *J R Soc Interface* **7** S435-S450
[3] Becker, J., A. Trugler, A. Jakab, U. Hohenester, and C. Sonnichsen 2010 *Plasmonics* **5** 161-167
[4] Chen, X., J. Tao, G. Zou, W. Su, Q.J. Zhang, and P. Wang 2010 *Chemphyschem* **11** 3599-3603
[5] Gunn, J.M., S.H. High, V.V. Lozovoy, and M. Dantus 2010 *J Phys Chem C* **114** 12375-12381
[6] Torres-Torres, C., N. Perea-Lopez, J.A. Reyes-Esqueda, L. Rodriguez-Fernandez, A. Crespo-Sosa, J.C. Cheang-Wong, and A. Oliver 2010 *Int J Nanomed* **5** 925-932
[7] Matsui, I. 2005 *J Chem Eng Jpn* **38** 535-546
[8] Xiao, J. and L. Qi 2011 *Nanoscale* **3** 1383-1396
[9] Millstone, J.E., S.J. Hurst, G.S. Métraux, J.I. Cutler, and C.A. Mirkin 2009 *Small* **5** 646-664
[10] Cortie, M.B. and A.M. McDonagh 2011 *Chem. Rev.* **111** 3713-3735
[11] Kundu, S., S. Panigrahi, S. Praharaj, S. Basu, S.K. Ghosh, A. Pal, and T. Pal 2007 *Nanotechnology* **18**
[12] Wang, D.W., Y. Liu, and T.Y. You 2010 *Crystengcomm* **12** 4028-4030
[13] Jayabalan, J., A. Singh, R. Chari, and S.M. Oak 2007 *Nanotechnology* **18**
[14] Aherne, D., D.M. Ledwith, M. Gara, and J.M. Kelly 2008 *Adv Funct Mater* **18** 2005-2016
[15] Métraux, G.S. and C.A. Mirkin 2005 *Adv. Mater.* **17** 412-415
[16] Bastys, V., I. Pastoriza-Santos, B. Rodríguez-González, R. Vaisnoras, and L.M. Liz-Marzán 2006 *Adv Funct Mater* **16** 766-773
[17] Yi, Z., X. Xu, X. Wu, C. Chen, X. Li, B. Luo, J. Luo, X. Jiang, W. Wu, Y. Yi, and Y. Tang 2013 *Appl. Phys. A* **110** 335-342
[18] Berta, M., B. Loppinet, D. Vlassopoulos, A. Askounis, V. Koutsos, I. Pastoriza-Santos, and L.M. Liz-Marzan 2012 *Polymer* **53** 5771-5778
[19] Salvador, M., B.A. MacLeod, A. Hess, A.P. Kulkarni, K. Munechika, J.I.L. Chen, and D.S. Ginger 2012 *Acs Nano* **6** 10024-10032
[20] Xue, C., X. Chen, S.J. Hurst, and C.A. Mirkin 2007 *Adv. Mater.* **19** 4071-4074
[21] Xue, C., Z. Li, and C.A. Mirkin 2005 *Small* **1** 513-516
[22] Xue, C. and C.A. Mirkin 2007 *Angew Chem Int Edit* **46** 2036-2038
[23] Boonstra, A.H. and T.N.M. Bernards 1989 *J. Non-Cryst. Solids* **108** 249-259
[24] Pilipavicius, J., I. Kazadojev, A. Beganskiene, A. Melninkaitis, V. Sirutkaitis, and A. Kareiva 2008 *Mater Sci-Medzg* **14** 283-287
[25] Mock, J.J., D.R. Smith, and S. Schultz 2003 *Nano Lett.* **3** 485-491

IOP Conf. Series: Materials Science and Engineering **77** (2015) 012007 doi:10.1088/1757-899X/77/1/012007

Characterization of poly(diphenylamine)-gold nanocomposites obtained by self-assembly

Ali Md Showkat, XuanThang Cao, Dong Woo Kim, Md Rafiqul Islam and Kwon Taek Lim

Department of Image Science & Engineering, Pukyong National University, Busan 608-737, Republic of Korea

E-mail: ktlim@pknu.ac.kr

Abstract. Composites of poly (diphenylamine) and gold nanoparticles were chemically prepared from solutions which containing methane sulfonic acid (MSA) with the addition of gold chloride trihydrate as the oxidant as a source for initiation of polymerization and a generation of Au nanoparticles. Poly (diphenylamine) decorated with noble-metal gold nanoparticles were synthesized successfully and exposure of the solutions to caused polymerization of diphenylamine to polymerize with gold salt. The nanocomposites were characterized for the structure, morphology and electronic properties through X-ray diffraction analysis (XRD), Fourier transform infrared spectroscopy (FT-IR), Transmission electron microscopy (TEM) and UV–visible spectroscopy.

1. Introduction

Composites are a special class of materials originating from combinations of two or more compounds by a suitable technique, which results in materials having unique physicochemical properties and large potential for application in diverse areas. Novel properties of composites can be derived from successful combination of the characteristics of the parent constituents into a single material. Conventional polymers usually serve as the matrix, which result in a special class of hybrid materials termed "polymeric nanocomposites". This composite material differs from the pure polymer with respect to some of the physical and chemical properties and hence it is useful for many applications in different field [1]. The syntheses of polymer metal composites, however, arouse great interest among researchers because of their potential use in technological applications and have attracted much attention recently due to their curious interesting properties [2, 3].

Polymer, metal nanocomposite has received considerable attention due to potential possibilities to create suitable materials for electrocatalysts, chemical sensors and microelectronic devices [4]. In addition, the exceptional uniform shape and the wide range of polymer microspheres further extend their application potential. The metal particles formed on the surface of the polymer microspheres were mainly noble metal particles like rhodium, silver and palladium. This particle not only combine the advantageous properties of metals and polymers but also exhibit many new characters that single-phase materials do not have. To prepare the nanoscale materials successfully, several approaches have been employed like sol-gel technique [5], physical mixing [6], in-situ chemical polymerization in the aqueous solution with the presence of polymer monomer and inorganic particles [7], sonochemical process [8], and γ-irradiation technique [9].

Poly (diphenylamine), (PDPA) composites and noble metal nanoparticles are currently of great research interest due to numerous applications arising from its good environmental stability and tunable electrical and optical properties [10]. Poly (diphenylamine) is polyaniline substituted derivatives and similar with polyaniline even this polyaniline has been studied exhaustively due to its unique properties and novel applications [11-14]. On the other hand, PDPA, an N-aryl substituted polyaniline, has been found to show many properties which are different from PANI and other N-substituted aniline derivatives [15, 16]. The utility of PDPA are known as sensor for alcohol [17], for modified electrode formation [18], as a glucose sensor [19], as a light emitting diodes [20], and as a lithium battery [21] have been established. Such studies have revealed that PDPA can form an effective alternative to PANI, and PDPA is known to have better solubility than PANI [22]. Recent studies [23,24] have attempted to incorporate newer properties into PDPA by grafting a nonconducting polymer onto its backbone. In fact, this creates the possibility of the creation of new polymeric materials with interesting characteristics. Moreover, this technique is easily controlled and adaptable. In the present work, synthesis of poly (diphenylamine)-gold nanocomposites by oxidative polymerization of diphenylamine from a solution containing methanesulfonic acid (MSA) and gold (III) chloride trihydrate as the oxidizing agent and characterized using various techniques.

2. Experimental

2. 1. Materials
DPA (diphenylamine), $HAuCl_4$ (gold chloride trihydrate), APS (ammonium persulfate), MSA (methanesulfonic acid) were obtained from Sigma-Aldrich. All chemicals were reagent grade and used as received. Solutions were prepared using Millipore water obtained from a MILLIPORE water purifying system.

2. 2. Preparation of poly (diphenylamine)-gold nanocomposites
A typical polymerization method for poly (diphenylamine)-gold nanocomposites were carried out from a solution of 0.05 mol.L^{-1} diphenylamine monomer and 0.025 mol. L^{-1} MSA in deionized water. To this mixture, (0.025 mol.L^{-1}) auric acid was added drop wise to the reaction mixture for about 2 hour under ice cold condition with constant stirring. The polymerization started with the slow addition of auric acid as indicated by the color change of the solution from bluish to the green form immediately. The poly (diphenylamine)-gold nanocomposite was filtered and washed with deionized water, methanol several times. Finally, the obtained product was dried under a vacuum dryer at room temperature for 48 hours.

2. 3. Characterization
Transmission electron microscopy images were obtained using a Jeol JEM-1230 electron microscope operated at 120 kV equipped with a Gatan Ultrascan 4000SP 4K X 4K CCD camera. TEM samples were prepared by placing a droplet of a colloid suspension in water on a Formvar carbon coated, 300 mesh copper grid (Ted Pella) and allowed to evaporate in air at room temperature. The X-ray diffraction (XRD) patterns were measured at room temperature with an X' Pert Philips materials research diffractometer using the CuKαR radiation. Infrared spectra were measured in the range of 400-4000 cm^{-1} on pellets made with KBR by means of a Perkin Elmer 1600 FTIR spectrophotometer, taking 32 scans at a resolution of 4 cm^{-1}. The optical absorption spectra were measured using a HP-8453 spectrophotometer.

3. Results and Discussion

3.1. TEM analysis of poly (diphenylamine)-gold nanocomposites

The nanocomposites of poly (diphenylamine) -gold was successfully synthesized from an aqueous solution of methanesulfonic acid using $HAuCl_4$ as oxidant by means of a self-assembly process. A typical TEM image of poly (diphenylamine)-gold nanocomposite product is shown in Figure 1. It can be seen that the poly (diphenylamine)-gold nanocomposites were quite uniform in size and morphology. Dark spots of Au nanoparticles on the surface of polymer can be noticed from TEM image with an average size of approximately 5-16 nm.

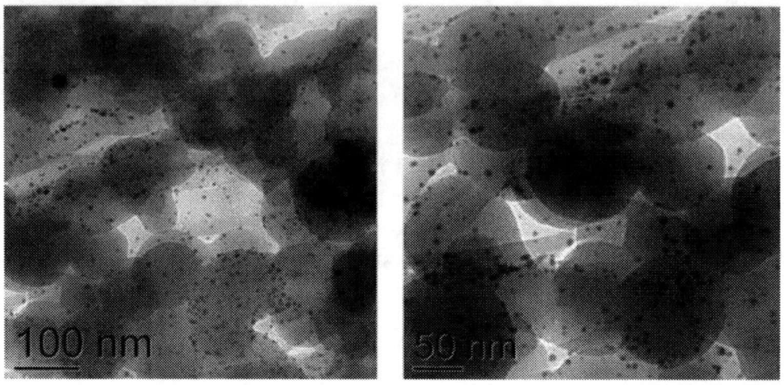

Figure 1. TEM images of poly(diphenylamine)-gold nanocomposites (a) and (b).

The elemental gold, reduced from $HAuCl_4$ in the reaction with diphenylamine, precipitated onto the self-assembled PDPA spheres possibly hindering the elongation process of spheres into tubes. The Cl^- anions, produced during reduction of the $HAuCl_4$ may be doping of the PDPA chain as well as doping with methanesulfonic acid. Firstly, the polymer of the composite, PDPA-MSA is amorphous, producing a characteristic ring in the diffraction pattern. Secondly, the Au nanoparticles can be seen in the form of dark spots superimposed on the polymer. Based on the above results, we can conclude that the poly (diphenylamine)-gold nanocomposites consist of amorphous PDPA spheres decorated with Au nanoparticles. This conclusion was further confirmed by XRD experiments.

3.2. XRD characterization of poly (diphenylamine)-gold nanocomposites

The XRD patterns of the self-assembly of polymerization of poly (diphenylamine)-gold nanocomposites prepared by oxidation with $HAuCl_4$. In case of PDPA, the broad peaks centered at $2\theta = 21 \sim 28°$ are observed in the nanocomposites (Figure 2), which are ascribed to the periodicity parallel and perpendicular to the polymer chains of PDPA similar to PANI [25]. In the diffraction pattern of PDPA-MSA/ Au nanocomposites of Figure 2, four additional sharp peaks at 38°, 45°, 64° and 78° revealed that represent Bragg's reflections from the (111), (200), (220) and (311) planes for the presence of gold (Au), respectively [26]. The XRD patterns confirmed presence of poly (diphenylamine)-gold nanocomposites.

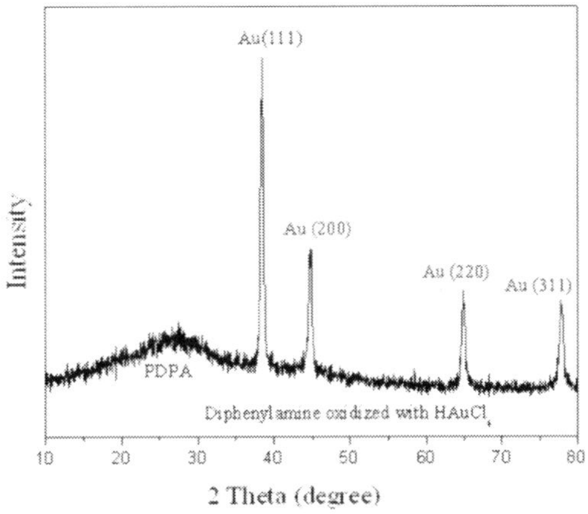

Figure 2. XRD diffraction pattern of poly (diphenylamine)-gold nanocomposites.

3.3. FTIR characterization of poly (diphenylamine)-gold nanocomposites

Figure 3 shows FTIR spectra of poly (diphenylamine)-gold nanocomposites. FTIR spectra of poly (diphenylamine)-gold peaks appear at 1592, 1502, 1303, 1285, 1170, 843, 580 cm^{-1}. The characteristic peaks at 1592 and 1502 cm^{-1} are attributed to the C=C stretching of quinoid phenyl and benzenoid phenyl rings, respectively [27]. The bands at 1303 can be assigned to the C-N modes of the secondary aromatic amine, the bands at 1170 cm^{-1} to the in-plane bending of C-H [28, 29] and the band at 843 cm^{-1} corresponds to the C-H out of plane bending vibration respectively [30]. A new peak at 1008 cm^{-1} related to a C-Cl vibration appears in the FTIR spectra of poly (diphenylamine)-gold is doped with the Cl^- from $HAuCl_4$.

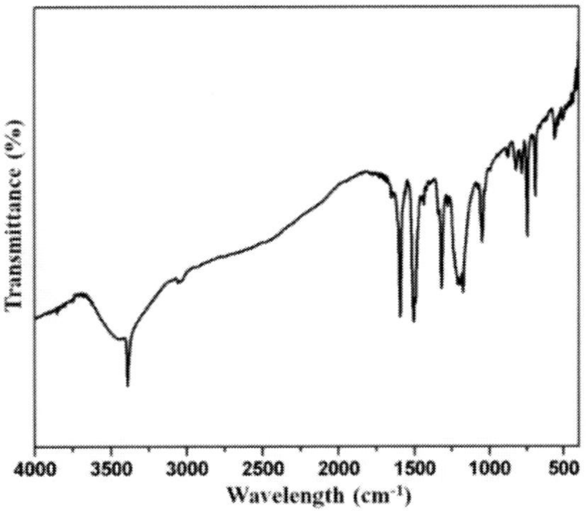

Figure 3. FTIR spectra of poly (diphenylamine)-gold nanocomposites.

The relatively broad peak at 3427 cm^{-1} can be attributed to the N-H stretching vibration mode. In addition, the characteristic bands of the parent PANI, the spectrum showed distinctive broad peaks of the 1005 ~ 1068 cm^{-1} assigned to the asymmetric and/or symmetric 0=S=0 stretching vibrations. The shifts to the lower energy of these peaks are directly related to the differences in electron density, which may originate from structural differences of these two materials or Au, as the electron density donator makes the electron cloud of PDPA more delocalized.

Table 1. Summarized for FTIR spectral data of poly (diphenylamine)-gold nanocomposites

Poly(diphenylamine)-gold nanocomposites	Peak Assignment
3427 cm^{-1}	N-H stretching vibration mode
1592 cm^{-1}	C=C stretching of quinoid phenyl rings
1502 cm^{-1}	C=C stretching of benzenoid phenyl rings
1303 cm^{-1}	C-N modes of the secondary aromatic amine
1170 cm^{-1}	in-plane bending of C-H
1008-1068 cm^{-1}	related to a C-Cl vibration appears in the spectra of PDPA-MSA/Au is doped with the Cl^- from $HAuCl_4$
843 cm^{-1}	C-H out of plane bending vibration
580 cm^{-1}	absorption of the $-SO_3H^-$ group

3.4. UV-Visible analysis

The UV-Vis absorption spectra of poly (diphenylamine)-gold nanocomposites obtained by chemically synthesized which shown in Figure 4. The characteristic bands of poly (diphenylamine) that appear at 310~370 nm are attributed to π-π* and polaron-π* transitions, respectively. The bands associated with the PDPA/Au products are located at 370 nm and 570 nm (region of interchain excitation). There is no

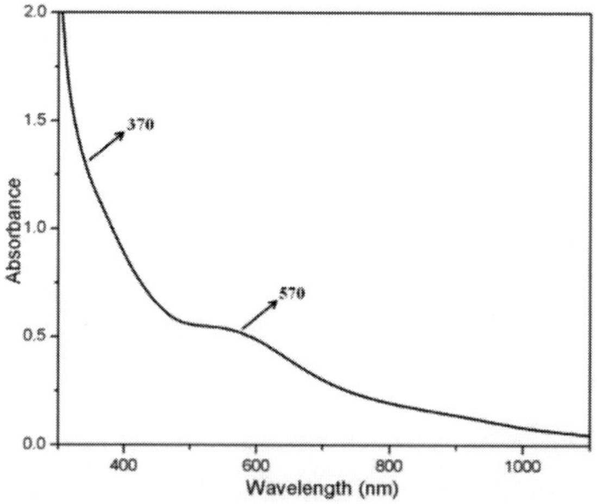

Figure 4. UV-Vis spectra of poly(diphenylamine)-gold nanocomposites.

variation in the position of the band at around 310 ~ 370 nm corresponds to the π-π* transition of the neutral form of PDPA. Conversely, the electronic spectra of composites shows a band at around 570 nm and inform that PDPA is present in the diphenosemiquinoamine form. This assignment is based on earlier reports on the doping of PANI with acids [31, 32]. PDPA-Au nanocomposites exhibit the surface plasmon resonance absorption band of Au at 520-570 nm.

This is because Au particles are incorporated into the polymer chain. The breadth of the 520 - 570 nm Plasmon resonance peak indicates a vary poly disperse Au nanoparticle solution [33]. Therefore, the wide band should also be attributed to the Plasmon resonance adsorption of Au particles, which provided a piece of evidence for the formation of Au particles. So, the broad absorption should originate from combined effect of the benzenoid to quinoid transition and plasmon resonance absorption of the gold particles.

4. Conclusions

We have described a new route for synthesis of poly (diphenylamine)-gold nanocomposites using $HAuCl_4$ as the oxidant in a solution of methanesulfonic acid. The presence of Au particles was confirmed by transmission electron. XRD study evidenced the presence of Au in the polymer sample in nanometer scale. FTIR spectra showed that the incorporation of gold seems to be easier with methanesulfonic acid. The synthesized nanocomposites can be taken for further application studies, depending upon their properties. Thus, the present work gives a simple and facile method of synthesis of poly (diphenylamine)-gold nanocomposites.

5. Acknowledgements

This work was supported by the BK21 plus program.

6. References

[1] Pfaendner R 2010 Nanocomposites: Industrial opportunity to challenge *Polymer Degradation and Stability* **95** (3): 369-373

[2] Xu P, Wang H, Lv R, Du Q, Zhong W, Yang Y 2006 Synthesis of TiO_2- SiO_2/polymer core-shell microspheres with a microphase- inversion method *J. Polym Sci Part A: Polym Chem* **44** 3911

[3] Grubbs R B 2005 Hybrid metal-polymer composites from functional block copolymers *J. Polym Sci Part A: Polym Chem.* **43** 4323

[4] Huang K, Zhang Y, Han D *et al* 2006 One step synthesis of 3D dendritic gold/polypyrrole nanocomposites via a self-assembly method *Nanotechnology* **17** (1): 283-288

[5] Jang S H, Han M G, Im S S 2000 Preparation and characterization of conductive polyaniline/ silica hybrid composites prepared by sol-gel processes *Synth Met.* **110** 17

[6] Akelah A, El-Deen N S, Hiltner A, Baer E, Moet A 1995 Organophilic rubber-montmorillonite nanocomposites *Mater lett.* **22** 97

[7] Ray S S, Biswas M 2000 Water-dispersible conducting nanocomposites of polyaniline and poly (N- vinylcarbazole) with nanodimensional zirconium dioxide *Synth Met.* **108** 231

[8] Kumar R V, Diamant Y, Gedanken A 2000 Sonochemical synthesis and characterization of nanometer-size transition metal oxides from metal acetates *Chem Mater* **12** 2301

[9] Yin Y, Xu X, Ge X, Xia C, Zhang Z 1998 Synthesis of cadmium sulfide nanoparticles *in situ* using γ-radiation *Chem. Commun* 1641-1642

[10] Doblhofer K, Rajeshwar K, Eds M D 1998 Handbook of conducting polymers; New York

[11] Jang J 2006 Conducting polymer nanomaterial and their applications *Adv. Polym. Sci.* **199**189

[12] Zhou H H, Ning X H, Li S L, Chen J H, Kuang Y F 2006 Synthesis of polyaniline- Silver

nanocomposite film by unsymmetrical square wave current method *Thin solid Films* **510**164

[13] Vivekchand S R C, Kam K C, Gundiah G, Govindaraj A, Cheetham A K, Rao C N R 2005 Electrical properties of inorganic nanowire polymer composites *J. Mater. Chem.* **15** 4922

[14] Feng, Mao C, Yang G, Hou W, Zhu J -J 2006 Polyaniline/Au composite hollow spheres: synthesis, characterization, and application to the detection of dopamine *Langmuir* **22** 4384

[15] Feng W, Bai X D, Lian Y Q, Liang J, Wang X G, Yoshono K 2003 Well aligned polyaniline/carbon- nanotube composite films grown by in-situ aniline polymerization *Carbon* **41** 1551

[16] Inzelt G 2002 Cyclic voltammetry of solid diphenylamine crystals immobilized on an electrode surface and in the presence of an aqueous solution *J Solid State Electrochem* **6** 265

[17] Athawale A, Milind V K 2000 Polyaniline and its substituted derivatives as sensor for aliphatic alcohols *Sens Actuators B* **67**173

[18] Santana H, Dias F C 2003 Characterization and properties of polydiphenylamine electrochemically modified by iodide species *Mater Chem Phys* **82** 882

[19] Santhosh P, Manesh K M, Uthayakumar S, Gopalan A I, Lee K-P 2009 Hollow spherical nanostructured polydiphenylamine for firect electrochemistry and glucose biosensor *Biosensors & Bioelectronics* **24** 2008

[20] Li C –Y, Wen T C, Guo T –F, Hou S –S 2008 A facile synthesis of sulfonated Poly (diphenylamine) and the application as a novel hole injection layer in polymer light emiting diodes *Polymer* **49** 957

[21] Gopalan A I, Lee K –P, Manesh K M, Santhosh P 2008 Poly (vinylidene fluoride)-polydiphenylamine composite electrospun membrane as high-performance polymer electrolyte for lithium batteries *Journal of Membrane Science* **318** 422

[22] Wen T C, Chen J B, Gopalan A 2002 Soluble and methane sulfonic acid doped Poly (diphenylamine)- synthesis and characterization *Mater Lett* **57**, 280

[23] Hua F, Ruckenstein E 2004 Fluorescence study of aggregation in water of PEO-grafted polydiphenylamine *Langmuir* **20** 3954

[24] Hua F, Ruckenstein E 2003 Water-soluble conducting polyethylene oxide-grafted polydiphenylamine synthesis through a " Graft Onto" process *Macromolecules* **36** 9971

[25] Zhang L, Wan M, Wei Y 2006 Nanoscaled polyaniline fibers prepared by ferric chloride as an oxidant *Macromol. Rapid Commun* **27** 366

[26] Feng X, Yang G, Xu Q, Hou W, Zhu J -J 2006 Self-assembly of polyaniline/Au composites: From nanofibers *Macromol. Rapid Commun* **27** 31

[27] Chen S –A, Lee H -T 1995 Structure and properties of poly~acrylic acid!-doped polyaniline *Macromolecules* **28** 2858

IOP Conf. Series: Materials Science and Engineering **77** (2015) 012007 doi:10.1088/1757-899X/77/1/012007

[28] Lee K P, Showkat A M, Gopalan A I, Kim S –H, Choi S –H 2005 Synthesis of Poly (diphenylamine) nanotubes in the channels of MCM-41 through self-assembly *Macromolecules* **38** 364

[29] Showkat A M, Lee K P, Gopalan A I, Kim S –H, Choi S –H 2006 Characterization and preparation of new multiwall carbon nanotube/ conducting polymer composites by in situ polymerization *Journal of Applied Polymer Science* **101** 3721

[30] Kim B –J, Oh S –G, Han M –G, Im S S 2000 Prepartion of polyaniline nanoparticlrs in micellar solutions as polymerization medium *Langmuir* **16** 5841

[31] Duke C B, Conwell E M, Paton A 1986 Localized molecular excitons in polyaniline. *Chem Phys Lett* **131** 82

[32] Tzou K, Gregory R V 1993 A method to prepare soluble polyaniline salt solutions- in situ doping of PANI base with organic dopants in polar solvents *Synth Met.* **53** 365

[33] Link S and El-Sayed M. A 1999 Spectral Properties and Relaxation Dynamics of Surface Plasmon Electronic Oscillations in Gold and Silver Nanodots and Nanorods *J. Phys. Chem. B* **103** 841

IOP Conf. Series: Materials Science and Engineering **77** (2015) 012008 doi:10.1088/1757-899X/77/1/012008

Synthesis and study of Au/TiO_2 and Au/CeO_2 nanocomposites for their physical properties and catalytic activity

S Chornaja[1], S Zhizhkuna[1], D Jankovica[2], D Karashanova[3], K Dubencovs[1], O Stepanova[1], V Kampars[1], G Poikane[1]

[1]Institute of Applied Chemistry, Riga Technical University, Latvia
[2]Institute of Inorganic Chemistry, Riga Technical University, Latvia
[3]Institute of Optical Materials and Technologies Bulgarian Academy of Sciences, Bulgaria

E-mail: Svetlana @ ktf.rtu.lv

Abstract. The effect of the gold nanoparticles size, gold load, nature of support and composites' specific surfaces area on the nanocomposites catalytic activity and selectivity was studied. It was demonstrated that glycerol conversion increased from 62% to 100% in the presence of the nanocomposites when Au particle size decreased from 26 to 1 nm. Glyceric acid was obtained as a main product with high selectivity.

1. Introduction

Supported gold catalysts are applicable in the hydrogenation processes [1], catalytic oxidations of CO [2, 3], alcohols [4, 5] and aldehydes [6]. There is also growing interest to apply such catalysts for water purification [7], air pollutions decreasing [8] as well as cars' exhausting gases emission decrease [9]. Glycerol is a bio-diesel manufacturing by-product, its amount reaching up to 10% of the total mass of the products mixture, thus making it important to find a solution for the glycerol utilization problem. The glycerol catalytic oxidation by oxygen demonstrates great potential in this regard. A number of important products can be obtained in this process. Platinum group metals are classical supported heterogeneous catalysts [4-6] used in liquid-phase oxidation process of alcohols and aldehydes, including glycerol. It is known [10-15], that supported gold catalysts can be used in the processes of glycerol catalytic oxidation by molecular oxygen. The advantage of gold based catalysts in comparison to the classic Pt and Pd systems is in the absence of the gold catalysts' surface deactivation by molecular oxygen in the process of oxidation [16, 17]. Gold catalyst can be used many times without regeneration required, so catalyst's life time becomes longer [18, 19]. Catalysts activity and selectivity depends significantly on the support nature, size of the gold particles on the surface, catalyst's synthesis conditions as well as the oxidation process parameters. Gold catalysts supported on carbon materials or metal oxides are mostly used [20 – 22]. Metal oxide-supported Au catalysts can be grouped into two categories, with respect to glycerol oxidation, which depend on the support material: gold catalysts with inert support materials, such as SiO_2, Al_2O_3, or MgO, and gold catalysts with active support materials, such as Fe_2O_3, TiO_2, CeO_2, ZrO_2, NiO_x, CoO_x. Oxygen is adsorbed in large quantities on active support materials in a molecular form, probably as a superoxide (O_2^-) species [23]. This explains the high surface mobility of adsorbed oxygen so that diffusion to the gold particles is not rate-limiting. It has been demonstrated [24] that supported gold catalysts are active in the basic environments only. The glycerol oxidation process has usually been studied at pH = 11-12, P_{O_2} = 3-10 atm and temperature of 298-373 K [11, 13, 24]. Glyceric acid is the currently only product that can be obtained with both high yield and selectivity. Authors of paper [11] report 100% selectivity by glyceric acid achieved in presence of 1%wtAu/C catalyst with glycerol conversion rate of 54-56% (P_{O_2} = 3 atm, 333 K, 3 h). There are works [10, 25] reporting about glycerol oxidation at the atmospheric oxygen pressure in presence of 1%wtAu/C or 1%wtAu/CeO_2 catalysts. Authors of the work [25] report that full glycerol conversion with selectivity by glyceric acid less than 53% has been achieved after two hours of oxidation in presence of 1%wtAu/C catalyst. Meanwhile, authors of the paper [10] report about glycerol conversion rate of just 38% with selectivity by glyceric acid 43% achieved after five hours of oxidation in presence of 1%wtAu/CeO_2. However, it is hard to find an appropriate catalyst and reaction conditions necessary to selectively obtain glycerol oxidation products other than glyceric acid.

This work is devoted to the synthesis and the study of the physical properties of Au/CeO_2 and Au/TiO_2 nanocomposites, as well as testing their catalytic activity in the reaction of glycerol oxidation and comparison of synthesized catalysts property to those of commercial Au/TiO_2 catalyst. The glycerol oxidation was studied at the atmospheric oxygen pressure.

2. Experimental

2.1. Catalyst preparation and characterisation

Two different types of nanocomposites, supported on TiO_2 or on CeO_2, were synthesed. TiO_2 nanoparticles were prepared by hydrolysis of titanium isopropoxide. The as-prepared powder consisted of anatase crystalline phase with crystallite size (d_{kr}) of 7-8 nm. The carrier synthesized was dispersed in ethanol and tetrachloroauric acid water solution diluted by C_2H_5OH (pH = 6-8) was added. After impregnating a carrier with tetrachloroauric acid, gold was reduced by formaldehyde or by hydroquinone ethanol solution (pH = 5-6). CeO_2 was prepared by the sol-gel combustion technique-ethylene glycol was used as a fuel and nitrate as oxidant. Cubic CeO_2 with d_{kr} 16 nm is obtained. The carrier obtained was dispersed in deionized H_2O and tetrachloroauric acid water solution (pH = 7) was added. After impregnating a carrier with tetrachloroauric acid, gold is reduced by formaldehyde water solution. Suspensions obtained after gold reducing were filtered, washed by water or ethanol and dried under the temperature 80-90 °C. Phase composition of prepared nanocomposites was determined by the X-ray diffraction analysis (D8 Advance, Bruker AXS). The crystallite size was calculated from broadening of diffraction maxima using Scherrer Equation (software Topas 3). Chemical analysis was performed by means of the S4 Pioneer X-ray Spectrometr (Bruker AXS). Specific surface area (SSA) of the powders was measured using HROM-3 chromatograph by BET method at temperature of liquid nitrogen. TEM measurements were carried out with by means of transmission electron microscope JEOL JEM 2100 at 200 kV accelerating voltage and at magnifications 40 000-200 000x. Histograms of Au particle size distribution were obtained by the counting of at least 200 particles from the micrographs, and the mean particle diameter (d_{Au}) was calculated as $d_{Au} = \sum(d_i \cdot n_i)\cdot\left(\sum n_i\right)^{-1}$, where n_i was the number of particles of specific diameter.

2.2. Glycerol oxidation

Oxidation of glycerol by molecular oxygen in the presence of supported gold catalysts was performed at atmospheric pressure in a thermostated glass batch reactor equipped with gas supply system and magnetic stirrer (2500 rpm). Reaction conditions, such as initial glycerol concentration c_0(glycerol) = 0.3 M, oxygen pressure $P(O_2)$ = 1 atm, NaOH initial concentration c_0(NaOH) = 1.5 M, glycerol and gold molar ratio n(glyc.)/n(Au) = 300 mol/mol, temperature 60 °C and oxidation time 1 hours were kept constant. Reaction mixture samples were filtered for separation of catalysts. The filtered samples were analyzed by high-performance chromatograph UFLC Shimadzu Nexera. Activity of the catalysts was estimated by glycerol conversion. Selectivity (mol%) of the products was calculated using the formula $selectivity = P_i \cdot 100\% \cdot \left(\sum P_i\right)^{-1}$, where P_i was the yield by a single product (mol%).

3. Results

Typical TEM microphotography and the corresponding histograms of synthesized nanocomposites are presented in figure 1. The morphology of the catalytic material, consisting of support and Au nanoparticles as an active phase, is visible (figures 1a,b,d,e). The support CeO_2 exhibits crystalline structure with less contrast. The support TiO_2 exhibits granular structure with less contrast. The metal nanoparticles possess higher contrast and sphere-like shape. In some regions of the sample they are coalesced, thus forming big aggregates. Very big Au particles with diameters about 30 - 50 nm are visible in some aggregates of the catalyst. They are not typical for the samples' studied structure. The corresponding size distribution histogram of Au nanoparticles in the 1wt%Au/CeO_2 and 5wt%Au/TiO_2 catalysts is shown in figure 1c, 1f. Electronic microscopy revealed that the average size of Au nanoparticles in 1wt%Au/CeO_2 catalyst is 8.7 nm and in 5wt%Au/TiO_2 catalyst – 9.9 nm.

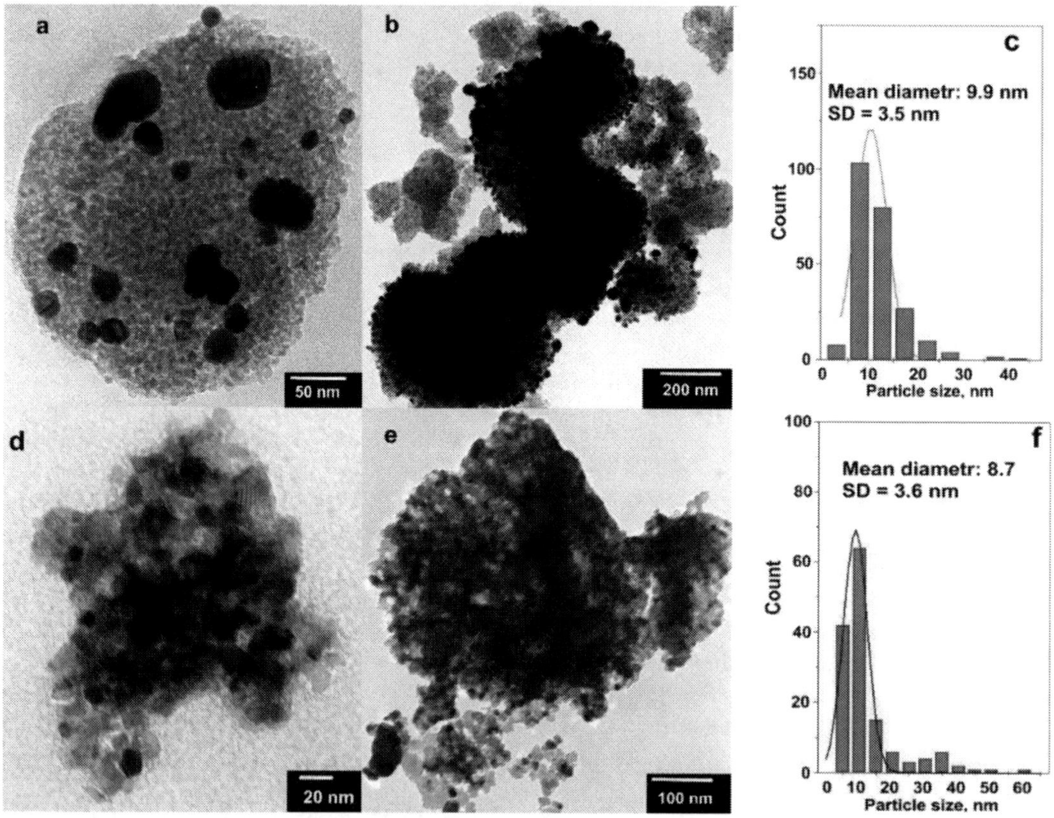

Figure 1. TEM images of Au nanocomposites with particle size distribution (a,b,c – 5wt%Au/TiO_2; d,e,f – 1wt%Au/CeO_2)

The prepared nanocomposites were tested in glycerol oxidation processes. The results of study of catalyst synthesis parameters effect on the catalyst's activity and selectivity, demonstrated that the Au/TiO_2 nanocomposites synthesized with hydrochinon as a reducer are inactive compared with the nanocomposites where formaldehyde is used. Therefore, all the gold nanocomposites used in later studies were synthesized using formaldehyde as a gold reducer. According to the data presented in the table 1 in all cases glycerol catalytic oxidation main product in the presence of the Au/CeO_2 or

Au/TiO_2 nanocomposites is the glyceric acid, by-products are – tartronic, lactic, glycolic, oxalic, acetic and formic acids. It is visible from the data presented in the table 1 that the size of gold particles defines catalytic activity and selectivity of nanocomposites in the process of glycerol oxidation. When Au particle size decreased from 26 to 1nm, glycerol conversion increased from 62 to 100%. In presence of Au/CeO_2 with Au particle size decreased from 9 to 1 nm the process selectivity by glyceric acid decreased from 52 to 44%. At the same time in the presence of catalyst Au/TiO_2 the process selectivity by glyceric acid practically did not depend on the size of Au particles. The results obtained in this work in presence of Au/CeO_2 are consistent with the data of works [11, 14, 26, 27]. According to the data presented in the mentioned works, the Au/C catalysts activity increases with the decrease of the gold particles size, although selectivity by glyceric acid decreases as well. For example, according to the data [14] with d_{Au} increased from 12 to 20 nm the activity of the 1wt% Au/C catalyst decreases 8 times but selectivity by glyceric acid increased from 69 to78% at the same time. The results obtained in this work in presence of Au/TiO_2 are consistent with the data of [12]. It is demonstrated in the work [12] that no obvious Au particles size impact on the catalyst activity and selectivity is found in case of glycerol oxidation in presence of 1wt%Au/TiO_2 catalyst.

Table 1. Glycerol oxidation over supported gold catalysts

Catalyst	TEM d_{Au}, (nm)	XRD $d_{kr.Au}$, (nm)	SSA, (m^2/g)	Glycerol conv., (mol%)	Selectivity, (mol%)						
					GLIC	TAR	LAC	GLY	OXA	ACE	FOR
1wt%Au/CeO_2	9	20-23	27	83	52	2	17	10	1	0	18
1.5wt%Au/CeO_2	1	29	27	100	44	12	17	4	4	13	6
2.7wt%Au/CeO_2	-	36	27	74	53	10	23	4	4	0	6
1.5wt%Au/TiO_2 (commer.)	2	2–5	-	93	62	16	16	2	6	0	0
				73[a]	69	15	6	1	5	4	0
2wt%Au/TiO_2	-	10-15	122	59	64	2	15	4	1	0	14
5wt%Au/TiO_2	10	10-15	117	78	62	3	17	6	1	0	11
5wt%Au/TiO_2	26	22-25	101	62	67	0	16	5	0	0	12

[a] – oxidation time is 0.5 h, GLIC – glyceric acid, TAR – tartronic acid, LAC – lactic acid, GLY – glycolic acid, OXA – oxalic acid, ACE – acetic acid, FOR – formic acid

From the data collected in table1 it could be concluded that Au/CeO_2 nanocomposites are more active but significantly less selective at the same time comparing to Au/TiO_2 nanocomposites having approximately same size of Au particles. For example, glycerol conversion equals to 83% (selectivity by glyceric acid is 52%) in presence of 1wt%Au/CeO_2 (d_{Au} = 9 nm), but glycerol conversion equals to 78% in presence of 5wt%Au/TiO_2 (d_{Au} = 10 nm) with selectivity by glyceric acid equals to 62%. The results obtained in this work are consistent with the data of [10]. It has been demonstrated in the work [10] that the 1wt%Au/CeO_2 catalyst is more active than 1wt%Au/TiO_2 catalyst in the oxidation of the glycerol with oxygen at the atmospheric pressure. However, the authors of the work [10] were capable to reach glycerol conversion of just 38% and 30% in five hours in presence of 1wt%Au/CeO_2 and 1wt%Au/TiO_2 catalysts respectively. Our work reports conditions for full glycerol conversion in presence of 1.5wt%Au/CeO_2 and 93% conversion in presence of 1.5wt%Au/TiO_2 in one hour (see table 1).

According to the data in table 1, it can be also concluded that there is no direct correlation between the catalysts activity and their specific surface area or gold content. Available reports on the subject are contradictive. The same conclusions (that there is no direct correlation between catalyst activity and specific surface area of the support) were also made in the works [10, 18] during the study of glycerol oxidation in the presence of Au based catalysts with different supports. The effect of the gold content is studied in the works [11, 13]. It is found in the work [11] that changing the gold content percentage from 0.25 to 1wt% does not affect d_{Au} (25 nm). However, authors explain the decrease of the Au/C catalyst activity with the decrease of the gold content by the decrease of the gold particles quantity per catalyst's surface unit. At the same time, work [13] demonstrates that with the increase of the gold content from 0.33 to 7.2wt% Au/C catalysts activity decreases, but the d_{Au} increases from 2.7 to 42.3 nm.

By comparing the data presented in table 1, it could be concluded that the most effective catalysts for glycerol oxidation to glyceric acid were commercial 1.5wt%Au/TiO_2 and 5wt%Au/TiO_2 (synthesized in this work) catalysts.

4. Conclusions.

Au/CeO_2 and Au/TiO_2 composites synthesized in this work as well as commercial 1.5wt%Au/TiO_2 are catalytically active in the process of glycerol oxidation under mild conditions. In comparison to the best results reported in literature, glyceric acid with high yield and selectivity is obtained at atmospheric pressure, using simple thermostated glass batch reactor. It was found that Au based catalysts activity depends from the size of gold nanoparticles, gold load and nature of the support.

References

[1] Perret N, Cárdenas-Lizana F, Keane M A 2011 *Catal. Commun.* **16** 159
[2] Li W-C, Comotti M, Schueth F 2006 *J. Catal.* **237(1)** 190
[3] Sakurai H, Akita T, Tsubota S, Kiuchi M, Harut M 2006 *Appl. Catal.*, A **291(1-2)** 179
[4] Parreira L A, Bogdanchikova N, A.Pestryakov, Zepeda T A, Tuzovskaya I, Farías M H, Gusevskaya E V 2011 *Appl. Catal.*, A **397** 145
[5] Wang H, Fan W, He Y, Wang J, Kondo J N, Tatsumi T 2013 *J. Catal.* **299** 10
[6] Biella S, Pratti L, Rossi M 2003 *J. Mol. Catal.* A: *Chem.* **197** 207
[7] Han Y F, Phonthammachai N, Ramesh K, Zhong Z, White T 2008 *Environ. Sci. Technol.* **42(3)** 908
[8] Gluhoi A C, Lin S D, Nieuwenhuys B E 2004 *Catal. Today* **90** 175
[9] Mellor J R, Palazov A N, Grigorova B S, Greyling J F, Reddy K, Letsoalo M P, Marsh J H 2002 *Catal. Today* **72** 145
[10] Demirel S, Kern P, Lucas M, Claus P 2007 *Catal. Today* **122** 292
[11] Carrettin S, McMorn P, Johnston P, Griffin K, Kiely C J, Hutchings G J 2003 *Phys. Chem. Chem. Phys.* **5** 1329
[12] Dimitratos N, Villa A, Bianchi C L, Prati L, Makkee 2006 M *Appl. Catal. A* **311** 185
[13] Demirel-Gülen S, Lucas M, Claus P 2005 *Catal. Today* **102-103** 166
[14] Ketchie W C, Fang Y, Wong MS, Murayama M, Davis R J 2007 *J. Catal.* **250** 94
[15] Ketchie W C, Murayama M, Davis R G 2007 *Top. Catal.* **44(1-2)** 307
[16] Dimitratos N, Messi C, Porta F, Prati L, Villa A 2006 *J. Mol. Catal.* A **256(1-2)** 21
[17] Gallezot P 1997 *Catal. Today* **37** 405
[18] Sobczak I, Jagodzinska K, Ziolek M 2010 *Catal. Today* **158** 121
[19] Gil S, Marchena M, Fernández C M, Sánchez-Silva L, Romero A, Valverde J L 2013 *Appl. Catal.*, A **450** 189
[20] Davis S E, Ide M S, Davis R J 2013 *Green Chem.* **15** 17
[21] Gil S, Cuenca N, Romero A, Valverde J L, Sánchez-Silva L 2014 *Appl. Catal.*, A **472** 11
[22] Purushothaman R K P , Haveren J van, Es D S van, Melián-Cabrera I, Meeldijk J D , Heeres H J 2014 *Appl. Catal.*, B **147** 92
[23] Liu H, Kozlov A I, Kozlova A P, Shido T, Asakura K, and Iwasawa Y 1999 *J. Catal.* **185** 252
[24] Carrettin S, McMorn P, Johnston P, Griffin K, Kiely C J, Hutchings G 2002 *J. Chem. Comm.* **7** 696
[25] Demirel S, Lehnert K, Lucas M, Claus P 2006 *Appl. Catal.* B 70**(1-4)** 637
[26] Porta F, Prati L 2004 *J. Catal.* **224** 397
[27] Dimitratos N, Lopez-Sanchez J A, Lennon D, Porta F, Prati L, Villa A 2006 *Catal. Lett.* **108(3-4)** 147

IOP Conf. Series: Materials Science and Engineering **77** (2015) 012009 doi:10.1088/1757-899X/77/1/012009

In-situ deposition of gold nanoparticles onto different substrates by chemical spray pyrolysis

I Oja Acik[1], N G Oyekoya[1], A Mere[1], A Katerski[1], V Mikli[2] and M Krunks[1]

[1] Laboratory of Thin Film Chemical Technologies, Department of Material Science, Tallinn University of Technology, Ehitajate tee 5, 19086, Estonia

2 Chair of Semiconductor Materials Technology, Department of Material Science, Tallinn University of Technology, Ehitajate tee 5, 19086, Estonia

E-mail: ilona.oja@ttu.ee

Abstract. Au-nanoparticles were deposited *in-situ* by chemical spray pyrolysis method onto glass and indium tin oxide (ITO) covered glass substrates in the temperature range of 260 to 400 °C. SEM studies showed that the number of Au-NPs increased and the particle size decreased with increasing the deposition temperature. The surface plasmon resonance effect of Au-NPs on glass substrate shifts from 550 to 520 nm with decreasing the mean crystallite size of Au-NPs from 49 to 24 nm. The surface plasmon resonance effect of Au-NPs on ITO substrate shift from 570 to 540 nm with decreasing the mean crystallite size of Au-NPs from 53 to 16 nm.

1. Introduction

Noble metal nanoparticles (NPs) such as gold, silver and copper have attracted researchers' interest due to their unique plasmonic properties and potential applications in optics, nanoelectronics, biomedicine, catalysis and photovoltaics [1-4]. Spectral position of plasmon resonance depends on the size and shape of nanoparticles, distance between them and permittivity of environment and metal [5]. Metal nanoparticles have been deposited by several methods such as electrodeposition [6], sputtering [7], spin-coating [8, 9] and spray pyrolysis [10-12].

Chemical spray pyrolysis (CSP) is a low-cost material deposition method; here the solution of appropriate precursor materials is deposited in the form of fine droplets onto the preheated substrate where the growth of material takes place. To obtain Au-NPs by spray pyrolysis either preformed nanoparticles are sprayed [13] or Au-NPs are generated via thermal decomposition of Au-precursor salts such as $HAuCl_4{\cdot}nH_2O$, $HAuCl_4$, $HAuCl_4{\cdot}3H_2O$ [10-12, 14,15]. The previous thermoanalytical study has shown that the thermal decomposition of $HAuCl_4{\cdot}3H_2O$ into pure gold and gaseous products takes place as a gradual decomposition process in the temperature range of 75-320 °C in air. Despite that, Au and AuCl were detected as the intermediate decomposition products at 240 °C [16].

Au-NPs deposited *in-situ* by chemical spray pyrolysis method have been grown directly onto glass and ITO (indium tin oxide) covered glass substrates by ultrasonic spray pyrolysis [11, 12] or deposited in combination with several metal oxide [11,12,14,15] and sulfides [10] matrixes. Moreover, it has been shown lately, that Au-NPs deposited *in-situ* by chemical spray pyrolysis method have enhanced the out-put characteristics of the solely sprayed thin film solar cell [17].

In this study we compare morphology, structural and optical properties of the Au-NPs deposited *in-situ* by chemical spray pyrolysis method onto bare glass and indium tin oxide covered glass substrates.

2. Experimental

2.1. Synthesis details

Au-nanoparticles (Au-NPs) were deposited *in-situ* by chemical spray pyrolysis method. Gold(III) chloride trihydrate ($HAuCl_4 \cdot 3H_2O$) was used as precursor for the synthesis of gold nanoparticles (Au-NPs). The $HAuCl_4 \cdot 3H_2O$ salt was dissolved in ethanol with the solution concentration of 0.005 mol/L. The spray solution was applied to the preheated glass and indium tin oxide (ITO) covered glass substrates at tin bath temperatures of 260, 300 or 400 °C in air, employing pulsed sprays, where each cycle consisted of 1 min of spraying followed by 1 min of pause. Two spray cycles were deposited with a spray rate of 3.0 mL min^{-1}.

2.2. Characterization

The Au-nanoparticles on glass and ITO substrates were characterised by their morphology, structure and optical transmittance spectra. The Au-NPs morphology and elemental composition was determined with the help of Zeiss HR FESEM Ultra 55 equipped with Bruker EDS system ESPRIT 1.8. An acceleration voltage for scanning electron microscopy (SEM) measurements was 4.0 kV and for EDX analysis 7.0 kV. The crystal structure of the samples was characterised by X-ray diffraction (XRD) patterns. XRD patterns were recorded by a Rigaku Ultima IV diffractometer with Cu Kα radiation (λ = 1.5406 Å, 40 kV at 40 mA) and using the silicon strip detector D/teX Ultra. The mean crystallite size of Au-NPs was calculated using the Scherrer formula. The total transmittance spectra of the Au-NPs were measured in the wavelength range of 300 –1000 nm on a Jasco V-670 UV–VIS–NIR spectrophotometer equipped with an integrating sphere.

3. Results and Discussion

3.1. Morphology and elemental composition

Figure 1 shows the SEM surface images of Au-NPs on glass substrates deposited in the temperature range of 260 and 400 °C. As seen the $HAuCl_4 \cdot 3H_2O$ solution deposited at 260 °C results in Au-NPs which tend to coalescence (Figure 1 A). According to the SEM image (Figure 1A) Au-NPs with diameters between 20 and 100 nm can be deposited at 260 °C. Increasing the deposition temperature to 300 °C results in Au-NPs (Figure 1B) with more uniform size distribution and Au-NPs with a size between 10 and 60 nm can be seen on the SEM image.

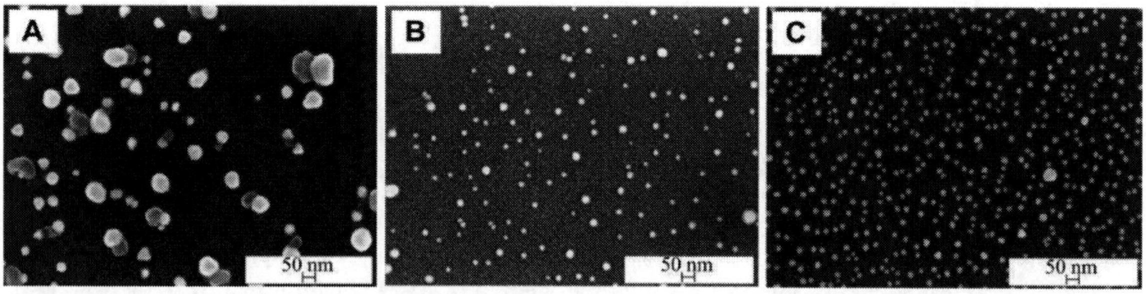

Figure 1. SEM images of the sprayed Au-NPs deposited by chemical spray pyrolysis method onto glass substrates at various deposition temperatures: (A) 260 °C, (B) 300 °C, (C) 400 °C. The magnification of the SEM images is 200 000 X.

Further increasing the deposition temperature to 400 °C results in more dense and uniform coverage of Au-NPs (Figure 1C). Au-NPs with a size in the range of 10 and 40 nm can be deposited at 400 °C onto glass substrates (Figure 1C).

In contrary to the Au-NPs deposited onto glass substrate, the Au-NPs grown onto ITO (Figure 2. Please note that the magnification of SEM images on Figure 2 is higher than those on Figure 1) are evenly distributed and separately placed irrespective of the deposition temperature. The Au-NPs with diameters between 20 and 60 nm were deposited at 260 °C (Figure 2 A). Increasing the deposition temperature narrows the size distribution and increases the number of Au-NPs grown on ITO substrate.

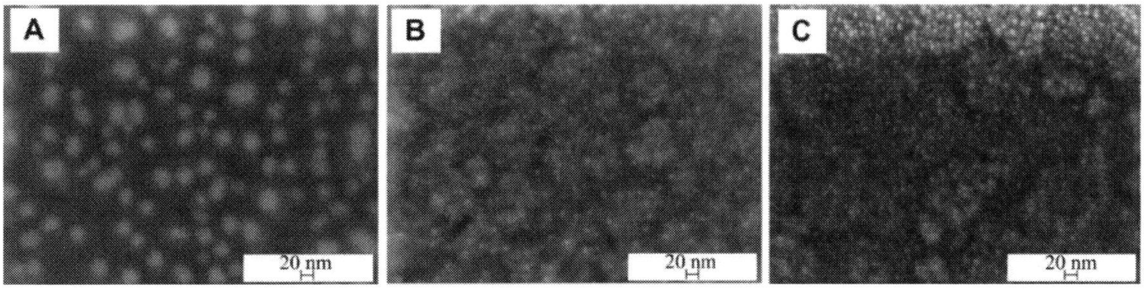

Figure 2. SEM images of the sprayed Au-NPs deposited by chemical spray pyrolysis method onto glass/ITO substrates at various deposition temperatures: (A) 260 °C, (B) 300 °C, (C) 400 °C. The magnification of the SEM images is 500 000 X.

In general, the results obtained are similar with the studies on Au-NPs deposited by ultrasonic spray pyrolysis on glass and ITO substrates, showing that the number of Au-NPs increased and the particle size decreased with increasing the deposition temperature [12].

EDX analysis confirms the formation of Au, irrespective of the deposition temperature. No chlorine residues were detected in the sample deposited at 260 °C, showing that Au-precursor is already decomposed at this temperature.

3.2. Structural properties

The XRD patterns of the Au-NPs deposited on glass (Figure 3) and ITO substrates (figure not shown) in the temperature range of 260 and 400 °C (Figure 3) reveal peaks at 2 theta values of 38.2 and 44.4 deg corresponding to the reflections from the (111) and (200) planes of metallic Au with cubic structure (PDF 00-004-0784). The XRD peaks characteristic of Au show lower intensity and become broader with increasing the decomposition temperature from 260 to 400 °C, referring to the lower amount of material and smaller crystallite size when depositing at higher temperatures. These results are in correspondence with the EDX data showing that the Au amount is decreasing with increasing the deposition temperature.

The (111) peak of the metallic Au in the XRD pattern of Au-NPs on glass and ITO substrates was used to calculate the mean crystallite size of the Au-NPs by the Scherrer formula. The mean crystallite size of the Au-NPs grown on glass substrate is decreasing from 49 to 24 nm and those on ITO substrate from 53 to 16 nm.

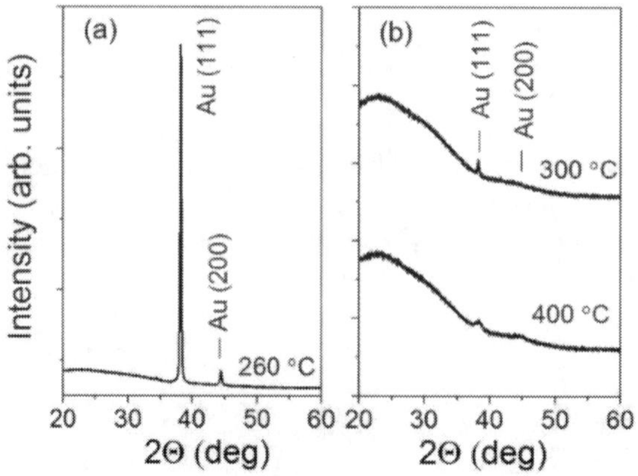

Figure 3. XRD patterns of the Au-nanoparticles deposited onto glass substrates at substrate temperatures of (a) 260 and (b) 300 and 400 °C.

3.1. *Optical properties*

The total transmittance spectra of the Au-NPs on glass and ITO substrates are shown in Figure 4 a and b, respectively. As seen, the overall transmittance is increasing with increasing the deposition temperature and is probably related with the decreased plasmonic scattering of the Au-NPs.

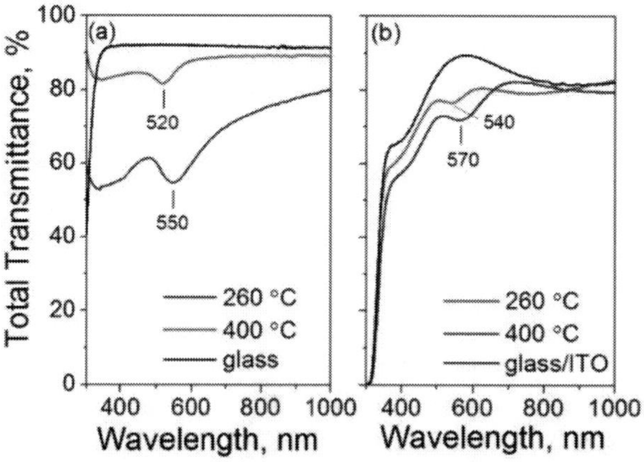

Figure 4. Total transmittance spectra of Au-NPs on (a) glass and (b) glass/ITO substrate deposited in the temperature range of 260 and 400 °C.

A broad band in the spectral region of 450 to 600 nm, clearly visible in the case of Au-NPs on glass substrate (Figure 4a), is associated with the surface plasmon resonance effect. This band is more pronounced when Au-NPs were deposited at lower temperatures. The surface plasmon resonance band maximum on glass substrate shifts from 550 to 520 nm with decreasing the mean crystallite size of the Au-NPs from 49 to 24 nm. In the case of Au-NPs on ITO substrate, the surface plasmon resonance band shifts from 570 to 540 nm with decreasing the mean crystallite size of the Au-NPs from 53 to 16 nm. According to the literature, the position of plasmon resonance depends on the size and shape of nanoparticles, distance between them and permittivity of environment and metal itself [5]. Thus, the different positions of the surface plasmon resonance effect on the glass and ITO substrates in the case of Au-NPs with similar size, viz. 49-24 nm and 53-16 nm, respectively, is related to the different

IOP Conf. Series: Materials Science and Engineering **77** (2015) 012009 doi:10.1088/1757-899X/77/1/012009

refractive index of the glass and ITO substrates. The refractive index of the ITO substrate (1.8-1.9) is higher than that of glass (1.4), thus the surface plasmon resonance effect in the presence of Au-NPs with similar sizes is red shifted due to the different substrates used.

4. Conclusions

Au-NPs were deposited onto glass and indium tin oxide covered glass substrates *in-situ* by chemical spray pyrolysis method. The Au-NPs on glass substrate tend to grow together when deposited at 260 °C, however further increasing the deposition temperature up to 400 °C results in non-aggregated and denser coverage of Au-NPs. The Au-NPs deposited onto ITO substrate are separately placed irrespective of the deposition temperature.

The mean crystallite size of the Au-NPs grown onto glass and ITO substrates is decreasing from 49 to 24 nm and from 53 to 16 nm, respectively, with increasing the deposition temperature from 260 to 400 °C. EDX studies coincide with the XRD data, showing the decrease in Au amount with increasing the deposition temperature.

The surface plasmon resonance effect of Au-NPs on glass and ITO substrate is shifted from 550 to 520 nm and from 570 to 540 nm, respectively, with decreasing the size of the Au-NPs.

5. Acknowledgements

This work is financially supported by the Estonian Ministry of Education and Research (IUT 19-4), TUT base financing project B24 and European Regional Development Fund through the projects: Centre of Excellence "Mesosystems: Theory and Applications" (MESO) TK114, 3.2.0101.11-0029 and project "Efficient plasmonic absorbers for solar cells" 3.2.1101.12-0023.

References

[1] Lue J T 2001 *J. Phys. Chem. Solids.* **62** 1599

[2] Bhattacharya R, Mukherjee P 2008 *Adv. Drug. Deliv. Rev.* **60** 1289

[3] Daniel M C, Astruc D 2004 *Chem. Rev.* **104** 293

[4] Atwater H A, Polman A 2010 *Nature Mater.* **3** 205

[5] Konishi Y, Tanabe I, Tatsuma T 2013 *Chem. Commun.* **49** 606

[6] El-Deab M S, Sotomura T, Ohsaka T 2006 *Electrochim Acta.* **52** 1792

[7] Kusnezh V, Danylov A, Il'chuk H, Petrus' R 2014 *Phys. Status Solidi A* **251** 1574

[8] Oja Acik I, Dolgov L, Krunks M, Mere A, Mikli V, Pikker S, Loot A, Sildos I 2014 *Thin Solid Films* **553** 144

[9] Buso D, Pacifico J, Martucci A, Mulvaney P 2007 *Adv. Funct. Mater.* **17** 347

[10] Katerski A, Kärber E, Oja Acik I, Dolgov L, Mere A, Sildos I, Mikli V, Krunks M 2014 *Nanoscale Res. Lett.* **9** 494

[11] Montero A M, Gennero M R, de Chialvo G, Chialvo A C 2009 J. Mater. Chem. **19** 3276

[12] De la Garza M, Lopez I, Gomez I 2013 *Adv. Mater. Sci. Eng.* 916908

[13] Wang W, Cassar K, Sheard S J, Dobson P J, Bishop P, Parkin I P, Hurst S 2009 *Nanotechn. Construct.* **3** 395

[14] Tarwal N L, Devan R S, Ma Y R, Patil R S, Karanjkar M M, Patil P S 2012 *Electrochim. Acta* **72** 32

[15] Oja Acik I, Oyekoya N G, Mere A, Loot A, Dolgov L, Mikli V, Krunks M, Sildos I 2014 *Surf. Coat. Tech.* submitted

[16] Otto K, Oja Acik I, Krunks M, Tõnsuaadu K, Mere A 2014 *J. Therm. Anal. Calorim.* **118** 1065

[17] Kärber A, Katerski A, Oja Acik I, Mikli V, Mere A, Sildos I, Krunks M 2014 *Beilstein J. Nanotech.* **5** 2398

IOP Conf. Series: Materials Science and Engineering **77** (2015) 012010 doi:10.1088/1757-899X/77/1/012010

Microwave-assisted synthesis and photocatalytic properties of sulphur and platinum modified TiO_2 nanofibers

R Drunka, J Grabis, Dz Jankovica, A Krumina, Dz Rasmane
Institute of Inorganic Chemistry, Riga Technical University, Latvia

E-mail: reinis_drunka@inbox.lv

Abstract. In the present work formation of active TiO_2 nanofibers in microwave synthesis and their modification with platinum were studied. Mixture of anatase and rutile nanopowder and 10M KOH solution were used as raw materials. Microwave assisted synthesis method permitted to obtain TiO_2 nanofibres with a diameter of 10nm and a specific surface area up to 40.2 m^2 / g. In order to modify TiO_2 nanofibers with platinum it was stirred in H_2PtCl_6 solution and illuminated with UV irradiation or reduced with sodium boronhydride. To modify titania with sulphur and prepare co-doped nanofibers platinum doped samples were extra treated in hydrogen sulphide atmosphere. Photocatalytic activity was determined by degradation of the methylene blue (MB) solution under UV and visible light irradiation. The obtained samples showed higher photocatalytic activity with respect to pure TiO_2 nanofibers. The doped TiO_2 nanofibers were appropriate for degradation of harmful organic compounds.

1. Introduction

The synthesis of nanofibers has become one of the most important research subjects in nanotechnology and various nanotubular materials have been produced during the last decades. TiO_2 is one of the most widely used photocatalyst for photocatalytic decomposition of organic compounds, splitting water and for dye sentized solar cells because of its high photocatalytic activity, excellent chemical stability, relatively low price, and low toxicity [1-2]. However, because of its wide band gap energy of ~3.2eV, it is only active in ultraviolet irradiation. As a consequence, significant efforts have been made to develop modified forms of TiO_2 that are active under visible light irradiation [3].
The microwave-assisted hydrothermal method has unique advantages of uniform and rapid heating in comparison with the conventional one. In addition, this method can significantly reduce the reaction time, leading to the fast crystallization and simplification of the preparation procedure [4]. It is established that photocatalytic activity of TiO_2 nanoparticles can be increased by doping with nitrogen, sulfur [5, 6], with such metal oxide as WO_3 [7]. One more promising dopant is platinum [8-10].The short reaction times provided by microwave synthesis make it ideal for rapid reaction scouting and optimization of reaction conditions. However, it is difficult to compare their effectivity for photocatalysis, because the conditions of preparation methods differ. The aim of the present work is to determine the characteristics of TiO_2 nanofibers doped with Pt and co-doped with platinum and sulfur, in dependence on doping agent content and to compare their photocatalytic activity.

IOP Conf. Series: Materials Science and Engineering **77** (2015) 012010 doi:10.1088/1757-899X/77/1/012010

2. Experimental

2.1. TiO_2 nanofiber preparation

The TiO_2 samples were prepared by a microwave method similar to hydrothermal method but in shorter reaction time. For this purpose, a certain amount of mixture of TiO_2 anatase and rutile nanopowder (Sigma–Aldrich, ≥99.7%, <25nm particle size) was dissolved in 700mL of 10M KOH at room temperature. This aqueous solution was poured in to a microwave vessel made from teflon. The microwave treatment was performed at 230°C for 40min by using Anton Paar Masterwave BTR microwave system. Solution was stirred with speed 700rpm during the reaction time. The as-obtained solution was then cooled until room temperature and left to precipitate TiO_2 particles. After this procedure KOH solution was decanted from the vessel and the obtained TiO_2 suspension was diluted with large amount of deionised water to decrease concentration of KOH. Washing and decanting procedures were repeated several times. Finally, a certain amounts 1M HCl was added to TiO_2 suspension to reduce pH to 7.0. The obtained solution was filtered by using a 1.0µm cellulose nitrate membrane filter. Particles were washed on the filter several times with deionised water and 96% ethanol. The powder was submitted to drying at 110°C for 24h. After drying TiO_2 nanofiber powder was stirred in 1M HCl solution for 24h to remove titanates. This procedure also allowed decreaseing the absorption process of MB on TiO_2 nanofibers surface.

2.2. TiO_2 nanofiber modification with platinum and sulphur

Two methods were used to modify TiO_2 nanofibers with platinum nanoparticles:
A) $H_2PtCl_6 \cdot 6H_2O$ was dissolved in 100mL of deionised water and stirred for 5min. TiO_2 nanofiber powder was added to the obtained solution and stirred for 30min. To modify TiO_2 nanofibers with platinum UV light source (FEK-56, 120W mercury high pressure UV lamp) was used. The solution was transferred into a quartz beaker and stirred under UV irradiation for 40min. The distance from the light source to the beaker was 11cm. Under UV irradiation $PtCl_6^{2-}$ anions were reduced and TiO_2 nanofibers were doped with platinum.
B) TiO_2 and $H_2PtCl_6 \cdot 6H_2O$ were dissolved in 60mL ethyl alcohol. Sodium borohydride was added and the suspension was stirred for 1h while $PtCl_6^{2-}$ anions were reduced and TiO_2 nanofibers were doped with platinum.
After the modifications all suspensions were filtered by using 1µm cellulose nitrate membrane filter. The obtained powders were submitted to drying at 110°C for 24h. After the drying process TiO_2 photocatalyst nanofibers were calcinated at 500°C for 2h to obtain anatase.
Co-doped with platinum and sulfur the TiO_2 nanofibers were prepared by platinum doped sample extra treatment at 380°C in H_2S flow for 2h. For H_2S gas synthesis the reaction between FeS and HCl in Kipp's apparatus was used.

2.3. Determination of photocatalysts activity

Photocatalytic properties of the obtained TiO_2 nanofibers were tested using the degradation of MB solution under UV and visible light irradiation. FEK-56, a 120W mercury high pressure UV lamp was used as the UV light source, and Philips Torndo 23W halogen lamp as the visible light source. The degradation process of the MB solution was controlled using the Janwey 6300 spectrophotometer. The absorption of the MB solution was measured at wave length of 662nm.0.0500g TiO_2 nanofiber powder was used for the degradation of 100mL MB(7.2mg/L).The smples were centrifuged before the measurements for 5min with a rotation speed of 10 000rpm.

IOP Conf. Series: Materials Science and Engineering **77** (2015) 012010 doi:10.1088/1757-899X/77/1/012010

3. Results and discussion

Characterizations of physical properties as morphology, crystallization, chemical content, surface areas were done by scanning electron microscopy (SEM), X-ray diffraction (XRD), X-ray fluorescence (XRF), BET surface area analysis methods, respectively. Specific surface area was obtained by using BET method and HROM 3 gas analyzer. Results of the specific surface area measurements and dopant content are shown in *Table 1.*

Table 1. *Comparison of specific surface area and dopant content in various TiO_2 samples*

Sample	***Specific surface area m^2/g***	***Platinum content %***	***Sulfur content %***
Pure TiO_2 nanopowder (anatase and rutile mixture) (commercial)	*45.0*	-	-
Pure TiO_2 nanofibers	*40.2*	-	-
C: TiO_2 nanofibers +Pt (Boronhydride reduction) after calcination in air at 500°C for 2h	*29.8*	*1.06*	-
D: TiO_2 nanofibers +Pt (Boronhydride reduction) after calcination in air at 500°C for 2h	*28.2*	*0.44*	-
G: TiO_2 nanofibers +Pt (UV reduction) after calcination in air at 500°C for 2h	*25.7*	*6.15*	-
H: TiO_2 nanofibers +Pt (UV reduction) after calcination in air at 500°C for 2h	*26.4*	*1.86*	-
L: TiO_2 nanofibers +Pt (Boronhydride reduction reduction) after calcination in H_2S flow at 380°C for 2h	*31.1*	*1.39*	*10.84*

In these experiments nanofibers with a specific surface area up to 40.2m^2/g were prepared and tested. Doped with Pt, TiO_2 nanofibers had a specific surface area 25.7 to 29.8m^2/g depending on Pt content. Co-doped with platinum and sulfur, TiO_2 nanofibers had a specific surface area 31.1m^2/g.
The chemical analysis was done by using an X-ray fluorescent spectrometer Bruker Pioneer S4. It was obtained that doped samples contain platinum from 0.44% to 6.15% and sulfur from 1.15% to 10.84%. SEM results show that samples were with nanofiber structure, and individual nanowires had a diameter of 10nm and lenght up to 15μm. Nanofibers were coated with nanosize Pt particles (Fig. 1).
The phase composition was determined by using an X-ray diffraction analysis with a D8 Advance, Bruker AXS system. Samples C, D, G and H were calcinated at 500°C for 2h to crystallize amorphous TiO_2 before the XRD analysis. XRD patterns of the prepared samples showed anatase and rutile maxima and Pt maxima for sample G with high content of Pt. No other shifts of the diffraction maxima were observed.

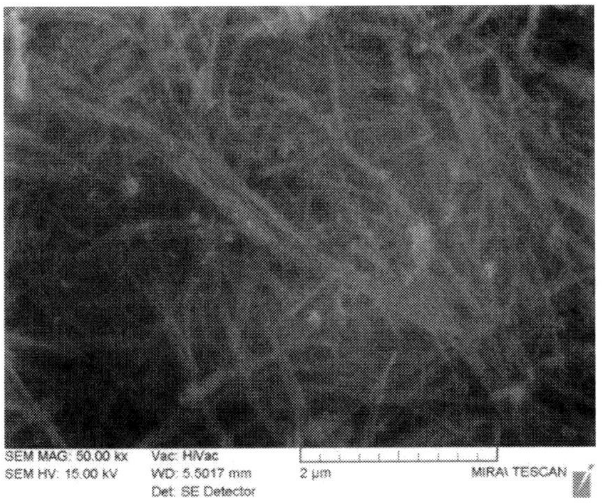

Figure 1. SEM micrograph of platinum modified TiO_2 nanofibers.

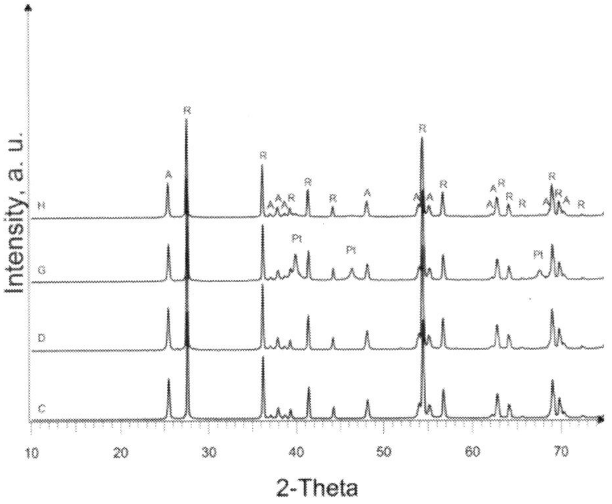

Figure 2. Results of the XRD analysis: A-anatase, Pt-platinum, R-rutile

Results of MB degradation under UV (Fig.3) and visible light (Fig 4) show that all prepared samples are active photo catalysts. Under UV irradiation up to 90% of MB was degraded in the 45min experiment and up to 50% under visible light in the 180min experiment. Sample L containing 1.39% Pt and 10.84% sulfur shows the highest activity under UV and also under visible light irradiation. Samples modified with platinum by the UV irradiation for H_2PtCl_6 reduction were with higher activity with respect to those that were prepared by using sodium boronhydride as reducing agent.

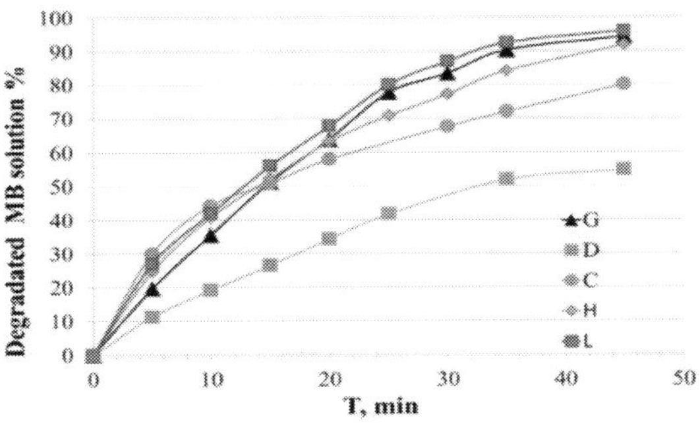

Figure 3. Degradation of MB solution under UV irradiation

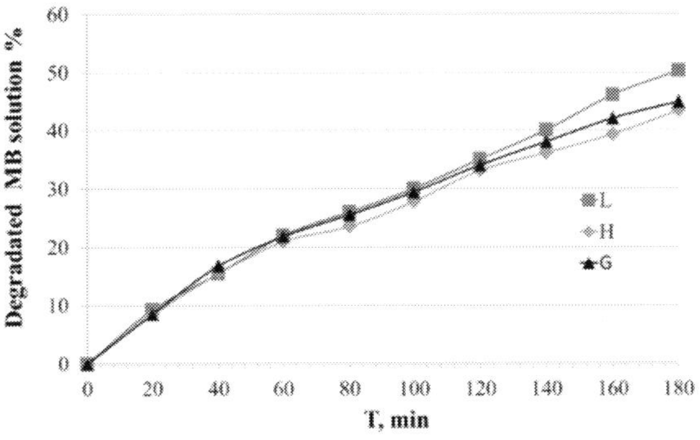

Figure 4. Degradation of MB solution under visible light irradiation

4. Conclusions

TiO_2 nanofibers were successfully fabricated through a simple microwave treatment of TiO_2 anatase and anatase/rutile mixture nanopowders. The obtained pure nanofibers, doped with platinum and co-doped with platinum and sulfur have specific surface area of up to 40.2m^2/g. Individual TiO_2 nanowires had a diameter of 10nm. Platinum and sulfur doping can effectively enhance the photoelectrocatalytic activity of TiO_2 under UV and visible light irradiation. The highest photocatalytic activity of doped TiO_2 nanofibers were shown by sample L containing 1.39% Pt and 10.84% sulfur. Modification with platinum and sulfur allows increase photocatalytic activity of TiO_2 nanofibers up to 40% under UV irradiation and up to 50% under visible light irradiation. Samples modified by using UV light source for $PtCl_6^{2-}$ anions reduction had a higher specific surface area and photocatalytic activity than samples modified using sodium boronhydride as reducting agent.

References

[1] Hashimoto K, Irie H, Fujishima A 2005*jpn. J. Appl. Phys.***12** (44) 8269
[2] Jitputti, J., Pavasupree, S., Suzuki, Y., Yoshikawa, S. 2008 *Jpn. J. Appl. Phys.* **47** 751
[3] LvJ, Sheng T, Su L, Xu G, Wang D, Zheng Z, Wu Y 2013 *Appl. Surf. Sci.* **284** 229
[4] Li L, Qin X, Wang G, Qi L, Du Q, Hu Z 2011 *Appl. Surf. Sci.* **257** 8006
[5] Sun L, Cai J, Wu Q, Huang P, Su Y, Lin C 2013 *El.chim.Acta* **108** 525
[6] Xiang Q, Yu J Jaroniec M 2011 Phys. Chem., **13**, 4853
[7] Chin Wei Lai, Srimala Sreekantan 2013 *El.chim.Acta* **87** 294
[8] Kim S, Hwang S,Choi W 2005 *J Phys Chem B.* **51** 109 24260
[9] Sui H L, Wang Z B, Yang M, Huo L, Gu D M, Yin G P 2014 *J. Pow. Sour.***255** 43
[10] Kim W, Tachikawa T, Kim H, Lakshminarasimhan N, Murugan P, Park H, Majima T, Choi W 2014 *Appl. Catal.B: Envi.* **147** 642

IOP Conf. Series: Materials Science and Engineering **77** (2015) 012011 doi:10.1088/1757-899X/77/1/012011

Hydrothermal Synthesis of Cobalt Ferrite Nanosized Powders

I Zalite[1], G Heidemane[1], L Kuznetsova[1], M Maiorov[2]

[1] Institute of Inorganic Chemistry of Riga Technical University, Latvia

[2] Institute of Physics of University of Latvia, Latvia

e-mail: ilmars@nki.lv

Abstract. Cobalt ferrite powders are synthesized by the co-precipitation technology, combined with the hydrothermal synthesis method and crystallite size, specific surface area (SSA), magnetic properties of synthesized products are obtained. All the synthesized ferrites are nanocrystalline single phase materials with crystallite size of 10-16 nm the SSA of 60±5 m^2/g and the calculated particle size of 20±2 nm. Synthesized Co ferrites are characterized by the saturation magnetization M_S of 59-60 emu/g, remanent magnetization M_r of 23 emu/g and coercivity H_c of 570-650 Oe.

1. Introduction

Recently ferrites have received an increased scientific and commercial interest. The most significant and popular usage of ferrites are in optics, electronics, mechanics and other technical fields [1]. Ferrites are also of great importance in medicine, for biomedical purposes and in chemical catalysis. More information appears about hyperthermia in scientific articles. With this method, ferrite nanoparticles are introduced in living organism and in controlled conditions nanopartices are transported to the cancer zones in organism, and in magnetic field with thermal treatment, cancer cells are eliminated [2].

The properties of nanomaterials vary from it morphology, particle size and microstructure and it is important to analyze methods of synthesis of nanoparticles, as well as obtain new metal oxides with new properties. Materials in the form of nanowires, nanosheets, etc., are especially important, because of their improved optical, electronic and magnetic properties allowing to extend application limits.

Ferrites with spinel structure are significant materials in development of several technological applications, where materials with high density and low porosity are necessary [3]. Ferrites, as most of ceramic materials, are obtained in solid phase reactions from different oxides. By improving nanotechnology, several liquid phase and gas phase synthesis methods have been developed – hydrolysis, hydrothermal synthesis [4], pyrolysis, microwave synthesis [1], co-precipitation method [5], sol-gel method [6], combustion [7] and plasma synthesis [8].

In this research the cobalt ferrite nanopowders synthesis have been prepared by hydrothermal method and their mechanical and magnetic properties have been studied.

IOP Conf. Series: Materials Science and Engineering **77** (2015) 012011 doi:10.1088/1757-899X/77/1/012011

2. Experimental

Synthesis of cobalt ferrite nanoparticles was performed by the co-precipitation technology, combined with the hydrothermal synthesis method. By the hydrothermal method cobalt ferrites were synthesized using reagent grade chemicals: $FeCl_3 \cdot 6H_2O$, urea, $Co(NO_3)_2 \cdot 6H_2O$, NaOH. Urea was hydrolyzed for 3 hours in the solution of $FeCl_3$ (molar ratio 3:1) at 70-75 °C. Cobalt nitrate was added to the cooled reaction mixture. The molar ratio $FeCl_3 \cdot 6H_2O$: $Co(NO_3)_2 \cdot 6H_2O$ was corresponding to the stoichiometry of the metal ions in ferrite. Stirring the suspension continuously with the 40% solution of NaOH, cobalt hydroxide is slowly precipitated until pH of the suspension is of 9-10, then inserted in the ultrasound bath for 20 min and after it treated for 24 h at 40 °C. The precipitate is washed by decanting with distilled water until the presence of Cl^- ion can not been registered. The volume of hydroxides mixture is reduced by decanting until 250 ml which are poured into the reaction vessel inserted in the autoclave. Then the obtained mixture of hydroxides was treated hydrothermally at different regimes (T = 200-250 °C, t = 1-3 h, p = 17-17,5 MPa). After hydrothermal treatment the formed precipitate is filtered by the water jet pump using the membrane filter, precipitate is washed by distilled water and dried at 40 °C.

Ferrite nanopowders were prepared for sintering as follows: 3 wt. % stearic acid is mixed mechanically with ferrite nanopowder sample for 1 h in planetary mills (400 rpm, vessel material ZrO_2, milling ball material ZrO_2) by using isopropanol as dispersing environment. After mixing samples were dried in an oven at 80 °C and sieved. For pressure-less sintering samples were pressed (200 MPa) in pellets with diameter of 12 mm and height of 4 – 6 mm. Stearic acid was burned out at 600 °C. Samples were sintered for 2 h at isothermal conditions in air atmosphere at the temperature range from (1000 – 1200) °C (10 °C/min).

All samples were analyzed by the X-ray diffractometer Advance 8 (Bruker AXS). Measuring parameters: CuK_α radiation (1.5418 Å), range (10 – 70) 2Θ°, increment 0.02 2Θ°, time per step 1 s, to reduce background effects the SolX detector was used. Crystallite size was determined by the Scherrer's equation. Magnetic properties of the synthesized ferrites were analyzed by vibrating sample magnetometry (VSM Lake Shore Cryotronics, Inc., model 7404 VSM). SSA was measured with the BET single point method. Measurements were made at the boiling point of liquid nitrogen (–196 °C), the adsorbing gas was Ar (~7 % Ar gas mixture in He gas).

3. Results and discussion

The characteristics of synthesized ferrite products are given in Table 1 and in Figures 2 – 3. It is found that all by hydrothermal synthesis obtained ferrites, except sample No 1 (synthesized 1 h at 200 °C - this sample shows also FeOOH phase), are nanocrystalline stoichiometric single phase materials (Fig. 2) with specific surface area of (55 – 63) m^2/g and calculated particle size of 20±2 nm (Fig. 3). The crystallite size in ferrite is of 10-16 nm. None of the obtained samples No 2-6 show other additional phases (commonly magnetite, maghemite, hematite or other metal oxides), which proves that obtained samples are of high purity.

With the increase of the temperature and the time of the hydrothermal treatment, the magnetic properties (saturation magnetization M_S, remanent magnetization M_r, emu/g and coercivity H_c, Oe) are also increasing (Table 1).

Slight weight changes (~ 1,5 mass %) and minor exo-effect in the temperature range of 250-350 °C has been found by the thermal analysis. The total change of the weight after treatment to the constant weight was ~ 7 mass %, therefore it is recomendated to calcinate the sample for 1 h at 400 °C. The crystallite size and SSA after calcinating change insignificantly (Table 2). After thermal treatment at higher temperatures the SSA of the ferrites has a tendency to decrease, while the crystallite size increases. This could be explained that particles are re-crystallizing and growing at higher temperatures, so the SSA is decreasing (Figure 4). With increase of temperature of thermal treatment, saturation magnetization of the ferrites increases (Figure 5).

Table 1. Characteristics of $CoFe_2O_4$ nanopowders prepared at different regimes.

No	MODE, °C/h	SSA, m^2/g	d_{50}, nm*	Crystallite size, nm	XRD phases	Magnetic properties		
						M_s, emu/g	M_r, emu/g	H_c, Oe
1	200/1	58	20	~10	$CoFe_2O_4$, FeOOH	13,2	2,6	105,1
2	200/3	59	19	13-14	$CoFe_2O_4$	-	-	-
3	230/1	63	18	10-13	$CoFe_2O_4$	50,0	10,2	494,0
4	230/3	55	21	15-16	$CoFe_2O_4$	58,9	17,8	643,4
5	250/1	61	19	12-13	$CoFe_2O_4$	57,3	17,3	565,5
6	250/3	62	18	10-12	$CoFe_2O_4$	59,8	16,8	574,2

*calculated from SSA

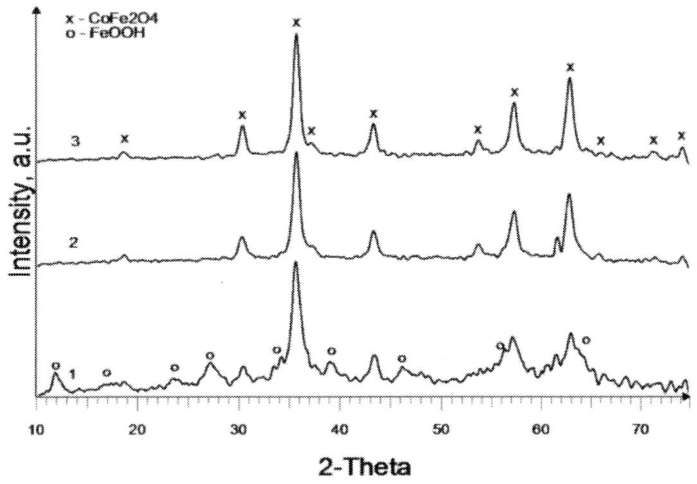

Figure 2. XRD patterns of the $CoFe_2O_4$ nanopowder prepared at:
1 – 200 °C, 1h; 2 – 230 °C, 1h; 3 – 250 °C, 1h.

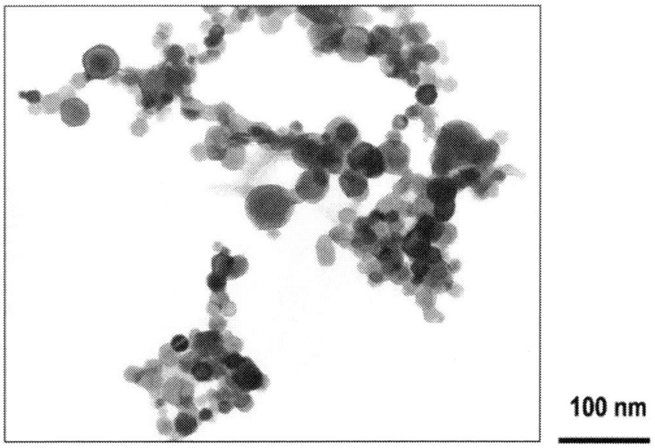

Figure 3. Electron microscopy images of $CoFe_2O_4$ obtained by the hydrothermal method.

IOP Conf. Series: Materials Science and Engineering 77 (2015) 012011 doi:10.1088/1757-899X/77/1/012011

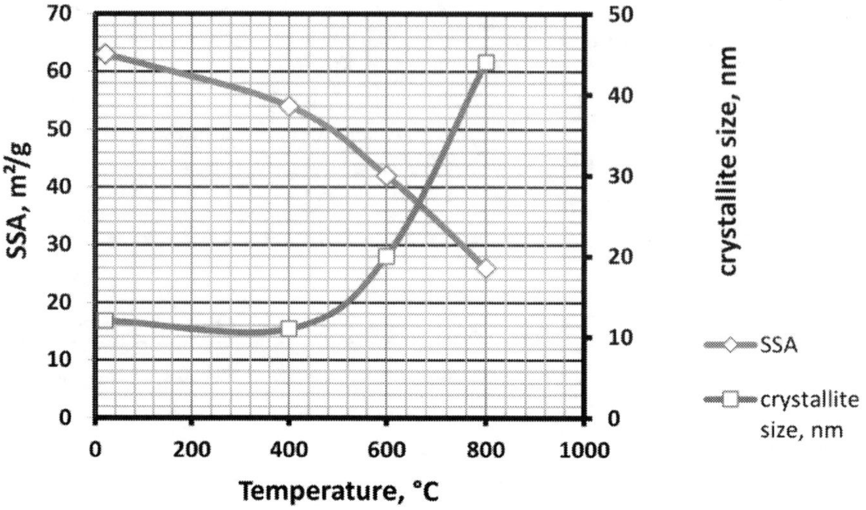

Figure 4. Specific surface area (SSA) and crystallite size of $CoFe_2O_4$ as functions of hydrothermal treatment temperature.

Table 2. Properties of synthesized $CoFe_2O_4$ nanopowders (sample No 3) after thermal treatment (2 h at different temperatures).

Heating temperature, °C	SSA, m²/g	d_{50}, nm*	Crystallite size, nm	XRD phases	Magnetic properties		
					M_s, emu/g	M_r, emu/g	H_c, Oe
Raw powder	63	18	10-13	$CoFe_2O_4$	50,0	10,2	494,0
400	54	21	10-11	$CoFe_2O_4$	50,1	12,6	390,1
600	40	29	19-20	$CoFe_2O_4$	62,8	22,4	757,0
800	26	44	43-45	$CoFe_2O_4$	71,6	28,9	875,8

*calculated from SSA

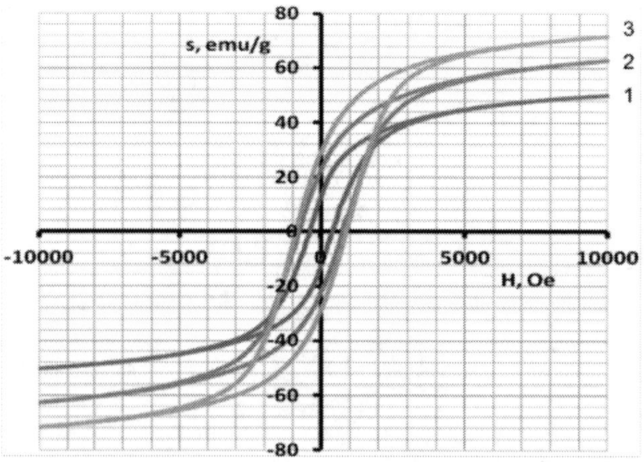

Figure 5. The magnetic properties of the sample No. 3 after thermal treatment at 400 (1), 600 (2) and 800 (3) °C.

After thermal treatment at 800 °C the magnetic saturation increased to 72 emu/g, which is close to the values of the standard bulk material (80 emu/g for $CoFe_2O_4$ [10]).

Sintering of nanosized $CoFe_2O_4$ powder was made at 1000, 1100 and 1200 °C. Some properties of the sintered ferrites are shown in Table 3. By increasing the sintering temperature, the size of nanograins and magnetization increases while coercivity decreases. In comparison to the ferrite nanopowders, obtained by hydrothermal method, the sintered materials have higher magnetization and lower coercivity, which could be explained with particle size and also crystallite size increase.

Table 3. Properties of $CoFe_2O_4$ (sample No 3) after 2h sintering.

Heating temperature, °C	Density, %	Open porosity, %	Crystallite size, nm	Magnetic properties		
				M_S, emu/g	M_r, emu/g	H_c, Oe
Raw powder	-	-	10-13	50,0	10,2	494,0
1000	81,3	14,2	75	-	-	-
1100	94,3	0,8	90-95	77,0	20,7	492,8
1200	95,0	0,1	150	81,3	14,1	168,6

4. Conclusions

Single phase cobalt ferrite nanopowders can be successfully synthesized by the hydrothermal synthesis method. The average particle size of the nanopowders obtained by this method is in the range of 20±2 nm (SSA of 60±5 m^2/g). The magnetic saturation values of these samples depend on hydrothermal synthesis temperature and time and reach up to 59-60 emu/g. After thermal treatment at 800 °C the magnetic saturation increased to 72 emu/g, which is close to the values of the standard bulk material (80 emu/g for $CoFe_2O_4$).

References

[1] Xue B, Liu R, Xu Z D, Zheng Y F 2008 *Chemistry Letters* **37** (10) 1058
[2] Fortin J P, Wilhelm C, Servais J, Menager C, Bacri J C F, Gazeau J 2007 *Chemical Society* **129** 2628
[3] Domenichini B, Caillot T 2003 *Acta Material* **51** 4815
[4] Millot N, Gallet S L, Aymes D, Bernard F, Grin Y 2007 *J. of Europ. Ceram. Soc.* **27** 921
[5] Ferreira T A S, Waerenborgh J C, Mendonça M H, Nunes M R, Costaa F M 2003 *Solid State Sciences* **5** 383
[6] Tang D Q, Zhang D J, Ai H 2006 *Chemistry Letters* **35** (11) 1238
[7] Zhongpo Z, Yue Z, Ziyu W, Wei W, Wufeng T, Jing S, Rui X 2008 *Applied Surface Science* **254** 6972
[8] Grabis J, Zalite I 2007 *Materials Science Forum* **555** 267
[9] Smit J, Wijn H P J 1959 In: Ferrites. Philips Technical Library. Eindhoven, Netherlands
[10] Cornell R M, Schwertmann U 2003 In: The Iron Oxides. Structure, Properties, Reactions, Occurrences and Uses. Wiley-VCH, Weinheim

IOP Conf. Series: Materials Science and Engineering **77** (2015) 012012 doi:10.1088/1757-899X/77/1/012012

ZnO Nanorods Grown Electrochemically on Different Metal Oxide Underlays

I Gromyko, T Dedova, M Krunks, V Syritski, A Mere, V Mikli, T Unt and I Oja Acik

Department of Material Science, Tallinn University of Technology, Ehitajate tee 5, 19086 Tallinn, Estonia

E-mail: inga.gromyko@gmail.com

Abstract. In this study we present results on electrochemically grown ZnO nanorods on different metal oxide underlays, such as ZnO seed layers with different morphologies, ZnS and TiO_2 compact thin films produced by spray pyrolysis on transparent conductive oxide (TCO) substrates. Also in this work we present results on ZnO nanorods directly deposited on some chosen TCO substrates. The relationship between nanorod formation and substrate properties were studied. All ZnO nanorod layers were grown electrochemically using $ZnCl_2$ aqueous solutions (c=0.2 mmol/L) at the bath temperature of 80 °C during one hour. The structural properties and morphology of metal oxide underlays and ZnO nanorods grown on them were studied by scanning electron microscopy (SEM), x-ray diffraction spectroscopy (XRD). Depending on the substrate morphology, ZnO rods with different dimension, orientation, shape and density were obtained. For instance, larger rods (d~200 nm, l~700 nm) were obtained on substrates, such as ITO/glass, FTO/glass and ZnO:In/ITO/glass. Smaller rods (d~60 nm, l~350 nm) were obtained on smooth, uniform and fine-grained underlays, such as ZnS and TiO_2.

1. Introduction

Zinc oxide (ZnO) is an n-type II-VI semiconductor with a direct band gap (3.37 eV) and large exciton binding energy of 60 meV [1-2]. ZnO has attracted research interest in recent years due to its potential applications in various nanodevices such as lasers and light emitting diodes [3], gas sensors [1, 4], field emission devices [1, 4] and solar cells [4, 5]. For hybrid organic/inorganic solar cells with "absorber layer/ZnO nanorod/blocking layer/TCO (transparent conductive oxide)" structure, it is highly important to synthesize ZnO nanorod/blocking layer/TCO structures. In order to increase the solar cell performance, it is desired to obtain high aspect ratio (l/d, where l – length and d – diameter) conductive ZnO nanorods. Among various deposition techniques, electrodeposition is a promising approach for growing ZnO nanorods, because it is simple, low-cost, low temperature and easily scalable to large-area deposition [6-7].

Morphology of the substrate has a particular significance in nanorods growth. A seed layer (usually ZnO thin film) is required prior to the electrodeposition of the ZnO nanorods to contribute the rods growth or improve their vertical alignment [1, 8, 9]. For some applications (emitting diodes, solar cells), it is required to grow ZnO nanorod directly on a transparent conductive oxide [10-11]. In this paper, we present a study on the growth of ZnO nanorods by electrodeposition on various substrates: 1) commercially available indium tin oxide (ITO) and fluorine doped tin oxide (FTO) coated glass substrates, 2) set of ZnO seed layers with various morphologies prepared by spray pyrolysis onto ITO/glass substrates and 3) ZnS and TiO_2 thin films deposited by spray pyrolysis on ITO/glass substrates. Herein we study relationship between the initial morphology of the substrate and final morphology of the ZnO nanorod layer. The morphology and structural properties of the substrates and ZnO nanorods deposited on them, respectively, are studied by means of high resolution scanning electron microscopy (SEM) and X-ray diffraction (XRD) techniques.

2. Experimental

2.1. Synthesis details

Prior the deposition, ITO and FTO covered glass substrates were washed thoroughly with soap, ethanol and sulphuric acid. The samples were rinsed with deionised water after each cleaning step. On some chosen ITO/glass and FTO/glass substrates, the ZnO nanorods were deposited directly onto the substrate. Commercially available ITO/glass substrates have been chosen for the deposition of metal oxide underlays such as ZnO seed layers, ZnS and TiO_2 thin films prepared by using spray pyrolysis method as described in details earlier [12-13]. Electrodeposition of ZnO nanorods was carried out potentiostatically in a three-electrode glass cell, where the TCO/glass was used as a working electrode, while a silver/silver chloride/3M KCl (Ag/AgCl/KCl) and a platinum (Pt) wire were used as reference and counter electrodes, respectively.

Solution of 0.2 mmol $ZnCl_2$ (Sigma-Aldrich), and 0.1 M KCl in amount of 50 ml was utilized as a supporting electrolyte. The growth temperature was kept at 80 °C using a temperature controlled circulating bath. Deposition time was fixed to 1 hour. Electrochemical deposition was done under -1.0 V potential, vs the reference electrode using the Radiometer Analytical potentiostat PGP201.

2.2. Characterization

The morphology of the substrates was studied by the high resolution scanning electron microscope Zeiss EVO-MA15 at the operating voltage of 10 kV. The crystal structure of the nanorods was characterized by using X-ray diffraction on a Rigaku Ultima IV diffractometer using CuKα radiation (λ=1.5406Å, 40 kV at 40 mA). Crystals density was evaluated from the SEM surface images, from area of 6.2 μm×4.7 μm.

3. Results and Discussion

3.1. Deposition of ZnO rods on TCO electrodes

In this study two different types of conducting electrodes were used: commercially available ITO and FTO coated glass substrates. The SEM images of TCO electrodes are shown in figure 1a (from ITO/glass) and in figure 1c (from FTO/glass). Corresponding morphologies of the electrodeposited ZnO layers directly grown on ITO and FTO electrodes are presented in figures 1b and 1d, respectively. As can be seen from the figures, the final shape of the ZnO nanorods differs depending on the substrate morphology. In case of ITO/glass, the rods are thinner (d~170 nm, l~700 nm) and show higher density (~7.4×10^8 cryst./cm^2) than those grown on FTO/glass (d~300 nm, l~600 nm, density ~6.5×10^8 cryst./cm^2). Such difference can be explained by the difference in the morphology of the substrate. Higher density of the grains on the FTO electrode (Figure 1c) provoke large amount of nuclei that can further coalescence and promote the lateral growth of the crystal. Similar behaviour has

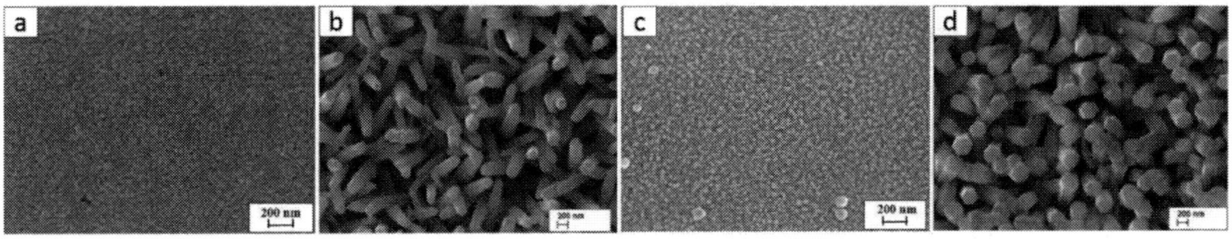

Figure 1. SEM images of: a) ITO/glass substrate, b) ZnO nanorods grown onto ITO/glass, c) FTO/glass substrate, d) ZnO nanorods grown onto FTO/glass.

already been reported for spray pyrolysis deposited ZnO nanorods on ITO/glass and FTO/glass substrates [14-15]. The resistivity of the substrates might be also a reason of such difference in final shape of the ZnO crystals. In our case, the ITO has lower resistivity (ρ~15-20 Ω·cm) than resistivity of the FTO susbtrate (ρ~150 Ω·cm). It was reported earlier by Kim et al., that conductive substrates greatly affect on the structural and optical properties of ZnO nanorods [16].

3.2. Deposition of ZnO nanorods on ZnO seed layers obtained by spray pyrolysis

In previous studies on ZnO thin films by spray pyrolysis [17-19], it was shown that temperature, concentration of the precursor solution (zinc acetate), dopant type and amount are the most important technological parameters that influence the morphology of the film. Here we prepared ZnO seed layers by spray pyrolysis on ITO/glass substrates using different technological parameters mentioned above. The morphology of ZnO seed layers are shown in figures 2a, 2c, 2e and 2g. The SEM images of ZnO nanostructures grown on those seed layers are presented in figures 2b, 2d, 2f and 2h, respectively.

Figure 2. SEM images of: a) ZnO seed layer SL-1 grown by spray of zinc acetate solution with c=0.05 mol/L at Ts=320 °C; b) ZnO nanorods on seed layer SL-1; c) ZnO seed layer SL-2 grown by spray of zinc acetate solution with c=0.1 mol/L at Ts=320 °C; d) ZnO nanorods on seed layer SL-2; e) ZnO seed layer SL-3 grown by spray of zinc acetate solution with c=0.1 mol/L at Ts=420 °C; f) ZnO nanorods on seed layer SL-3; g) ZnO seed layer doped with In (3 at.%) SL-4 grown by spray of zinc acetate solution with c=0.2 mol/L at Ts=440 °C, (ZnO:In); h) ZnO nanorods on seed layer SL-4.

Seed layer SL-1 is compact and flat, composed of densely packed fine grains. As a result, ZnO nanorods grown on SL-1 have a diameter of 80 nm and length of ca. 270 nm, and show uniform dense coverage (figure 2b). SL-2 seed layer from solution with higher concentration at the same temperature have highly structured surface composed of the flake-like grains with different sizes (figure 2c). Because of such nonuniformity in the dimensions of the grains on this substrate, the sizes of the ZnO rods vary greatly. ZnO nanorod layers compose of smaller crystals (d~80 nm) as well as of much larger crystals (d~350 nm) with length of ca. 300 nm (figure 2d). Seed layer SL-3 is also composed of grains with different shapes including platelets and nanoneedles (figure 2e). The ZnO nanostructures grown on this seed layer are mixture of fat and thin compactly standing ZnO crystals (d~150-300 nm, l~150 nm) (figure 2f). The ZnO nanorods have been deposited also onto ZnO:In seed layer. SEM image for SL-4 seed layer is presented in figure 2g and for the resultant ZnO nanorod crystals in figure 2h. Rods have uniform dimensions (d~200 nm, l~700 nm) comparable to those grown on bare ITO/glass layer. Rods grown on ZnO:In are markedly larger than those grown on undoped ZnO layers, that might be explained with difference in conductivity of ZnO:In and ZnO films [20-21].

XRD patterns of ZnO seed layers and of samples with ZnO nanorod grown on those seed layers are presented in figure 3. All the samples show the reflections at 2θ of 31.73°, 34.36° 36.21° and 62.75° corresponding to the (100), (002), (101) and (103) planes of ZnO (PDF 01-080-0074) [20].

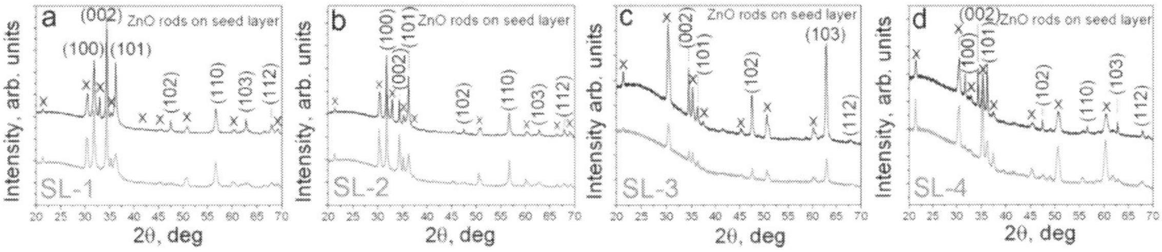

Figure 3. XRD patterns of the nanorods grown on different ZnO seed layers: a) SL-1; b) SL-2; c) SL-3; d) SL-4; x – reflections from ITO/glass substrate.

For the SL-1 seed layer, the $I_{(002)}/I_{(100)}$~1 and $I_{(002)}/I_{(101)}$~4 (figure 3a). As in ZnO powder reference [20] the $I_{(002)}/I_{(100)}$~0.7 and $I_{(002)}/I_{(101)}$~0.4, thus the crystallites in SL-1 film are orientated along the (002) plane parallel to the substrate. The $I_{(100)}/I_{(002)}$~1.9 is characteristic of the SL-2 ZnO seed layer, in ZnO powder $I_{(100)}/I_{(002)}$~1.4, thus the orientation is along the (100) plane (figure 3b). For the SL-3 ZnO seed layer the $I_{(103)}/I_{(002)}$~2 , in ZnO powder $I_{(103)}/I_{(002)}$~0.7, thus the crystallites are orientated along the (103) plane (figure 3c). According to XRD pattern of ZnO:In film (figure 3d), the intensity of the (002) reflection is negligible and the intensity of the (101) is the strongest, the crystallites do not have preferred orientation. Observed result is in a good correspondence with the results of earlier studies [21]. Therefore, according to the XRD patterns presented in figure 3, all the ZnO underlays show different crystallographic orientations. Herein are presented intensities ratios of resultant ZnO nanostructured layers: $I_{(002)}/I_{(101)}$~1.6, $I_{(100)}/I_{(101)}$~1.2, $I_{(103)}/I_{(002)}$~2.2, $I_{(002)}/I_{(101)}$~1 (figures 3a, b, c, d, respectively). Comparing them with the ZnO powder references [20] where the $I_{(002)}/I_{(101)}$~0.4, $I_{(100)}/I_{(101)}$~0.56, $I_{(103)}/I_{(002)}$~0.7, $I_{(002)}/I_{(101)}$~0.4 it seems that for all resultant ZnO nanostructured layers intensities ratios changed. Orientation along the (002) plane is characteristic of all electrodeposited ZnO layers except of ZnO rods on seed layer SL-2 (figure 3b). It could be that differences in seed layer thicknesses (SL-2 is deposited from more concentrated solution than SL-1 and thus SL-2 layer is probably thicker [15, 22]) or surface morphologies (compare figures 2a and 2c) are responsible for such kind behavior. XRD study shows that the ZnO rods grow preferably along the (002) plane parallel to the substrate independent of the orientation of corresponding ZnO seed layer.

3.3. Deposition of ZnO nanorods on ZnS and TiO_2 underlays grown by spray on ITO/glass substrates

For some purposes (dye-synthesized solar cell etc.) it is important to have a less conductive thin barrier layer between TCO and ZnO nanorods, therefore it is important to find out technological parameters in order to grow nanorods on such types of the substrate. Here we prepared ZnS and TiO_2 thin films by spray pyrolysis using knowledge of the earlier studies on ZnS [23] and TiO_2 films deposition by spray [24]. Both ZnS and TiO_2 films have smooth morphologies according to SEM images (figures 4a and 4c). Both substrates resulted in very uniform and dense coverage of well-shaped hexagonal ZnO nanorods. The dimension of the rods grown onto ZnS are slightly larger (d~150 nm, l~450 nm), than those grown on TiO_2, (d~60 nm and l~370 nm). The difference in rods dimensions could be explained by different surface roughness on ZnS and TiO_2 underlays.

Figure 4. SEM images of the a) ZnS underlay; b) ZnO nanorods deposited on underlay (a); c) TiO_2 underlay; d) ZnO nanorods deposited on underlay (c).

4. Conclusions

Herein we presented a study on the growth of ZnO nanorods by electrodeposition on various substrates: commercially available ITO, FTO substrates on glass, metal oxide underlays such as ZnO seed layers, ZnS and TiO_2 thin films prepared by using spray pyrolysis. The underlays with highly structured surface composed of grains with different sizes provoke the growth of ZnO crystals with different sizes and shape. ZnO rods were grown on smooth, flat, uniform and fine-grained substrates, such as ITO, FTO, ZnS and TiO_2 thin films, ZnO seed layer obtained at 320 °C and c=0.05 mol/L and indium-doped ZnO seed layer. The conductivity of the substrates might also affect the shape of resultant ZnO crystals. Electrically more conductive substrates- such as ITO, FTO and indium-doped ZnO seed layer provoke the growth of ZnO rods with larger dimensions. Thus, dimensions, shape and density of the rods depends strongly on the properties of the substrate or seed layer used, the orientation of the seed layer has minor effect.

5. Acknowledgements

This work is financially supported by the Estonian Ministry of Education and Research (IUT 19-4), TUT base financing project B24 and European Regional Development Fund through the projects: Centre of Excellence "Mesosystems: Theory and Applications" (MESO) TK114, 3.2.0101.11-0029 and project "Efficient plasmonic absorbers for solar cells" 3.2.1101.12-0023 and the Estonian Research Council grant PUT150.

References

[1] Yi G-C, Wang C and Park W I 2005 *Semicond. Sci. Technol.* **20** 22
[2] Wang Z L 2004 *J. Phys.: Condens. Matter* **16** 829
[3] Choi Y-S, Kang J-W, Hwang D-K and Park S-J 2010 *IEEE Trans. Electr. Dev.* **57** 26
[4] Weintraub B, Zhou B, Li Y and Deng Y 2010 *Nanoscale* **2** 1573
[5] Yin Z, Wu S, Zhou X, Huang X, Zhang Q, Boey F and Zhang H 2010 *Small* **6** 307
[6] Pauporte T, Bataille G, Joulaud L and Vermersch F J 2010 *J. Phys. Chem.* C **114** 194
[7] Kim H, Moon J Y and Lee H S 2011 *Electron. Mater. Lett.* **7** 59–62
[8] Kim D C, Kong B H and Cho H K 2008 *J. Mater. Sci. Mater. Electron.* **19** 760
[9] Yin Y T, Que W X and Kam C H 2010 *J. Sol-Gel Sci. Technol.* **53** 605
[10] Lee H K, Kim M S and Yu J S 2011 *Nanotechnology* **22** 445602
[11] Masyda Y and Kato K 2009 *Cryst. Growth Des.* **9** 3083
[12] Dedova T, Krunks M, Oja Acik I, Klauson D, Volobujeva O and Mere A 2013 *Mater. Chem. Phys.* **141** 69
[13] Krunks M, Dedova T and Oja Acik I 2006 *Thin Solid Films* **515** 1157
[14] Dedova T, Oja Acik I, Krunks M, Mikli V and Volobujeva O 2012 *Thin Solid Films* **520** 4650
[15] Dedova T, Volobujeva O, Krunks M, Mikli V, Gromyko I, Katerski A and Mere A 2013 Growth of Zinc Oxide Rods on FTO Electrodes by Spray Pyrolysis *Int. Conf. on Functional Materials and Nanotechnologies (Tartu, Estonia, 21–24 April 2013) IOP Conf. Series: Mater. Sci. Eng. vol 49* (Bristol: IOP Publishing) pp. 1-4
[16] Kim H, Moon J Y and Lee H S 2009 *Electron. Mater. Lett.* **5** 135
[17] Krunks M and Mellikov E 1995 *Thin Solid Films* **270** 33
[18] Dedova T, Klauson J, Badre C, Pauporte Th. Nisumaa R, Mere A, Volobujeva O and Krunks M 2008 *Phys. Status Solidi* A **205** 2355
[19] Kriisa M, Karber E, Krunks M, Mikli V, Unt T, Kukk M and Mere A 2014 *Thin Solid Films* **555** 87
[20] International Centre for Diffraction Data (ICDD), PDF-2 Release 2008
[21] Kriisa M, Krunks M, Karber E, Kukk M, Mikli V and Mere A 2013 *J. Nanomater.* **2013** 1
[22] Shinde V R, Gujar T P and Lokhande C D 2006 *Sensors and Actuators B: Chemical* **120** 551
[23] Dedova T, Krunks M, Gromyko I, Mikli V, Sildos I and Unt T 2014 *Phys. Status Solidi* A **211** 514
[24] Oja Acik I, Junolainen A, Mikli V, Danilson M and Krunks M 2009 *Appl. Surf. Sci.* **256** 1391

IOP Conf. Series: Materials Science and Engineering **77** (2015) 012013 doi:10.1088/1757-899X/77/1/012013

Properties of PP/MWCNT-COOH /PP composites made by melt mixing versus solution cast /melt mixing methods

I Reinholds[1], Z Roja[1], J Zicans[2], R Merijs Meri[2], J Bitenieks[2]

[1] Faculty of Chemistry of University of Latvia, 48 Valdemara Str., Riga, Latvia
[2] Institute of Polymer Materials, Riga Technical University, 14 Azenes Str. 14, Riga, Latvia

E-mail: ingars.reinholds@lu.lv

Abstract. An approach on improvement of the properties of polypropylene / carbon nanotube (PP/CNT) composites is reported. PP blend compositions with carboxylic acid functionalized multi-walled carbon nanotubes (MWCNT-COOH) at filler content 1.0 wt.% were researched. One part of the composites was manufactured by direct thermoplastic mixing PP with the filler, but the other one was made from solution casted masterbatch with the following thermoplastic mixing. An increase of mechanical properties (Young's modulus, storage modulus and tensile strength), compared to an increase of glass transition temperature indicated a reinforcement effect of CNTs on PP matrix, determined from the tensile tests and differential mechanical analysis (DMA), while the elongation was reduced, compared to PP matrix. By differential scanning calorimetry (DSC) analysis, the effect of nanofiller on the reorganization of PP crystallites was observed. A noticeable enhanced effect on increase of the crystallization temperature was indicated for masterbatch manufactured composite. An increase of thermal stability was also observed, compared to pristine PP and the composite made by direct thermoplastic mixing PP with the filler.

1. Introduction

As one of the most used thermoplastic polymers, isotactic polypropylene, has many benefits in industrial applications due to its low cost, recyclability, noteworthy mechanical properties and the thermal resistance determined by its tacticity. From this point, it is a prospective material for manufacturing reinforced nanocomposites with novel properties. Single and multi-walled (SWCNT, MWCNT) carbon allotropic nanofillers and their modified forms have many benefits on the improvement of mechanical, electrical, thermal and other properties of polymer materials [1-2]. For sufficient enhancement of PP nanocomposite properties due to the reinforcing effect, the carbon nanofiller should be well dispersed in PP matrix [2-3]. Melt mixing is the most common method used for manufacturing composites of PP with macro and nano fillers [3]. The physical properties of such materials depend directly on the content of nanofiller, as an increase of its content can lead to reduction of performance due to agglomeration. The Van der Waals interaction forces of filler are too strong to be disrupted by the mechanical mixing [4]. The aim of the present work was to investigate the physical properties of composites made from isotactic PP and the carboxyl group functionalized MWCNTs by the direct thermoplastic mixing and by the formation of blends from PP/CNT masterbatches made by solution mixing and sonification both the components in o-xylene and to study extensively the dispersion of filler particles by the change of mechanical and thermal properties.

IOP Conf. Series: Materials Science and Engineering **77** (2015) 012013 doi:10.1088/1757-899X/77/1/012013

2. Materials and investigation methods

2.1. Materials

As the matrix, isotactic PP of trademark Moplen HP456J supplied by LyondellBasell Industries (density 0.905 g/cm^3, melt flow index 3.0 g/10 min) was used. Industrial grade carboxylic group functionalized multi-walled carbon nanotubes (MWCNT-COOH) of trademark TNIMC4 supplied by Chengdu Organic Chemicals (carboxylic acid content 1.55 wt.%, density 2.1 g/cm^3) synthesized by chemical vapor deposition (CVD) were applied as fillers. Reagent grade solutions of o-xylene (Alfa Aesar) and methanol (Penta) were used for masterbatch preparation.

The sample description for creation PP/1 wt.% MWCNT-COOH composite from masterbatch (composite B) is as follows: PP properly weighted using analytical weights was added to 100 mL of o-xylene in Erlenmeyer flask. The solution was heated in glycerol bath equipped with temperature measurement for 30 minutes at 125 °C with 1,000 rpm. In the second flask, MWCNT-COOH was added to the second portion of 100 mL of o-xylene followed by the mixing at 1,000 rpm for 15 minutes at room temperature and for 30 minutes in an ultrasound bath (ULTRASONS 3000515, 50/60 Hz, 4000 W). The solution of filler dispersion in o-xylene was heated at 125 °C for about 5 minutes and added to the mixture of PP/ solution followed by heating up to 10 minutes at 130 °C. The hot mixture was sonificated for 5 min in ultrasound bath followed by the heating for 5 min at 130 °C, repeated by twice. The following mixture was poured by portions in 500 mL of cold methanol, followed by a rapid stirring for 5 minutes. The masterbatch was filtered by membrane filter (Millipore VVLP, 0.1 μm) and desiccated at 80 °C for 24 h. Then in was crashed to the powder and added to PP followed by thermoplastic mixing of a total duration of 6 min on twin roll-mills at 190 °C, which was also used for the direct mixing PP with the filler (composite A). The plates for all compositions were obtained by compression molding at 180 °C and a pressure of 5 MPa with a transverse cross section 60 × 0.5 mm.

2.2. Methods

Differential scanning calorimetric measurements were carried out with Mettler Toledo DSC 1/200W under a nitrogen atmosphere. DSC scans were performed in a temperature range of 25 to 200 °C at a constant heating rate of 10 °C/min. The first heating cycle was used to eliminate prior thermal history. The cooling and the second heating scans were analyzed. The average weight range of samples was 10 mg. Results were calculated by applied software Mettler Graph Ware TA72 PS. 5.

Thermogravimetric analysis (TGA1/SF) was carried out by Mettler TG-50 equipment at temperature range from 25 to 600 °C at a heating speed of 10 °C/min under a nitrogen atmosphere.

Tensile tests of composite specimens were performed at 23±2 °C on a Tinius Olsen H1KS UTM at a deformation speed of 1 mm/min for determination of Young's modulus (E) and at a deformation speed of 50 mm/min for stress-strain characteristics, e.g. stress at yield (σ_Y), tensile strength (σ_B) and deformation at break (ε_B) in accordance to ISO 527-2 standard. At least 8 parallel specimens were tested.

Dynamic mechanical analysis (DMA) was done by Mettler DMA/SDTA861e equipment within the temperature range from -40 to 100 °C at a heating speed of 10 °C/min, using nitrogen as a carrier gas.

3. Results and discussion

The average results of stress-strain properties for pristine PP and PP/MWCNT-COOH composites with the filler content 1 wt.% measured at room temperature are compared in Table 1. The addition of carbon nanotubes by direct mixing in PP matrix and by manufacturing through the masterbatch technique affect an increase of the Young's modulus by 9.4 % and by 22 % compared to pristine PP, indicating an increase of MWCNT distribution in polymer matrix for the last ones.

However, the elongation at break decreases as well as the stress at break for composites, while the yield stress increases by 14 % for masterbatch manufactured composite, indicating an increase of the nanofillers distribution in PP matrix, acting as the reinforcing filler (Table 1).

Table 1. Mechanical properties of PP and PP/MWCNT-COOH composites.

Sample	E (MPa)	σ_Y (MPa)	σ_B (MPa)	ε_B (%)
PP	1380±100	34.5±0.5	42±2	700±100
Composite A	1510±100	33.5±0.5	26.5±0.5	19±8
Composite B	1690±100	39.5±0.5	37±2	16±6

E – the Young's modulus, σ_Y – yield stress, σ_B – stress at break, σ_B – elongation at break

The investigation of DSC cooling scans indicate a notable increase of PP crystallization temperature by 6 and by 11 °C for composites A and B compared to pristine PP due to the CNTs nucleating effect (Figure 1a) with an enhanced behavior for masterbatch compounded composite, that also indicates a reduction in melting temperature, due to an decrease of the arrangement of crystallites in the PP matrix in the presence of distributed CNTs (Figure 1b). The degree of crystallinity for B composite increased from 42 % for pristine PP and A composite to 48 % for B composition.

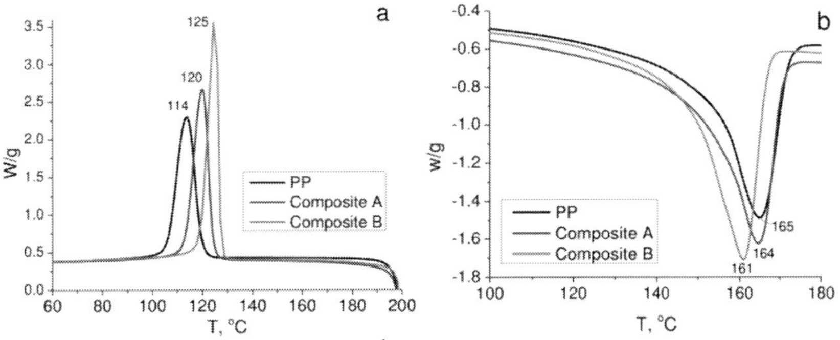

Figure 1. DSC curves of cooling (a) and the 2nd heating scan (b) for pristine PP, PP/MWCNT-COOH composites (filler content 1 wt.%) made by direct thermoplastic mixing (A) and made from masterbatch (B).

The measurements of storage modulus (E′) by DMA indicate an increase of storage modulus of PP in the presence of the filler at higher extent for masterbatch manufactured composite (Figure 2a).

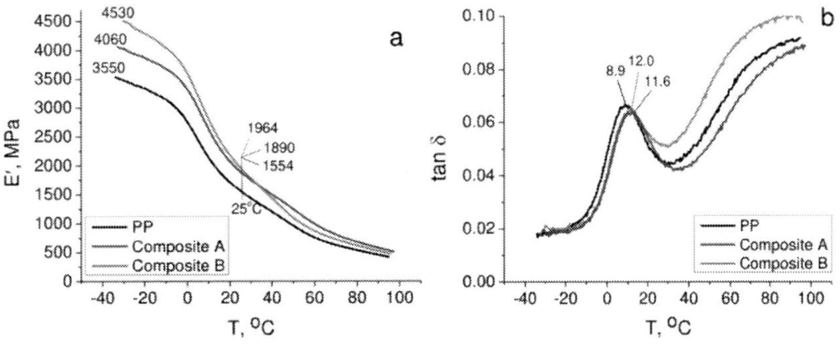

Figure 2. DTA curves of the storage modulus (a) and the mechanical loss factor (b) for pristine PP, PP/MWCNT-COOH composites (filler content 1 wt.%) made by direct thermoplastic mixing (A) and made from masterbatch (B).

It should be noted that MWCNT nanofiller affects an increase of storage modulus at lower temperatures by 1.14 and by 1.27 times for A and B composites compared to pristine PP, affecting an increase of crack resistance of PP matrix that is beneficial for this material [2]. However, the storage modulus increase also at room temperature by 1.26 and by 1.21 times due to the reinforced effect of the nanofiller dispersion in PP also indicated by the shift of the mechanical loss factor curves to higher temperatures and an increase of Ps glass transition temperatures in the presence of nanofillers.

In Figure 3a the profiles of TGA curves related to the materials obtained are shown. There is notable a shift of the curves of PP/MWCNT-COOH nanocomposites to higher values of degradation temperature in relation to those corresponding to pristine PP. However, the masterbatch made composite may include PP fragments of reduced PP crystallinity of lower thermal stability, affecting the small peak in Figure 3b, corresponding to the first derivative of TGA. Meanwhile, the filler particles affect an increase of T_{max} values by 6 °C and 16 °C for composites A and B, indicating an increase of the filler distribution in PP matrix for masterbatch manufactured composites (Figure 3b).

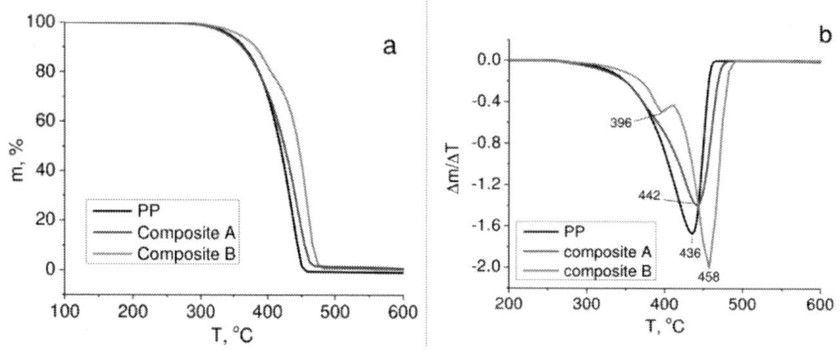

Figure 3. TGA curves (a) and the first derivatives (b) for pristine PP, PP/MWCNT-COOH composites (filler content 1 wt.%) made by direct thermoplastic mixing (A) and made from masterbatch (B).

4. Conclusions

An experimental investigation on the effect of functionalized MWCNT filler interaction with PP matrix at MWCNT-COOH content equal to 1 wt.% was done. The method of manufacturing nanocomposites by a complex solution mixing and thermoplastic mixing method was done, showing an enhanced effect of mechanical and thermal properties for made nanocomposites compared to pristine PP and even to composite made by the direct thermoplastic mixing. The changes of structure indicated an increase of crystallization temperature that could indicate a nucleating effect of filler nanoparticles also affecting a reduced thermal destruction compared to pristine PP. The future experiments should be continued in order to better explain the changes of composite morphology especially at higher content of nanofiller in contrast to manufacturing method evaluation of composites.

Acknowledgements: This work was supported by The National Research Program of Latvia in Materials Science No.6.

References

[1] Breuer O Sundararaj U 2004 *Polymer composites* **25** 630-645
[2] Bikaris D 2010 *Materials* **3** 2884-2946
[3] Han Z D Fino A 2011 *Progress in Polymer Science* **36** 914-944
[4] Zhorin V A Muhina L L Razumovskaya I V 2008 *Polym Deg Stabil* **93** 952-967

IOP Conf. Series: Materials Science and Engineering **77** (2015) 012014 doi:10.1088/1757-899X/77/1/012014

Synthesis and properties of magnetic iron oxide/platinum nanocomposites

V Serga[1], M Maiorov[2], L Kulikova[1], A Krumina[1] and D Karashanova[3]

[1] Institute of Inorganic Chemistry, Riga Technical University, 34 Miera str., Salaspils, LV-2169, Latvia
[2] Institute of Physics, University of Latvia, 32 Miera str., Salaspils, LV-2169, Latvia
[3]Institute of Optical Materials and Technologies, Bulgarian Academy of Sciences, 109 Acad. G.Bonchev str., 1113 Sofia, Bulgaria

E-mail: vera_serga@inbox.lv

Abstract. Iron oxide/platinum nanocomposites have been synthesized by the extractive-pyrolytic method (EPM) involving gradual decomposition of iron capronate and n-trioctylammonium hexachloroplatinate initially produced by solvent extraction. The content of platinum in the composites was 1.2 wt%, 2.4 wt% and 4.8 wt%. Phase composition, morphology and magnetic properties of the produced materials were investigated. XRD analysis and magnetic measurements show that the magnetic phase (magnetite Fe_3O_4) dominates in a carrier sample produced by the pyrolysis of iron carboxylate, but hematite α-Fe_2O_3 exists there as an admixture. Referring to the TEM results, the produced composites contain ultra-disperse platinum particles on the carrier, and the mean size of these varies from 3 nm to 9 nm.

1. Introduction

Magnetic composite nanoparticles on the base of ferromagnetic iron oxides combined with platinum are of particular interest for applications in catalysis [1, 2], sensor materials [3], and in biomedicine [4].

Commercial carriers [1, 5] as well as carriers produced by traditional methods, e.g., by coprecipitation of iron salts in an alkali medium [2-4], are used at the production of platinum nanoparticles on magnetic oxides (γ-Fe_2O_3 or Fe_3O_4). Platinum nanoparticles are produced by metal vapour synthesis [1], by colloid-deposition route [2], by the reduction of platinum ions from aqueous solutions using $NaBH_4$ as a reducing agent [3] or by γ-ray irradiation [5].

In earlier studies [6-8], platinum-containing composites were produced by the extractive- pyrolytic method (EPM) involving thermal decomposition of platinum-containing n-trioctylammonium salt (n-trioctylammonium hexachloroplatinate $[(C_8H_{17})_3NH]_2PtCl_6$) on various commercial and plasma processed oxide carriers. Therefore, of certain interest was the use the EPM not only for the covering with metal nanoparticles, but also for the production of oxide carrier.

In the reported study, iron oxide/platinum nanopowders were produced by the EPM, and the phase composition, morphology and magnetic properties of the produced materials were investigated.

IOP Conf. Series: Materials Science and Engineering **77** (2015) 012014 doi:10.1088/1757-899X/77/1/012014

2. Experimental

The synthesis of iron oxide/platinum nanocomposites was begun with the production of a carrier. Iron (III) was extracted from a nitrate aqueous solution by caproic acid (HR - $C_6H_{12}O_2$), with no diluent, by adding a solution of sodium hydroxide in stoichiometric quantities. Sodium ions were removed by 3-fold contact of the extract with a freshly produced aqueous solution of iron nitrate. To prepare iron capronate (FeR_3) as a paste, an excess of the extractant was removed in vacuum. FeR_3 was thermally decomposed by heating until a temperature of 350 ^{0}C and annealing during 30 min. Further on, a platinum-containing extract (precursor) was prepared using a solution of n-trioctylamine in toluene as an extractant, following the procedure described in [8].

The platinum-containing composites were produced following the previously developed procedure: the carrier was impregnated by a platinum-containing precursor (C_{Pt} = 0.4 M at the production of composites with the 2.4 wt% and 4.8 wt% metal content, and C_{Pt} = 0.03 M at the production of composites with the 1.2 wt% metal content), then drying followed by pyrolysis [6-8].

The phase composition of the produced carrier and Pt-containing composites was characterized by an X-ray diffraction method using a diffractometer D8 Advance (Bruker Corporation) with CuK_α radiation (λ = 1.5418Å). The mean size of platinum crystallites was defined from the (111) peak width by the Scherrer method. TEM measurements were made using the JEOL JEM 2100 operating at 200 kV. The specific surface area (SSA) of the carrier was measured by the BET method at the temperature of liquid nitrogen using the HROM-3 chromatograph. Magnetic measurements were carried out at room temperature using a vibrating sample magnetometer (Lake Shore Cryotronics, Inc., model 7404) with a maximum magnetic field of 1 T. The temperature dependence of sample magnetization was measured within a range of temperatures from room to 680 ^{0}C.

3. Results and discussion

The XRD phase analysis of the products of iron capronate pyrolysis has shown (Fig. 1, curve 1) the dominance of the magnetic phase – magnetite Fe_3O_4 and/or maghemite γ-Fe_2O_3 with a non-magnetic admixture of hematite α-Fe_2O_3 (PDF ICDD 01-071-5088).

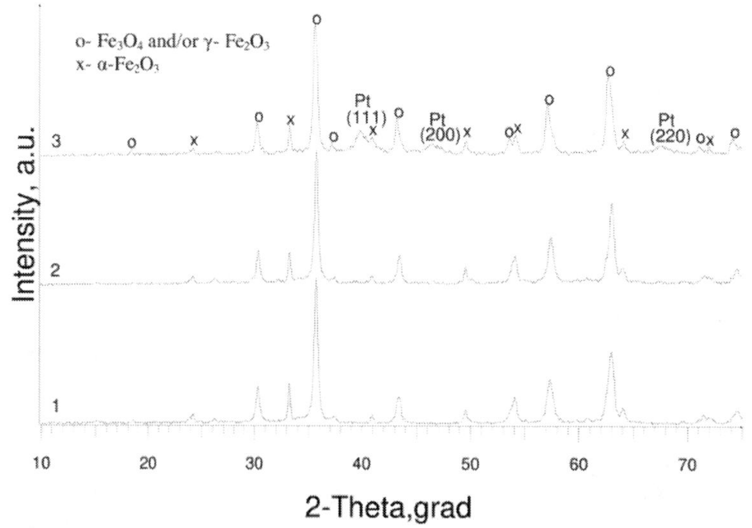

Figure 1. XRD patterns of samples: 1 – iron oxide carrier; 2 – iron oxide/1.2 wt% Pt and iron oxide/2.4 wt% Pt composites; 3 – iron oxide/4.8 wt% Pt composite.

The similarity of the crystal lattices of Fe_3O_4 and γ- Fe_2O_3 does not allow to conclude on which magnetic phase is present in the sample. Both oxides have the structure of spinel and similar crystal lattice parameter values: for Fe_3O_4 a_0 = 8.378 Å (PDF ICDD 01-071-6336) and for γ- Fe_2O_3 a_0 = 8.361 Å (PDF ICDD 00-071-6450).

The produced iron oxide powder (SSA = 39 m^2/g) was later used as a carrier for the production of platinum-containing composites. When preparing a sample with the 4.8 wt% Pt-content, during the pyrolysis of a platinum-containing precursor, platinum particles (PDF ICDD 00-004-0802) of a mean size of 8 nm (Fig. 1, curve 3) were found to form on the carrier; but with the 1.2 wt% and 2.4wt% Pt-content in the samples, amorphous metal particles were found (Fig. 1, curve 2).

With reference to the results of the magnetic measurements, the iron oxide powder (the carrier) exhibited ferromagnetic properties and had the coercivity H_c = 11 кA/m and the spontaneous magnetization $\sigma \approx 60$ $A{\cdot}m^2/kg$ (Fig. 2).

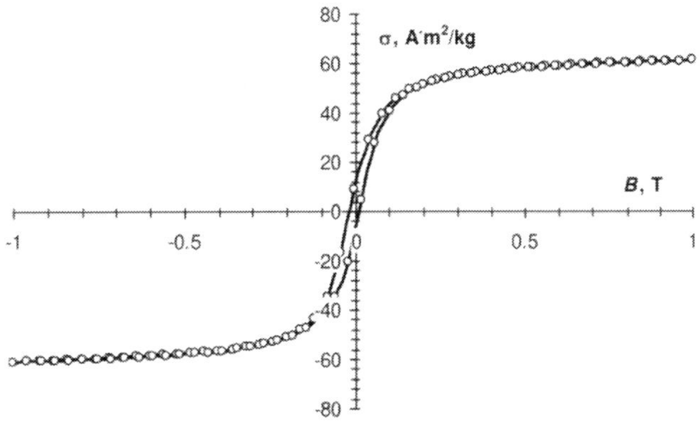

Figure 2. Magnetization loop of the carrier.

The temperature dependence of the sample magnetization showed (Fig. 3) that the Curie temperature of the magnetic material under study was 580 ^{0}C that corresponds to the Curie temperature of bulk magnetite (the Curie temperature of bulk maghemite is 675^{0}C) [9]. This fact gives grounds to suggest that the basic magnetic phase of the sample is magnetite, which can include only a small amount of maghemite admixture.

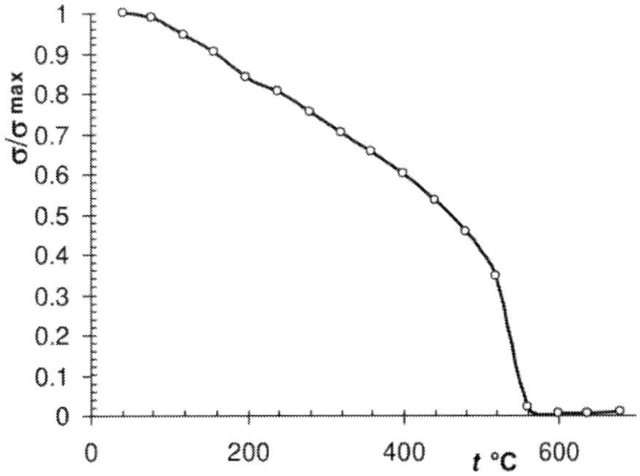

Figure 3. Temperature dependence of the carrier magnetization under the 1 T magnetic field.

With reference to the XRD data (Fig. 1), after covering with platinum, the phase composition of the carrier remained the same. This was also confirmed by the results of magnetic measurements: the H_c value did not practically change, σ decreased a little to 50 A·m^2/kg, which seems to be determined by the presence of platinum in the sample. The temperature dependence of the sample composite magnetization (the platinum content 4.8 wt%) was similar to the dependence plotted for the carrier (Fig. 3).
TEM investigations have revealed (Fig. 4) that the Pt-nanoparticles in the produced composites have a spherical shape and are immobilized on the carrier surface.

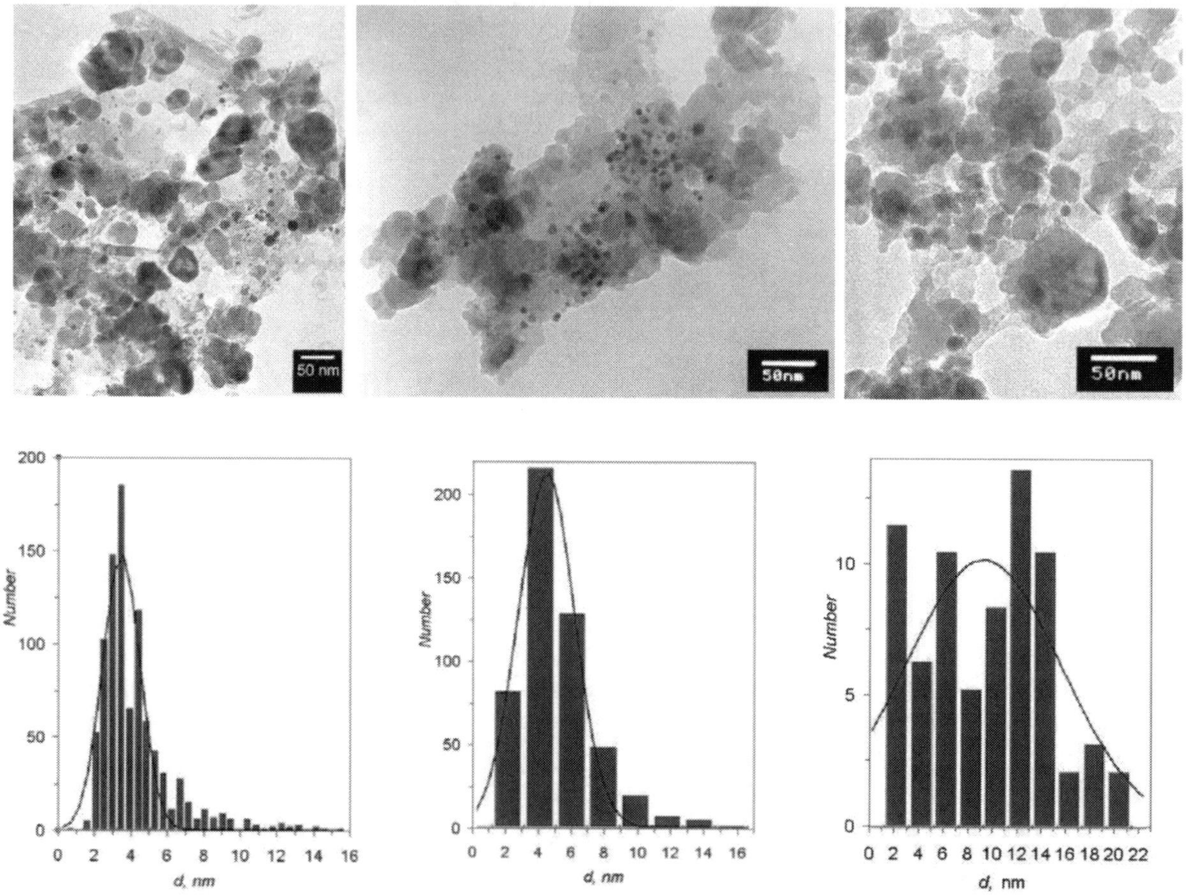

a **b** **c**

Figure 4. TEM images and size distribution of platinum nanoparticles in composites: a – iron oxide/4.8 wt% Pt; b – iron oxide/2.4 wt% Pt; c – iron oxide/1.2 wt% Pt.

A comparative analysis of the data listed in Table 1 evidences that the mean size of the particles in sample 1 (TEM data, Fig. 4a) is smaller that the mean size of the crystallites (XRD data). The X-ray amorphous platinum particles in sample composites 2 and 3 (TEM data, Figs. 4 b, c) have a mean diameter of 4 and 9 nm, respectively. The difference of the XRD and TEM data obtained for samples 1 and 3 is probably caused by the presence of a large amount of small-sized particles being amorphous (Figs. 4a, c).

Table 1. Platinum particle size in iron oxide/Pt composites from XRD and TEM data

Sample No.	Pt content in composite, wt%	Pt crystallite mean size from XRD, nm	Pt particle mean diameter from TEM, nm
1	4.8	8	3
2	2.4	amorphous	4
3	1.2	amorphous	9

The catalytic properties of the produced composites with the 2.4 wt% and 4.8 wt% platinum content were studied in the reaction of glycerol oxidation by molecular oxygen in alkaline aqueous solutions and it has been shown that composites exhibit catalytic activity [10].

4. Conclusions

The performed investigations have demonstrated the dominance of the magnetic phase – magnetite Fe_3O_4 – in the sample carrier produced by iron carboxylate pyrolysis, but hematite α-Fe_2O_3 is present there as an admixture. During the production of metal-containing composites by the EPM, the phase composition of the carrier does not practically change, and the mean size of the platinum particles in the composites varies from 3 nm to 9 nm.

References

[1] Evangelisti C, Aronica L-A, Botavina M, Martra G, Battocchio C and Polzonetti G 2013 *J. Mol. Catal. A: Chemical* **366** 288
[2] Li S, Liu G, Lian H, Jia M, Zhao G, Jiang D and Zhang W 2008 *Catal. Commun.* **9** 1045
[3] Liu Y, Zhou J, Gong J, Wu W-P, Bao N, Pan Z-Q and Gu H-Y 2013 *Electrochim. Acta* **111** 876
[4] Lu A-H, Salabas E L and Schuth F 2007 *Angew. Chem. Int. Ed.* **46** 1222
[5] Seino S, Kinoshita T, Otome Y, Maki T, Nakagawa T, Okitsu K, Mizukoshi Y, Nakayama T, Sekino T, Niihara K and Yamamoto T A 2004 *Scripta Mater.* **51** 467
[6] Serga V, Kulikova L, Cvetkov A and Krumina A 2012 *Mater. Sci. Eng.: IOP Conf. Series* **38** 012062
[7] Ivanov I, Petrova P, Georgiev V, Batakliev T, Karakirova Y, Serga V, Kulikova L, Eliyas A and Rakovsky S 2013 *Catal. Lett.* **143** 942
[8] Palcevskis E, Kulikova L, Serga V, Cvetkovs A, Chornaja S, Sproge E and Dubencovs K 2012 *J. Serb. Chem. Soc.* **77** 1799
[9] Mee C 1967 *The Physics of Magnetic Recording* (Moscow, Energy) (in Russian)
[10] Sproge E, Chornaja S, Dubencovs K, Zhizhkun S, Kampars V, Serga V, Kulikova L and Palcevskis E 2013 *J. Serb. Chem. Soc.* **78** 1359

IOP Conf. Series: Materials Science and Engineering **77** (2015) 012015 doi:10.1088/1757-899X/77/1/012015

Production of nano-ceramic coatings on titanium implants

A A Fomin[1], I V Rodionov[1], M A Fomina[1], N V Petrova[2]

[1] Yuri Gagarin State Technical University of Saratov, Saratov, Russia
[2] Saratov State University, Saratov, Russia

E-mail: afominalex@rambler.ru

Abstract. Composite titania coatings modified with hydroxyapatite nanoparticles were obtained on intraosseous implants fabricated from commercially pure titanium and titanium alloy Ti-2.5Al-5Mo-5V. The present study aims to identify consistency changes of morphological characteristics and physico-mechanical properties of titanium items coatings obtained by oxidation during induction heat treatment and modification with colloidal hydroxyapatite nanoparticles. The influence of temperature between 600 and 1200 °C and duration of thermal modification from 1 to 300 s was studied. It was established that high hardness about 6.7±1.9 GPa for nanocrystalline TiO_2 coatings and 19.2±0.6 GPa for nano-ceramic "TiO_2+HAp" coatings is reached at 1000 °C and 120 s.

1. Introduction

In medical practice, metals, particularly titanium and its medical alloys (Ti-2.5Al-5Mo-5V, Ti-4Al-6V, etc.), are widely used when endoprostheses and dental implants are manufactured [1]. It is especially important to produce biocompatible materials and coatings in order to improve osseointegration on the surface of these medical items [2]. The metal substrate of such implants has resistance to mechanical loads of distributed type. When installed with an interference fit into the prepared bone bed there is significant shear force causing coating delamination and biocompatibility reduction. Under such extreme conditions the surface layer mechanical characteristics (a combination of hardness and sufficient elasticity) require special attention. Moreover, the biocompatible coating should have high morphological heterogeneity of the microstructure, combined with the homogeneity of nanostructure [3, 4].

The surface structural modification is usually performed by plasma spraying, sol-gel deposition, oxidation, etc. [5, 6]. A characteristic feature of these methods is high energy consumption and cost of materials, complex technological sequence, relatively long production of the required phase state, lower mechanical strength combined with high porosity, limited or lacking possible formation of nanoscale elements. It is an important problem of obtaining of mechanically strong functional coatings having high biocompatibility on the surface of metallic materials.

Electrical equipment that generates high frequency eddy currents is used to ensure the necessary physico-chemical effects on the well-conductive materials and various composites. Hence the aim of this work is to develop a technology enabling to produce on titanium biocompatible mechanically durable oxide coatings with increased morphology of micro- and nanostructure using a new method of induction heat treatment (IHT) and surface modification by nano-ceramic bioactive materials, such as hydroxyapatite (HAp) nanoparticles.

IOP Conf. Series: Materials Science and Engineering **77** (2015) 012015 doi:10.1088/1757-899X/77/1/012015

2. Methodology

2.1. Preparation of coatings

Titanium samples are cylinders fabricated from commercially pure (CP) titanium and titanium alloy Ti-2.5Al-5Mo-5V. Their surface is subjected to machining and micro-texturizing treatment comprising abrasive air-blasting by corundum abrasive with dispersion of 200...300 μm. Titanium samples are subjected to ultrasonic cleaning in aqueous surfactants and ethanol.

The surface of the prepared samples was oxidized by IHT in an oxygen-containing environment. An IHT thermal cycle comprises intensive heating up to the required temperature, exposition at the same temperature from 1 to 300 s and still air cooling.

The next stage included modification with colloidal HAp nanoparticles and final IHT of composite structure for less than 300 s. Using atomization the microdroplets of colloid solution are entered into the air stream and further this mixture is directed onto the surface of the sample with an oxide matrix coating.

The IHT effect at 600...1200 °C on the performance of micro- and nanostructure of the coatings and mechanical properties was defined. IHT regimes were given double numbers: the first number indicated the temperature of titanium substrate, the second one marked the heat treatment duration in seconds. For example, IHT regime "1000-001" designates the temperature of 1000 °C and the exposure duration of 1 s (transition to stationary thermal treatment).

2.2. Coating characterization

To analyze the surface morphology of the coatings scanning electron microscopy (SEM) was applied. SEM of the coatings is performed on MIRA II LMU (HV: 30 kV). The microstructure of coatings was studied at magnification from ×1000 to ×20000 and nanostructure – from ×100000 to ×200000.

Hardness of the coatings were analyzed by nanoindentation using a mechanical properties tester Nanovea Ergonomic Workstation with low (100 mN) load applied to Berkovich indenter.

3. Results and discussion

3.1. Surface morphology

The structure of the titania (TiO_2, rutile [3]) coatings is determined by intensive oxidation due to thermophysical effect of high-frequency currents in titanium substrate (Figure 1a).

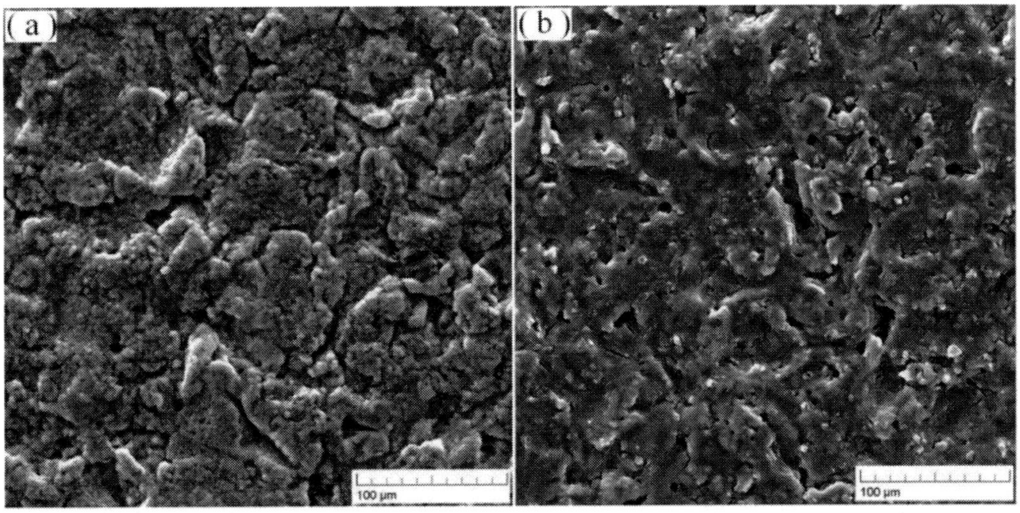

Figure 1. Microstructure of titania (a) and nano-ceramic (b) coatings (×1000).

The shape of rutile nanocrystals is associated with the IHT temperature, so at 600…800 °C rounded, needle-like and plate oxide structures are formed [4]. On the surface of titanium alloy Ti-2.5Al-5Mo-5V mainly rounded crystals are formed (Figure 2a). When the IHT temperature reaches 800 °C, the crystals grow as well (Figure 2b). When IHT exceeds 1000 °C and 150...180 s the number of coating defects increases causing plate splitting and the formation of prismatic crystals. When the temperature and IHT duration increase the crystals grow to 150…300 nm. At the temperature above 1000 °C and exposure duration more than 120 s the oxide coating becomes quite thick as there appear numerous fractures and its spontaneous destruction is observed.

Figure 2. Nanostructure of titania coatings obtained on IHT regime: a – "800-120" on titanium alloy Ti-2.5Al-5Mo-5V (×133000); b – "800-030" on CP-Ti (×100000).

The matrix of this composite structure is formed by titania crystals with the relief elements uniformly modified with a thin layer of HAp nanoparticles having an average size of 30...70 nm (Figure 1b, 3).

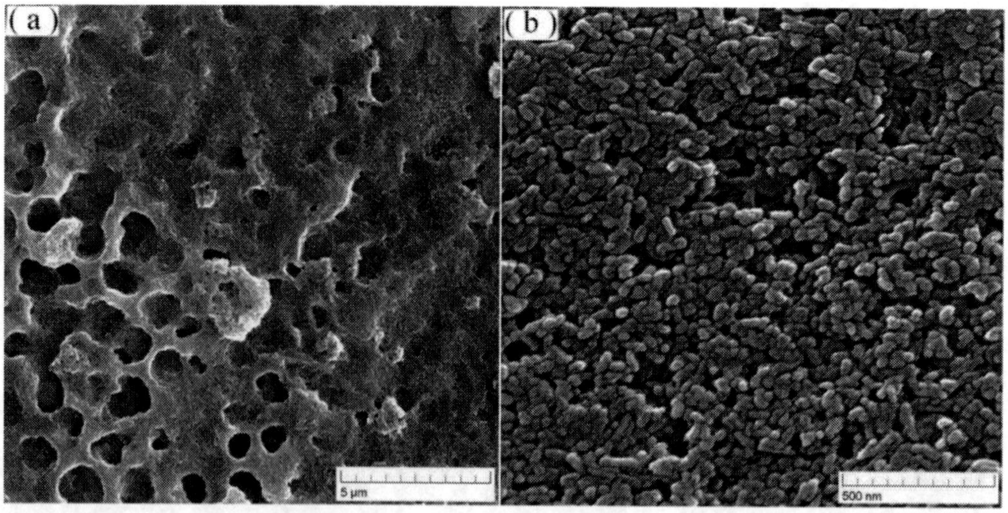

Figure 3. Porous structure of nano-ceramic TiO_2+HAp coatings obtained on IHT regime: a – "1000-030" (microstructure, ×20000); b – "600-120" (nanostructure, ×200000).

At this modification HAp nanoparticles can penetrate into pores, micro- or nanochannels, and fractures. Changing the saturation of oxide matrix with the filler mechanical properties of the coatings, hardness in particular, and other biocompatibility qualities can be controlled.

3.2. Hardness

Oxide TiO_2 and nano-ceramic TiO_2+HAp coatings are characterized by enhanced hardness (Figure 4). In general, the required hardness values characterize these coatings as high strength, and the margin of hardness compared to cortical bone is 8...12 times higher [3]. This parameter completely ensures the integrity of biocompatible coating during the installation with the interference fit into the prepared bone bed and further functioning under the physiologically normal loads.

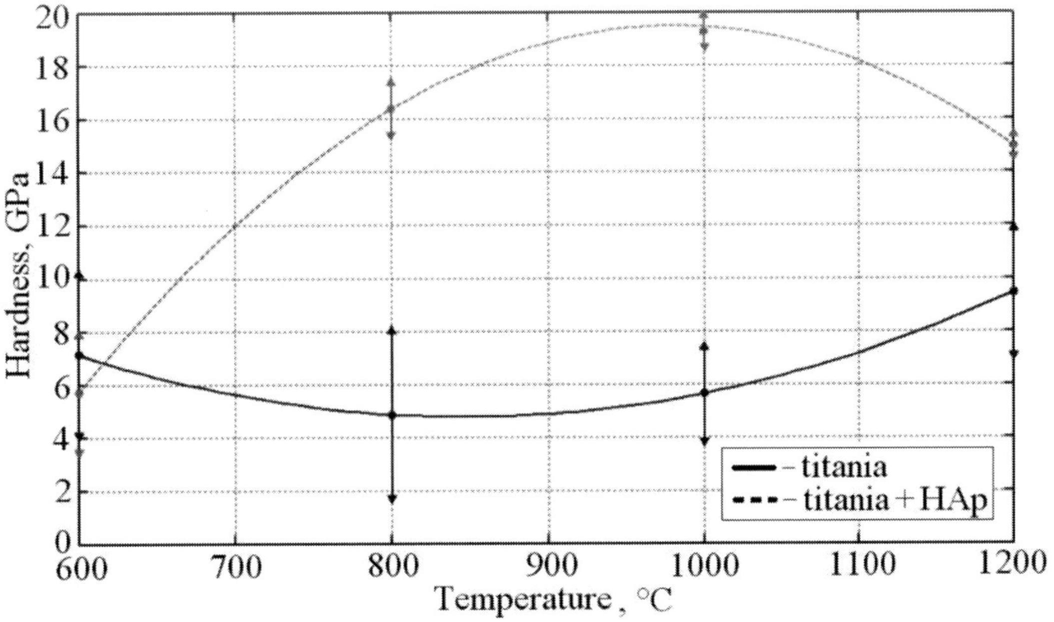

Figure 4. Hardness of nanocrystalline TiO_2 and TiO_2+HAp coatings.

The hardness of coatings modified with HAp nanoparticles at IHT of 600 °C is characterized by triple increase to 6 GPa compared to the titanium substrate. When the IHT temperature increases to 800...1200 °C the hardness value is highest and equals 15.0...19.2 GPa. Highest hardness is reached at the chosen duration of thermal treatment about 120 s.

4. Conclusions

The surface structure of titanium and titanium alloy Ti-2.5Al-5Mo-5V after IHT is characterized by high morphological heterogeneity and mechanical properties. Titania coatings are formed of rounded, needle-like, plate, and prismatic nanocrystals. The optimal combination of nanostructure parameters and high hardness 4.9...9.5 GPa of thin-coatings is achieved at IHT regimes "800-030" and "800-120". To obtain a thick-layer coating (≥ 10 μm) with high hardness on the implant surface it is required to form a porous structure of prismatic crystals at IHT regimes "1000-120", "1200-030" and "1200-120".

Oxidation at IHT and subsequent modification with HAp provide the formation of mechanically strong composite structure consisting of porous titania matrix and nanocrystalline bioceramic filler. It was established that porous titania coatings modified with HAp nanoparticles and formed by heating from 800 to 1200 °C and holding for at least 30 s have high bioactivity and hardness of 15.0...19.2 GPa.

Acknowledgments
The research was carried out with financial support: grant the Russian Foundation for Basic Research RFBR No. 13-03-00898 “a” and project No. 1189 in the framework of the basic part of the state educational task for institutions of higher education, under the jurisdiction of the Ministry of Education of the Russian Federation in the field of scientific activity.

References
[1] Paital S and Dahotre N 2009 *Mater. Sci. Eng.* R **66** 1
[2] Dorozhkin S V 2010 *Biomaterials* **31** 1465
[3] Fomin A A, Steinhauer A B, Rodionov I V, Fomina M A, Zakharevich A M, Skaptsov A A, Gribov A N and Karsakova Ya D 2014 *J. Frict. Wear* **35** 32
[4] Fomin A A, Steinhauer A B, Rodionov I V, Fomina M A and Zakharevich A M 2013 *Tech. Phys. Lett.* **39** 969
[5] Sung Y-M, Shin Y-K and Ryu J-J 2007 *Nanotechnology* **18** 065602
[6] Lee B-H and Koshizaki N 2008 *Nanotechnology* **19** 415303

IOP Conf. Series: Materials Science and Engineering 77 (2015) 012016 doi:10.1088/1757-899X/77/1/012016

Highly Porous Wood Based Carbon Materials for Supercapacitors

A Volperts[1,6], G Dobele[1], A Zhurinsh[1], D Vervikishko[2], E Shkolnikov[2], J Ozolinsh[3], N Mironova-Ulmane[4] and I Sildos[5]

[1]Latvian State Institute of Wood Chemistry, 27 Dzerbenes St., 1006 Riga, Latvia, E-mail: quizzus@marshal.lv
[2]Scientific Association for High Temperatures, Russian Academy of Sciences, 13/2 Izhorskaya St., 125412 Moscow, Russia
[3]Riga Technical University, Faculty of Material Science and Applied Chemistry 14/24 Azenes Street, LV-1048, Riga, Latvia
[4]Institute of Solid State Physics, University of Latvia 8 Kengaraga street, Riga, LV-1063, Latvia
[5]Institute of Physics, University of Tartu, Ravila 14c, 50411, Tartu, Estonia

Abstract: Wood based activated carbons synthesized by two-stage thermocatalytical synthesis with NaOH activator were studied and used as supercapacitor electrodes (sulphuric acid electrolyte). Porous structure and electrochemical properties of carbons vs synthesis conditions were assessed. It was found that there are correlations between carbons synthesis variables, their porosity and supercapacitors functional characteristics. At the temperature 600°C and activator/precursor ratio 1.25 porosity decreased, however energy capacitance of supercapacitor increased calculating on elementary cell mass.

1. Introduction

The existing methods of energy storage lag behind the intensity of modern power systems development, thus it is of high importance to provide the solutions meeting these demands. One of the possible solutions to level peak loads in the grid, to suppress fluctuation in Diesel-generators and to recuperate breaking energy for vehicles is the application of electrochemical capacitors with double electric layer – so called supercapacitors (SC) [1-4]. Activated carbons (AC) are universally accepted as a main component of SC electrodes since they are stable in various media, have highly developed porosity, sufficient conductivity and are relatively cheap [5-7]. AC properties depend on precursor type, its modification as well as activation conditions [8-10].

Power capacitance of carbon electrode is mainly influenced by dipsersity, elemental composition and micro- and mesopores size distribution of AC. Electrodes should provide high capacitance, minimal energy drop with power increase and stable operation at high number of charges-discharges cycles [11].

[6] To whom any correspondence should be addressed.

IOP Conf. Series: Materials Science and Engineering **77** (2015) 012016 doi:10.1088/1757-899X/77/1/012016

Alkali thermochemical activation (KOH, NaOH) is the most widespread method for production of highly effective AC from various precursors including wood and its components [12-17]. For the biomass based AC the universally accepted pore model is slit-like [18,19], where primary pores are represented by the gaps between carbon layers [20]. AC particles are the more or less structured 3D graphene structures combinations. Soprtion methods and Raman spectroscopy are the most popular non-destructive technique for the characterization of carbon-containing materials [21, 22]

In the process of charcoal alkali activation size and volume of pores increase on the expense of carbon layer thickness decrease due to carbon oxidation and volatile products formation. These reactions start on the graphenes periphery, between boundary carbon atoms and molten alkali [12]. With the increase of alkali/carbon ratio, treatment temperature and time, porosity and, accordingly, specific surface of AC increase. Variations of these parameters let to SC with high capacitance - 320-370 F g^{-1} [23,24] in water and 170-195 F g^{-1} [25] in aprotic electrolytes.

Numerous literature sources testify that activator amount is the deciding factor for the porous structure formation [26-28]. There is a widespread opinion that AC pore volume and specific surface increase is the necessary prerequisite for SC electrical characteristics increase. For some types of precursors this dependency exists indeed, however for others there is a maximum capacitance value at the some definite AC specific surface value [29].

Our research is devoted to study of capacitance characteristics of SC, providing the minimum parasitic electrolyte mass, altering porous structure of wood based alkali activated carbons.

2. Experimental

2.1. Activated carbons synthesis

AC samples were synthesized by two stage thermal treatment. At the first stage wood chips (birch, 0.2-0.4 mm fraction) were pretreated in nitrogen atmosphere at 600°C for 150 min (4°C min^{-1} rate). At the second stage the obtained charcoal was impregnated with 50 w% NaOH water solution, ratio of alkali/charcoal was varied from 1 to 3.7. The mixture was activated at 600-800°C in the stream of argon for 120 min, washed with dionised water, HCl, and water again up to pH 5, and then dried overnight at 105°C. Ash content of the synthesized AC was 0.1-0.4%. The main variables in the experiment were activation temperature and activator/charcoal ratio (further denoted as K).

2.2. SC assembly

AC was first mixed with ethanol, then a binder - water suspension of PTFE F-4D (10 wt% on dry electrode mass) was added, and the resulting mixture was calandered in rollers. Electrodes were dried and then impregnated with 4.9M water solution of H_2SO_4. Cellulosic separator NKK TF 4030 (10 μm) was used to make a suprecapacitor (SC). The electrode area was 4.15 cm^2. Foil from thermally expanded graphite (200 μm) was used as a current collector. Assembled SC were pressed under 10 atm in a dry box. SC capacitance was assessed using potentiostate ELLINS 30-S at 1 V for 5 minutes. Inner resistance was calculated by voltage drop at current transient in the beginning of SC discharge. Capacities shown are calculated on a dry mass of electrode, without electrolyte.

2.3. Porous structure assessment

Porous structure was assessed by N_2 adsorption at 77 K (Kelvin 1042, Costech Instruments). Pores volumes were calculated using Dubinin-Radushkevich and Brunauer–Emmet–Teller (BET) theories using N_2 sorption isotherms [21].

2.4. Raman scattering

Raman spectra of carbon material were measured in back-scattering geometry at room temperature through 50 x microscope objective using micro-Raman spectrometer equipped with argon laser (514.5 nm, max cw power Pex=10 mW). The spectral signal was dispersed by the 2400 grooves/mm grating onto Peltier-cooled (-60 C) CCD detector. Frequency shifts in the Raman spectra were calibrated using $CaWO_4$ as a reference.

3. Results and discussion

Raman spectroscopy was used to compare structures of charcoal and activated carbon (AC) synthesized at 700°C and K=2. The Raman-active vibration numbering for natural single-crystal graphite exhibits a single Raman peak at around 1580 cm^{-1}, called the G peak. This peak is associated with the in-plane C–C stretching mode of the sp^2 hybridized carbon atoms. For polycrystalline graphite, depending on the size of the crystallites, a second peak at 1350 cm^{-1} appears namely the disorder or D peak. If the long-range order of the crystalline material is lost and the carbon phase becomes glassy, both the G and the D peaks broaden [30]. Three Raman scattering peaks at around 1364, 1596 and 2325 cm^{-1} are observed. The peak at 1580 cm–1 (G band) is attributed to an e_{2g} mode of graphite and is related to the vibration of sp^2 – bonded carbon atoms [22]. Note that in a perfect graphite crystal the first-order vibrational mode of the d band is forbidden because of the selection rules. Decreasing particle size or bending of the lattice fringes may activate this band [31]. We can see this phenomenon: after activation of carbon intensity of Raman scattering D band increases with decreasing particle size of carbon (figure 1 and figure 2). Stretch vibrations of N_2 in the ambient might cause the sharp (narrow) peak at about 2325 cm^{-1} [31].

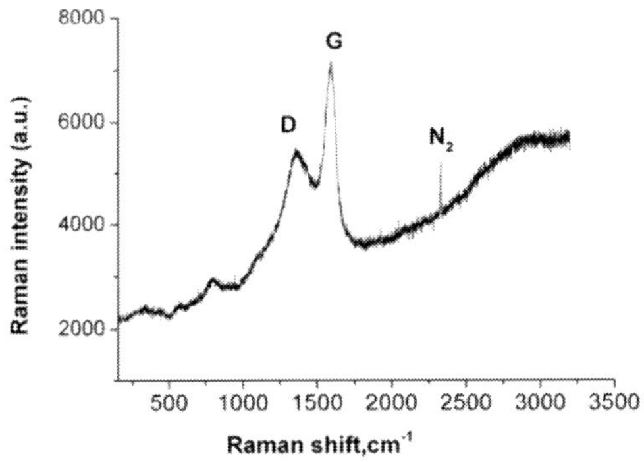

Figure 1. Micro-Raman scattering spectrum of wood charcoal

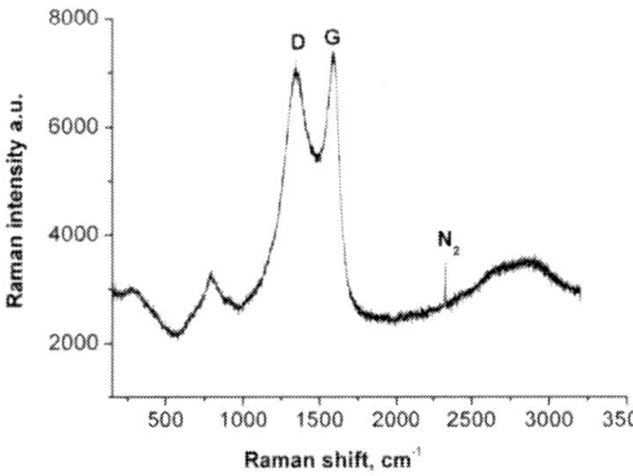

Figure 2. Micro-Raman scattering spectrum of activated carbon after activation (activation temperature 700°C, K=2)

Structural characteristics of AC synthesized at activation temperature 700°C and K variation from 1.7 to 3.7 were calculated using N_2 sorption isotherms. With increase of alkali addition total specific area and pore volume increase as well, however micropores volume changes negligibly (figure 3).

When these AC were used for electrodes in SC alkali addition rate variation manifested itself in increase of total volume of adsorbed electrolyte and specific capacitance (310 F g^{-1}) practically stays the same. Probably the minimal micropores volume is required to form double electric layer. There can be another factor negating positive effect of total porosity increase – it was reported [28] that change of surface groups' content can be such a factor, since they take part in redox reactions and lead to pseudocapacitance effect.

Activation temperature is another important parameter in AC synthesis. Characteristics of activated carbons synthesized in temperature range 600-850°C and K=2 are shown in figure 4. Specific surface area and pore volume reach maximum at activation temperature 700°C and diminishe after that. The same trend can be observed for SC specific capacitance – 330 F g^{-1} at 600°C and 180 F g^{-1} at 850°C.

Total adsorbed electrolyte volume significantly increases with the increase of activation temperature. It should be noted that inner resistance practically does not change with the increase of activation temperature up to 800°C and corresponding volume of adsorbed electrolyte (figure 4). At higher temperature resistance somewhat increases, probably this can be explained by the increase of larger pores volume. The same trend was observed in the work [23], where decrease of capacitance was attributed to decrease of surface C-O groups, which influences pseudocapacitance (reversible faradaic redox reactions).

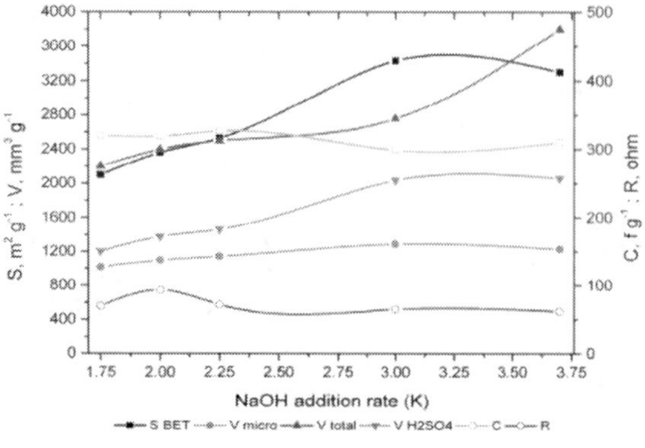

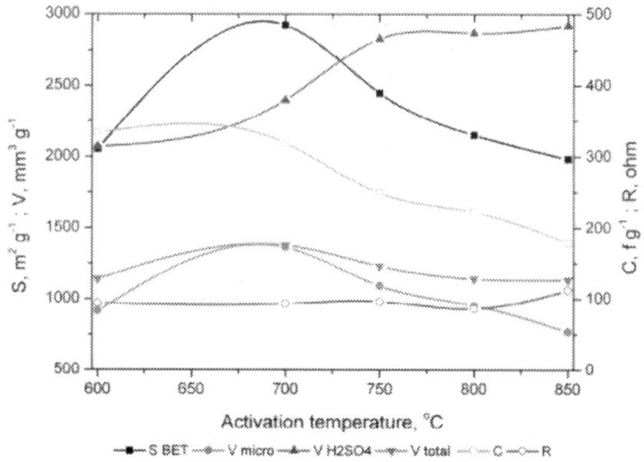

Figure 3. Dependence of activated carbons porous structure and supercapacitor characteristics from NaOH addition rate at activation temperature 700°C

Figure 4. Dependence of activated carbons porous structure and supercapacitor characteristics from activation temperature at NaOH addition rate K=2

S_{BET} – specific surface in $m^2\ g^{-1}$, V_{micro} – micropores volume, $m^3\ g^{-1}$, V_{total} – total pore volume, $mm^3\ g^{-1}$, V_{H2SO4} – volume of adsorbed electrolyte, $mm^3\ g^{-1}$, C – specific capacitance, F g^{-1}, R – resistance, ohm*10^{-3}

Thus in our case lower activation temperature positively influences specific capacitance of SC, however it should be noted that activation temperature lower then 600°C is not feasible since thermodynamical boundary of the main reactions is 580°C [33].

Results of the experiments devoted to assessment of the lowest beneficial K value for AC synthesis at 600°C are shown in figure 5.

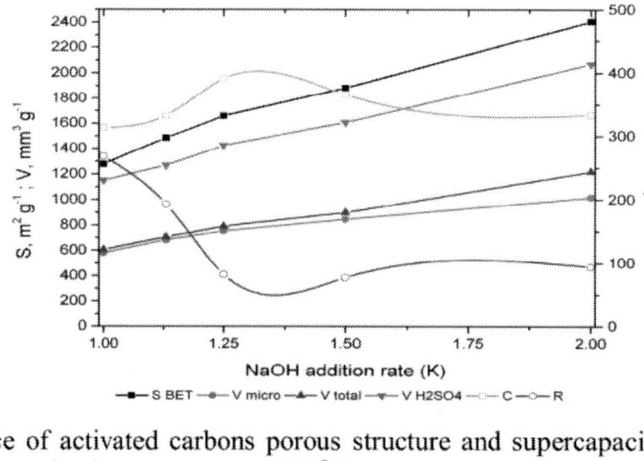

Figure 5. Dependence of activated carbons porous structure and supercapacitor characteristics from NaOH addition rate at activation temperature 600 °C

S_{BET} – specific surface in $m^2\ g^{-1}$, V_{micro} – micropores volume, $mm^3\ g^{-1}$, V_{total} – total pore volume, $mm^3\ g^{-1}$, V_{H2SO4} – volume of adsorbed electrolyte, $mm^3\ g^{-1}$, C – specific capacitance, F g^{-1}, R – resistance, milliohm

Increase of activator addition rate from 1 to 2 leads to increase of total and micropore volume, though the latter is pronounced to the lesser extent. At K=1.25 SC specific capacitance reaches its maximum, which is accompanied with significant decrease of resistance. This value of resistance, which is the sum of ohmic losses in solid phase, electrolyte resistance and contact resistances, does not change with the subsequent increase of K from 1.25 to 2. In or case decrease of charcoal activation temperature and activator addition rate led to 30% increase of SC specific capacitance calculating on elementary cell mass.

Thus electrode pores should provide transport and availability of ions in the process of double electric layer charge and discharge. The total volume of electrolyte adsorbed in macro-, meso- and micropores in our experiment (figure 5) is 1400 mm^3 g^{-1}. Obviously there is a minimal micropores volume (in this case around 750 mm^3 g^{-1}) required to form double electric layer, and meso- and macropores provide transport function for electrolyte ions. These pores do not contribute into electric double layer formation, and additional electrolyte mass negatively influences on the SC specific characteristics calculating on cell or whole device.

4. Conclusions

Porous structure and electrochemical properties of hardwood based activated carbons obtained by thermocatalytical NaOH activation were assessed. It was found that the effective double layer in the supercapacitor cell electrodes (sulfuric acid electrolyte) and maximum capacitance 330 F g^{-1} can be achieved at micropores volume 750 mm^3 g^{-1} and specific surface area 1600 m^2 g^{-1}. These parameters can be achieved at activation temperature 600°C and activator/charcoal ratio 1.25 and with the increase of carbon porosity electrical capacitance decreases.

5. Refences

[1] Yu A, Chabot V and Zhang J 2013 *Electrochemical Supercapacitors for Energy Storage and Delivery: Fundamentals and Applications* (Boca Raton: CRC Press)
[2] Beguin F, Frackowiak E and Lu M 2013 *Supercapacitors: Materials, Systems and Applications* (Weinheim, Wiley-VCH)
[3] Jayalakshmi M and Balasubramanian K 2008 *International Journal of Electrochemical Science* **3** 1196
[4] Burke A 2007 *Electrochimica Acta* **53** 1083
[5] Lipka S M and Swartz C R 2007 *Energeia* **18** 1
[6] Frackowiak E 2006 *J. Braz. Chem. Soc.* **17** 1074
[7] Simon P and Gogotsi Y 2008 *Nature Materials* **7** 845
[8] Xu B, Chen Y, Wei G, Cao G, Zhang H and Yang Y 2010 *Materials Chemistry and Physics.* **124** 504
[9] Linares-Solano A, Lillo-Rodenas M A, Marko-Lozar J P, Kunowsky M and Romero-Anaya A J 2012 *International Journal of Energy, Environment and Economics* **20** 59
[10] Тамаркина Y V 2010 *Scientific works of Donetsk National Technological University, Chemistry and Chemical technology series* **14** 70
[11] Conway B E 1999 *Electrochemical supercapacitors – scientific fundamentals and technological applications* (Norwell: Kluwer Academic publishers)
[12] Ioannidou O and Zabaniotou A 2007 *Renew Sustain Energ Rev* **11** 1966
[13] Marsh H and Rodríguez Reinoso F 2005 *Activated Carbon Hardcover* (Amsterdam: Elsevier)
[14] Kalinicheva O A, Bogdanovich N I and Dobele G V 2008 *Forest Journal* **2** 117
[15] Lillo-Ro´denas M A, Marco-Lozar J P, Cazorla-Amoro´s D and Linares-Solano A 2007 *J. Anal. Appl. Pyrol.* **80** 166
[16] Mikova N M, Ivanov I P, Parfenov V A, Fetisova O Y, Chesnokova N V 2011 Journal of Siberian Federal University. Chemistry **4** 356
[17] Dobele G, .Jakab E, Zoltan S, Volperts A, .Zhurins A and Telysheva G 2013 *J. Anal. And Appl. Pyrol* **103** 173
[18] Fryer J R 1981 *Carbon* **19** 431
[19] Dubinin M 1979 *News of USSR Sceince Academy. Chemical series* **8** 1691
[20] Kaneko K, Ishii C, Ruike M and Kuwabara H 1992 *Carbon* 1992 30 1075
[21] Bansal R and Goyal M 2005 *Activated carbons Adsorption* (Boca Raton: CRP Press)

[22] Ferrari A C, Meyer J C, Scardaci V, Casiraghi C, Lazzeri S, Mauri F, Piscanec S, Jiang D, Novoselov A K, Roth S and, Geim A K 2006 *Phys. Rev. Lett.* **97**, 187401
[23] Bleda-Martinez M J, Macia-Agullo J A, Lozano-Castello D, Morallon E , Cazorla-Amoros D and Linares-Solano A 2005 *Carbon* **43** 2677
[24] Tamarkina Y V, Shendrik T G, Kucherenko V A and Khabarova T V 2012 Journal of Siberian Federal University **5** 24
[25] Liao W-C, Liao F-S, Tsai C-T and Yang Y-P 2012 *China Steel Technical Report* **25** 36
[26] Zhang Z, Cui M, Lai Y, Li J and Liu Y 2009 *J. Cent. South Univ. Technol.* **16** 91
[27] Lota G, Centeno T A, Frackowiak E and Stoeckli F 2008 *Electrochimica Acta* **53** 2210
[28] Cuhadaroglu D and Uygun O A 2008 *African Journal of Biotechnology* **7** 3703
[29] Lozano-Castello D, Marko-Lozar J P, Bleda-Martinez M J, Montilla F, Morallon E, Linares-Solano A and Cazorla-Amoros D 2013 *Tanso* **256** 41
[30] Cheng Y H , Tay B K, Lau S P, Shi X, Qiao X L, Chen J G, Wu Y P and Xie C S 2001 *Appl. Phys. A* **73** 341
[31] Klinke C, Kurt R and Bonard J-M 2002 *J. Phys. Chem. B*, **106** 11191
[32] Lillo-Rodenas M A, Cazorla-Amoros D and Linares-Solano A 2003 *Carbon* 41 267

Acknowledgments
Support of ERAF 2OP/2.11.10/14/APIA/VIAA/050 and Latvian Science Council Grant No 666/2014 are gratefully acknowledged.

IOP Conf. Series: Materials Science and Engineering **77** (2015) 012017 doi:10.1088/1757-899X/77/1/012017

Variations of the parameters of PbSe nanolayers with change of their technology

A Pashaev[1], O Davarashvili[2], M Enukashvili[2], Z Akhvlediani[2,3], R Gulyaev[2] and M A Dzagania[2]

[1]National Aviation Academy, Bina 25km, Baku AZ-1045, Azerbaijan
[2]Iv. Javakhishvili Tbilisi State University, Chavchavadze 1, Tbilisi 0128, Georgia
[3]E. Andronikashvili Institute of Physics, Tamarashvili 6, Tbilisi 0186, Georgia

E-mail: omardavar@yahoo.com

Abstract. For designing IR photodetectors based on strained PbSe layers, the optimization of the technology of molecular epitaxy with a "hot-wall" has been performed with the aim of fabrication of the layers with combining high values of residual deformation and thickness. By investigating the characteristics of layers such as the thickness, tangential lattice constant, deformation and x-ray diffraction line halfwidth, the best conditions for growing the layers of optimal thickness of 50-80nm and the highest deformation $1 \cdot 10^{-2}$ were obtained. The width of the forbidden gap in the obtained layers increased by 0.15eV at T=300K as compared with unstrained PbSe layers.

1. Introduction

Recently we have proposed to use the strained layers of PbSe grown on KCl substrates with a great lattice constant for designing the high-sensitivity high-temperature IR photodetectors [1]. This objective can be achieved through the realization of effective "negative" pressure in PbSe layers when the forbidden gap width increases in them and, when doped with impurities Cr, In and Yb having variable valency, then levels of impurities shift into the depth of the forbidden gap. In this case, the impurities can significantly compensate electrically active nonstoichiometric defects. So the concentration of current carriers in the PbSe layers can be decreased by a factor of 4 or 5 [2]. The optimal thickness of epitaxial PbSe layers for sufficient absorption of IR radiation, from one hand, and for the shift of the levels of the impurities in the forbidden gap, on the other hand, is estimated as 50-80nm [3]. Formation, structure of other IV-VI semiconductor nanolayers and their possible applications were discussed in [4-6].

Hence, it is important to establish the process conditions for fabrication of epitaxial PbSe layers for realization in them the maximum residual deformation at the layer thickness of 50-80nm.

2. Variations of experimental conditions

For fabrication of epitaxial PbSe layers by molecular epitaxy a "hot-wall", the source of epitaxy, polycrystalline PbSe, was placed at the bottom of the quartz ampoule. Colliding with the hot wall, the evaporated molecules reached the KCl (100) substrate placed on a stainless mask at the open end of the

IOP Conf. Series: Materials Science and Engineering **77** (2015) 012017 doi:10.1088/1757-899X/77/1/012017

quartz ampoule practically without losses. Optimal process conditions were established by varying the following conditions: growth temperature (temperature of the KCl substrate), and the growth rate and duration, and analyzing the obtained parameters: layer thickness, tangential lattice constant, deformation and x-ray diffraction line halfwidth. The growth rate, the rate of bringing the evaporated molecules to the substrate, was varied by varying the source temperature and the distance between the mask with the substrate and the open end of the quartz ampoule. Three cases were studied: 1) the distance between the substrate and the end of the ampoule was 12nm at the epitaxy source temperature 470°C ; 2) the distance between the substrate and the end of ampoule was 2.5mm at the epitaxy source temperature of 470°C; 3) the distance between the substrate and the end of the ampoule was 1.0mm at the epitaxy source temperature of 510°C.

No.	Layer	Epitaxy source temperature, °C	Substrate temperature, °C	Growth duration t,s	Layer thickness d,nm	Overall growth rate V, nm/s	Growth rate stage, nm/s	Tangential lattice constant a, Å	Deformation $\varepsilon = (a_1 - a_{PbSe}) / a_{PbSe} \cdot 10^2$	x-ray diffraction line halfwidth Δ, min
1	SL-537	470	280	90	159	1.8	5.0	6.142(1)	0.27	21
2	SL-541	470	300	80	95	1.2	4.3	6.162(0)	0.59	19
3	SL-555	470	300	75	73	1.0	4.9	6.177(4)	0.83	20
4	SL-562	470	300	67	56	0.8	6.0	6.188(5)	1.00	20
5	SL-577	470	240	30	182	6.0	7.8	6.149(9)	0.37	22
6	SL-581	470	240	10	35	3.5	12.5	6.210(1)	1.38	26
7	SL-602	510	240	12	226	19	22.6	6.139(2)	0.20	31
8	SL-605	510	240	20	440	22	23.9	6.132(6)	0.10	36

3. Results and discussion

In all cases, the "delay" of epitaxy was observed: this was the first stage of formation and merging of islands, proceeding at low rate. The thickness of layers was close to the sensitivity of its detection by the x-ray method and made up 5-10nm. At the second stage, the two –dimensional growth continued at a higher rate. The "delay" in the first case made up 60s, the second case – 8s, and the third case – 2s [7].

Table 1. Data on the conditions of growth, tangential lattice constants and deformation of epitaxial PbSe layers on KCl substrates.

In the Table 1 are listed the data on the layers selected for illustration of the abovementioned cases. The layers thickness was measured by the x-ray method by measuring the radiation from two reflecting planes - from a clean substrate and a substrate with a deposited layer.

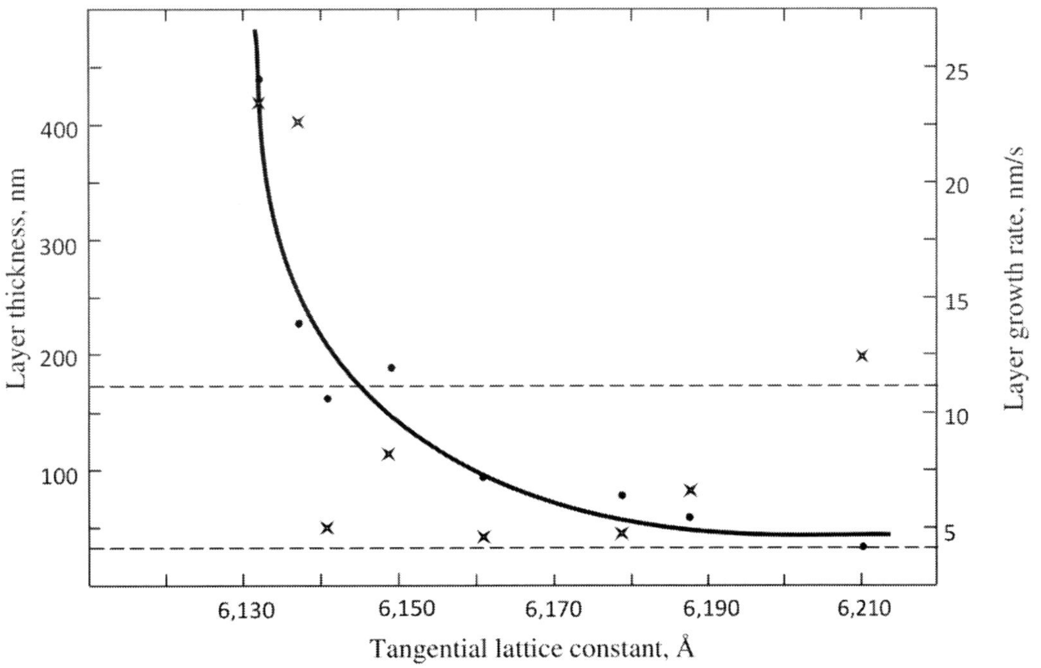

Fig.1. Relationship between the thickness, growth rate and tangential lattice constant for epitaxial PbSe layers : • - thickness of the layers, x - growth rate.

The tangential lattice constant of the layer was determined by a symmetrical circuit and by diffraction patterns of θ -2θ scanning at the reflection from plane (400). Radiation $C_{ok\alpha}$ (λ = 1789Å) was investigated. Deformation ε was determined as $\varepsilon = (a_l - a_{PbSe}) / a_{PbSe}$ in respect to a nonstrained PbSe layer (monocrystal). The halfwidth of the x-ray diffraction line was also determined by diffraction patterns at reflection from plane (400).

In Fig.1 is shown the relationship among the thickness, the growth rate and the tangential lattice constant of layers.

Based on the data given in the Table and Fig.1, we can draw the first general conclusions. The tangential lattice constant of the layers increase with decreasing thickness. The lattice constant range from 6.140 to 6.190Å was achieved at the growth rate 4-8nm/s (the region between dotted lines in Fig.1) and corresponding layers thickness from 55 to 180nm. The lattice constant range from 6.126 to 6.140Å was realized at the layers, thickness from 200 to 450nm and more at the growth rate of 20-25nm/s. The lattice constant 6.210Å was realized at the layers thickness of 35nm and the rate of cooling 12.5nm/s.

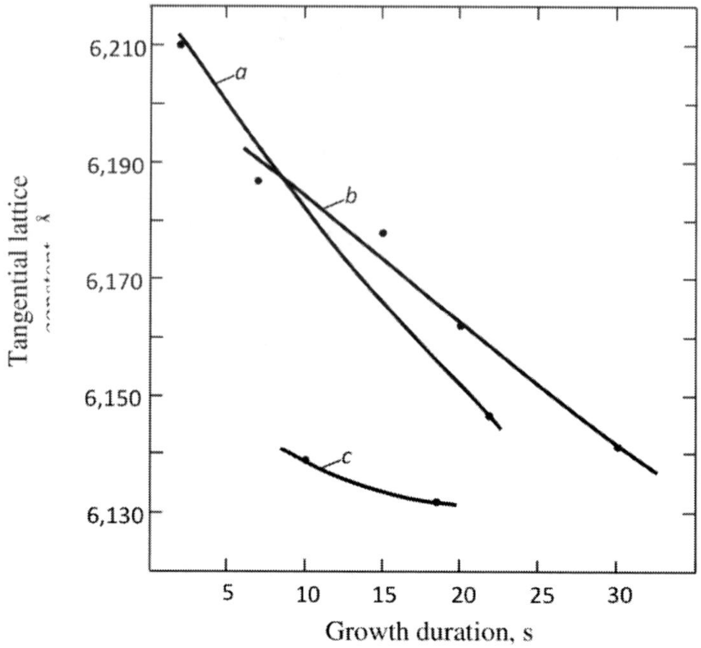

Fig.2 Dependence of the tangential lattice constant of epitaxial PbSe layers at growth duration : a) l = 2,5 mm, T_e = 470°C ; b) l = 12mm, T_e = 470°C ; c) l = 0,5mm, T_e = 510°C.

According to works [8,9], in addition to these data, the sizes of subgrains in the layers, their number and the disorientation among them, and also the critical thickness at which the number of dislocations grows and they were accelerated were assessed.

For the layers thickness 55-180nm, the size subgrains made up 50-70nm, their average disorientation was $(1.2\cdot10^{-3})'$ their number was equal, to $2\cdot10^{4}$, and the corresponding critical thickness was ~ 70nm. For the layers thickness from 200 to 450nm, the size of subgrains was ≥ 100nm, their average disorientation was $(3.6\cdot10^{-4})'$, the corresponding critical thickness being ~ 15nm. For the layer 35nm thick, the critical thickness was 80nm at the average disorientation of subgrains $(8\cdot10^{-4})'$ - their number making up $3\cdot10^{4}$. The value of deformation ε of the layer 35nm thick was equal to $1.38\cdot10^{-2}$.

By concrete examples, it is seen how the growth duration of layers affects their parameters (Fig. 2). First of all, this effect was revealed in the row of layers 1-4 (Table) when the lattice constant increased from 6.142 to 6.188Å after, the growth time decreased from 30 to 7' and at close growth rates of 1-2nm/s at the first stage. The obtained lattice constants can be explained by that, at the epitaxy temperature of 300°C, nonstochiometric defects managed to migrate with the growth process and, the shorter the growth duration, more efficiently they accommodated, or more precicely being annigilated, in dislocation nuclei, they impeded the energy transfer of elastic energy of the layers to deformations.

The growth duration determines the level of parameters of the layers in the case of different growth rates both the overall and at the second stage, the lattice constant of layer SL-577 was higher than that of layer SL-537. In this case, the growth rate for the latter layer made up 22s, while for the latter one -30s higher.

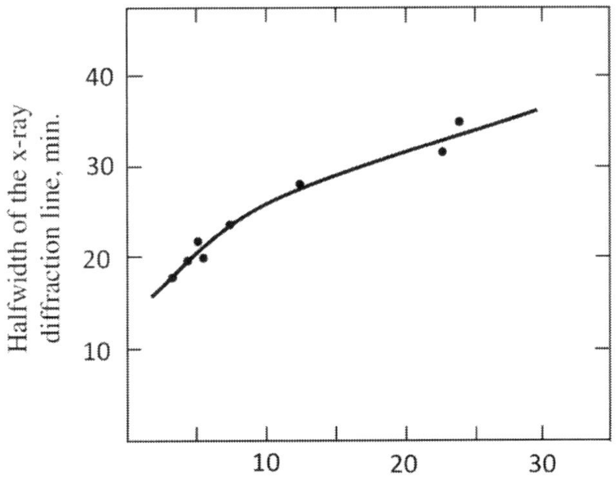

Fig.3. Dependence of the halfwidth of x-ray diffraction pattern (reflecting plane (400)) on the growth rate of epitaxial PbSe layers.

At higher growth rate than that of the lattice constant of 6.210Å was achieved for layer SL-581 at short growth duration -2s. This means that, despite low temperature of epitaxy (240°C), for layers SL-577 and SL-581, high lattice constants were achieved at short growth duration. Despite that, for layers SL-602 and SL-605, at the second stage the growth duration made up 10 and 18s, respectively, at the growth rate ≥ 20nm/s and the thickness of layers 220 and 440nm, respectively larger subgrains ≥ 100nm with higher disorientation $(3\cdot10^{-4})'$ were formed. At critical thickness ~15nm, relaxation of strain at subgrain boundaries increased, and hence the lattice constants were not high. As is shown in Fig.3, in this case, the halfwidth of the x-ray diffraction line was equal to 31′ and 36′, respectively, which exceeded those for SL-577 and SL-562 (26′ and 20′, respectively).

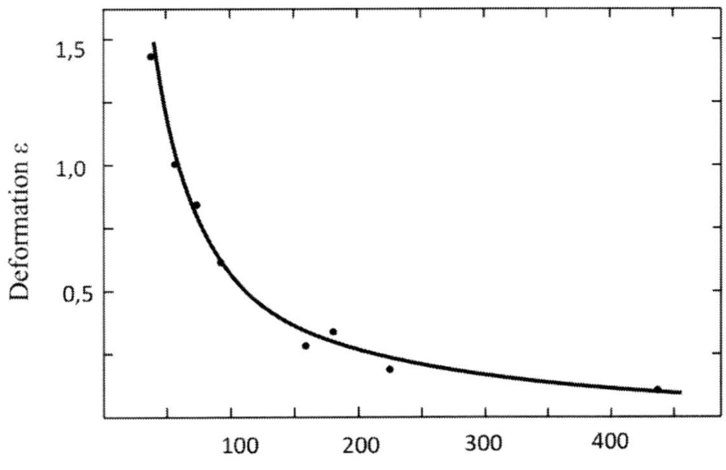

Fig.4. Deformation $\varepsilon = \frac{a_l - a_{PbSn}}{a_{PbSn}} \cdot 10^2$ of epitaxial PbSe layers dependence at their thickness

Thus, the high growth rate causes additional relaxation of strain, which is due to an increase in the elastic energy of layers at high values of thickness as well.

IOP Conf. Series: Materials Science and Engineering **77** (2015) 012017 doi:10.1088/1757-899X/77/1/012017

On the other hand, in layers SL-541 and SL-577, at close growth duration at the second stage, 20 and 22s, respectively, and comparable x-ray diffraction line halfwidth of 19 and 22′, respectively, the constant lattice of layer SL-541 (6.162Å) was higher than that of layer SL-577 (6.149Å). This can be explained by lower growth temperature of layer SL-577 and by close critical thickness of these layers ~ 70nm, i.e. by low tendency to relaxation of strain at subgrain boundaries.
In Fig.4 is shown the thickness dependence of deformation in PbSe layers. in the thickness range from 55 to 80nm, the deformation made up $0.7\cdot10^{-2}$ -$1\cdot10^{-2}$.
Under such deformation in the layers, the forbidden gap width increased by 0.15eV as compared to the unstrained layer [10, 11]. The epitaxial PbSe layers with 50-80nm thick with the lattice constant 6.140-6.190Å and the deformation of the level to $1\cdot10^{-2}$ can serve as the basis for designing of IR photodetectors and lasers.

4. Conclusions

If we follow the character of relaxation of strain with decreasing lattice constant as a whole, we will see that, in the case of layer with the lattice constant 6.210Å, critical thickness >100nm and significantly finer subgrains (30nm), the substrate – layer interface. For layers 1-5 (Table) with the lattice constant 6.140 – 6.190Å, the size of subgrains 40-70nm and critical thickness 50-70nm, only minor part of the relaxation of strain takes place at subgrain boundaries, and it generally occurs at the substrate – layer interface as well. For the layers with lattice constants 6.140 – 6.190Å , and the size of subgrains >100nm and the critical thickness ~ 15nm, major part of the relaxation of strain occurs at the subgrain boundaries. Solid solutions PbSnSe have considerable reserves for realization of the dielectric state when the levels of impurities shift deeply into the forbidden gap and the concentration of current carriers can decrease by at least a factor of 7. Without a doubt, this opens up new possibilities for designing high – sensitivity high –temperature IR photodetectors.

References

[1] Pashaev A, Davarashvili O, Enukashvili M, Akhvlediani Z, Gulyaev R, Bychkova L and Zlomanov V 2013 *Nano Studies 8 253*

[2] Pashaev A, Davarashvili O, Enukashvili M, Bychkova L, Dzagania M and Zlomanov 2014 *Herald Azer.Eng. Acad. 6 49*

[3] Pashaev A, Davarashvili O, Enukashvili M, Akhvlediani Z, Gulyaev R and Zlomanov V 2013 *Bull Azer. Nat. Acad. Sci. 33 31*

[4] Prabahar S, Surynaraynan N, Rajasekar K and Srikanth S 2009 Chalcogenide Zett. 6 203

[5] Abbas N, Al-Fawade E and Alatya S 2013 *J.Mat.Sci. Eng. A3 82*

[6] Cruz S, Tellez G, Rosas U, Perez R, Merino R, Lima L and Moreno O 2013 *J.Mat.Sci.Eng. A3 621*

[7] Pashaev A, Davarashvili O, Enukashvili M, Akhvlediani Z, Gulyaev R, Bychkova L and .Zlomanov V 2013 *J.Mat. Sci. and Eng. A3 235*

[8] Pashaev A, Davarashvili O, Enukashvili M and Dzagania M 2014 *Bull Azer. Nat. Acad. Sci. 38 31*

[9] Pashaev A, Davarashvili O, Enukashvili M, Akhvlediani Z, Gulyaev R and Zlomanov V 2013 *J.Mat. Sci. and Eng. A3 117*

[10] Pashaev A, Davarashvili O, Enukashvili M, Akhvlediani Z, Gulyaev R and.Zlomanov V 2014 *IJEIT 3, 318*

[11] Pashaev A, Davarashvili O, Enukashvili M, Akhvlediani Z, Bychkova L and Gulyaev R 2014 *Advanced Materials Research 1025 831*

IOP Conf. Series: Materials Science and Engineering **77** (2015) 012018 doi:10.1088/1757-899X/77/1/012018

Dark relaxation of holographic gratings in azobenzene and chalcogenide films

A Ozols, P Augustovs, K Kenins, V Kokars, E Zarins, K Traskovskis and D Saharov

Faculty of Material Science and Applied Chemistry, Riga Technical University, Riga, Latvia

E-mail: aozols@latnet.lv

Abstract. Studies of transmission holographic grating (HG) relaxation (recorded with *s-s, p-p, s-p, L-L, L-R* laser beam polarizations) have been carried out in organic molecular azobenzene glassy films W-50 and W-75, and in chalcogenide glassy a-As_2S_3 film. Results previously obtained with azobenzene molecular glassy K-RJ-9 films also are used for comparison. HG with the periods of 0.50, 2.0 and 8.6 μm were studied in these films. It is found that recording efficiency in the same sample depends on both recording polarizations and HG period. The HG stability in the course of relaxation also depends on these factors. Anticorrelation has been observed of HG recording efficiency and stability. Recording was efficient with *p-p* or *L-R* polarizations at 2.0 μm HG period whereas it was stable with *s-s* *L-L* polarizations at 8.6μm HG period. As a rule, HG stability was higher in the samples with higher glass transition temperature. The HG dark (relaxational) self-enhancement of HG was observed in the case of W-75 (*L-R* polarizations, 0.50μm; *s-s* polarizations, 2.0μm) and in previously studied K-RJ-9 (*p-p* polarizations, 8.6μm) samples. Relaxational changes in diffraction efficiency are explained in terms of post-recording mass transfer under the influence of photoinduced spatially periodic mechanical stress.

1. Introduction

Relaxation processes in disordered materials after the optical irradiation are much less studied than photo induced processes [1]. However, they are important both from fundamental point of view as well from the point of view of applications. As mentioned by O.N.Oliveira, Jr, J.Kumar, L.Li and S.K.Tripathy [2] with respect to organic amorphous films "Perhaps the most important unexplained problem is the post writing evolution which can be due to the mechanical relaxation that has not been treated in any of the models". Relaxation studies of optical recording in information storage materials are practically significant to establish the possible information storage time. They are also important for production of holographic optical elements. Previously we have experimentally investigated the relaxation of holographic gratings (HG) in fresh amorphous chalcogenide As_2S_3 films [4]. A strong relaxational (dark) self-enhancement (RSE) of holographic gratings up to 18 times was found. After that no strong RSE was found in other materials, to our knowledge. Are a-As_2S_3 films unique in this respect? Recently we also have explored the relaxation of HG in azobenzene molecular K-RJ-9 [(5,5,5-triphenylpentyl 4-((4-(bis(5,5,5-triphenyl-pentyl)amino) phenyl) diazenyl) benzoate] molecular glassy film [1]. RSE of HG was found there. Interestingly, RSE of HG was more pronounced in reflection mode readout rather than in the transmission mode readout.

In this paper, we have experimentally studied the dark relaxation of transmission HG in two azobenzene molecular glassy films synthesized in our faculty and in one amorphous chalcogenide film varying recording beam polarizations and HG period. It was found that these factors indeed influence the relaxation. Besides, the glass transition temperature of the sample was found to have a strong effect. RSE of HG was observed in some cases but not as strong as in [4]. The observed HG relaxation properties are discussed in terms of post writing mass transfer initiated by mechanical stress periodically distributed in space after the HG recording. Previously reported results [1] partially also are included in discussion for generality.

2. Experiments and results

Studies of transmission holographic grating relaxation (recorded with *s-s, p-p, s-p, L-L, L-R* laser beam polarizations) have been carried out in organic molecular azobenzene glassy films W-50, W-75, and in a chalcogenide glassy a-As_2S_3 film (the latter provided by Dr.J.Teteris from the Institute of Solid State Physics, University of Latvia). The precise chemical notation of the films and the chemical structure of molecules of azobenzene films are given in Figs.1 and 2. Films were spin-coated onto the glass substrates. Amorphous chalcogenide As_2S_3 film was thermally evapourated onto the glass substrate.Its bandgap was E_g=2.39 eV, corresponding to λ=519 nm. Glass transition temperature T_g, long wave length absorption band maximum location , λ_{max} and thickness of the samples, d are given in the Table 1.There, data for the sample K-RJ-9 studied previously in [1] are also included for comparison and for further discussion.

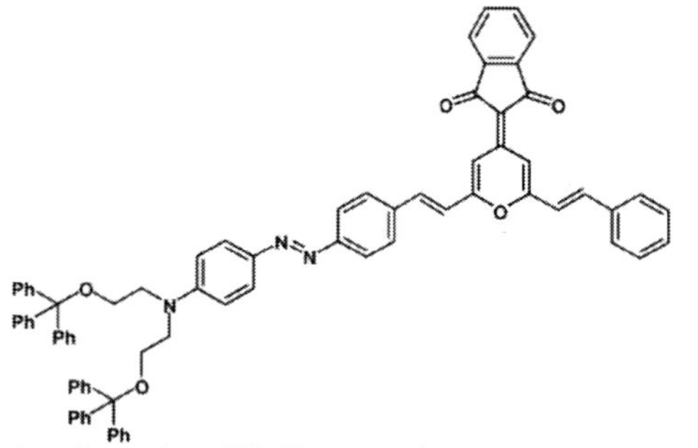

Fig.1. Sample W-50 –azobenzene molecular glassy film: 2-(2-(-4-(-(4-(bis(2-(trityloxy)ethyl) amino) phenyl)diazenyl)styryl)-6-styryl-4H-pyran-4-ylidene) -2H-indene-1,3-dione.
Donor group is on the left side from the azobenzene, acceptor group is on the right side.

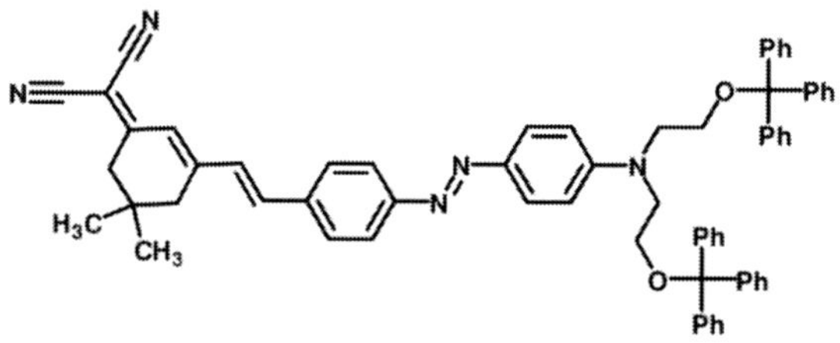

Fig.2. Sample W-75 (also IWK-2D) – azobenzene molecular glassy film: 2-(3-(4-((4-(bis(2-(trityloxy)ethyl) amino)phenyl)diazenyl) styryl)-5,5-dimethylcyclohex-2-enylidene).
Donor group is on the right side from the azobenzene but the acceptor group is on the left side.

Table 1. Characteristics of studied samples

Sample	W-50	a-As_2S_3	W-75	K-RJ-9
T_g , ^{0}C	176	187	105	66
λ_{max} , nm	478	519	504	457
d ,µm	1.5	12	3	2

HG with the periods (Λ) of 0.50, 2.0 and 8.6 µm were studied in these films. Recording wavelength was 532 nm and the readout wavelength was 632.8 nm. Recording light intensity was 1.76 W/cm^2, the recording beam powers were equal. The main experimental setup is shown in Fig.3.

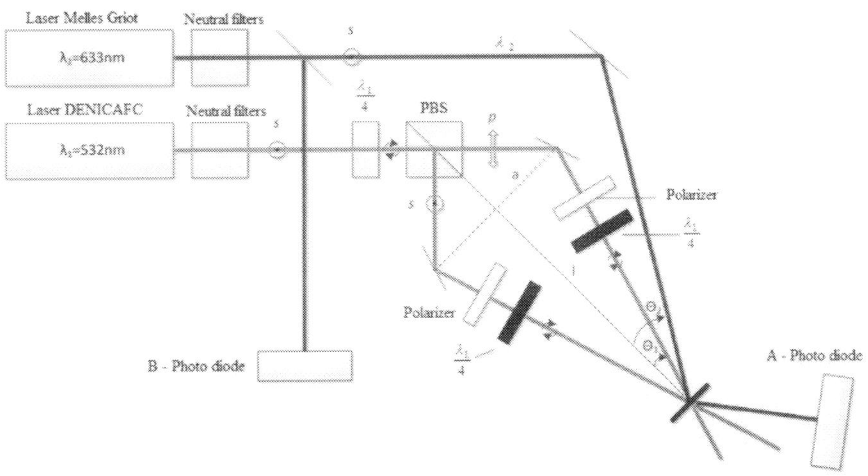

Fig.3. The main experimental setup for HG recording and readout in the case of circular recording beam polarizations. Other wave plates were inserted in the case of other polarizations. A-photo diode was measuring the diffracted beam power, the B-photo diode was measuring the incident beam power. Their ratio gave the diffraction efficiency.

The experimental procedure was as follows. First, HG were recorded until the maximum diffraction efficiency (DE) or (in the case of samples W-50 and a-As_2S_3) until the maximum self-diffraction efficiency [5] was reached. Then DE at 632.8 nm ($DE_{initial}$) was measured in all cases and the relative DE ξ = $DE/DE_{initial}$ was monitored periodically during the relaxation time in the dark. The experimental relative DE relaxation time dependences are presented in Figs.4-8. Polarizations C-1 and C-2 in Figs. 4,5 corresponded to *L-L* and *L-R*, respectively, but they were slightly elliptical for technical reasons.

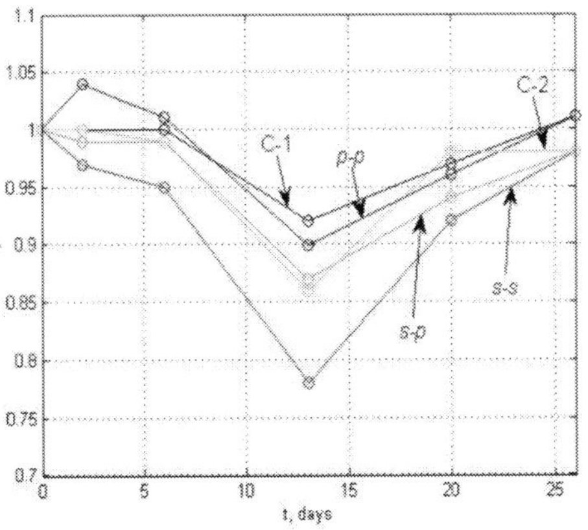

Fig.4. Sample W-50. Relative DE relaxation time dependences for different recording polarizations. Λ = 2.0μm.

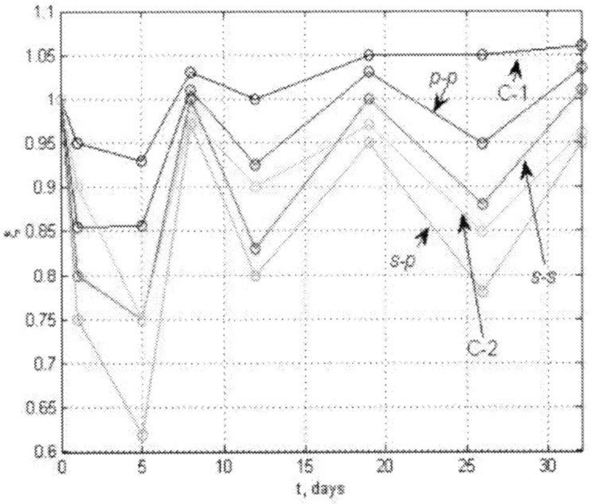

Fig.5. Sample a-As_2S_3. Relative DE relaxation time dependences for different recording polarizations. Λ = 2.0μm.

It is seen from Figs. 4-8 that relaxation curves were changing non-monotonically, with oscillations.DE relaxation behaviour depended on recording beam polarizations. In some cases of the sample W-75 RSE was observed. DE relaxation time dependences for sample K-RJ-9 has been presented in our previous study [1].

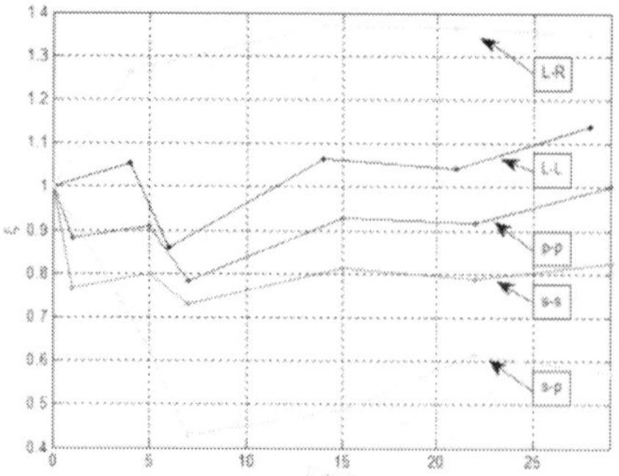

Fig.6. Sample W-75. Relative DE relaxation time dependences for different recording polarizations. $\Lambda = 0.50\mu m$.

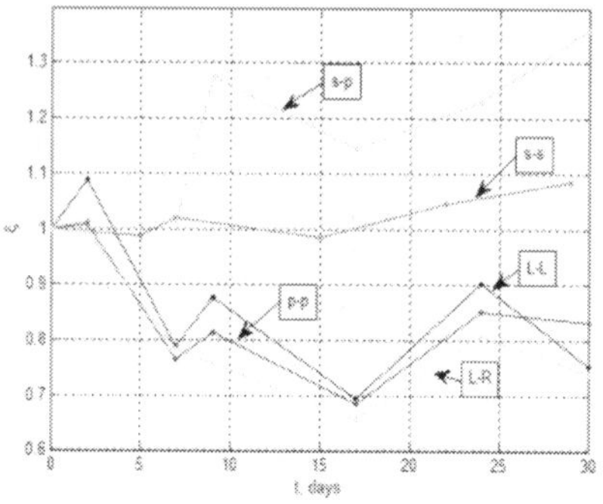

Fig.7. Sample W-75. Relative DE relaxation time dependences for different recording polarizations. $\Lambda = 2.0\mu m$.

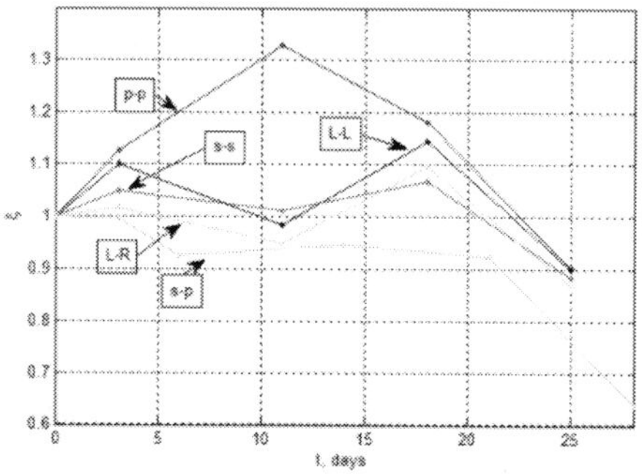

Fig.8. Sample W-75. Relative DE relaxation time dependences for different recording polarizations. $\Lambda = 8.6$ μm.

The obtained results are summarized in Tables 2-7. There, the initial DE, the mean square deviation σ of relative DE and the extremal relative DE deviation $\Delta\xi_{extr} = \xi_{extr}$-1 are given. The mean square deviation was calculated according to the formula

$$\sigma = \sqrt{\frac{1}{N}\sum_{n=1}^{N}(\xi_n - 1)^2}$$

where N is the number of experimental points for each DE relaxation curve.

Table 2. DE relaxation parameters of the sample W-50 at $\Lambda = 2.0$ μm.

	s-s	*p-p*	*s-p*	*L-L*	*L-R*
$DE_{initial}$, %	4.0	10.1	5.2	7.0	17.0
σ	0.15	0.055	0.066	0.041	0.066
$\Delta\xi_{extr}$	-0.22	-0.10	-0.13	-0.08	-0.14

Table 3. DE relaxation parameters of the sample a-As_2S_3 at $\Lambda = 2.0$ μm.

	s-s	*p-p*	*s-p*	*L-L*	*L-R*
$DE_{initial}$, %	17.0	17.0	5.7	15.0	13.0
σ	0.15	0.088	0.21	0.051	0.12
$\Delta\xi_{extr}$	-0.25	-0.15	-0.37	-0.07	-0.15

Table 4. DE relaxation parameters of the sample W-75 at $\Lambda = 0.50$ μm.

	s-s	*p-p*	*s-p*	*L-L*	*L-R*
$DE_{initial}$, %	0.085	1.1	0.0070	0.093	0.14
σ	0.22	0.12	0.46	0.10	0.33
$\Delta\xi_{extr}$	-0.27	-0.21	-0.55	0.13	0.37

Table 5. DE relaxation parameters of the sample W-75 at $\Lambda = 2.0$ μm.

	s-s	*p-p*	*s-p*	*L-L*	*L-R*
$DE_{initial}$, %	0.81	9.9	0.70	8.0	24.0
σ	0.10	0.23	0.23	0.27	0.25
$\Delta\xi_{extr}$	0.09	-0.31	0.35	-0.30	-0.33

Table 6. DE relaxation parameters of the sample W-75 at $\Lambda = 8.6$ μm.

	s-s	*p-p*	*s-p*	*L-L*	*L-R*
$DE_{initial}$, %	1.63	4.7	0.74	4.4	8.5
σ	0.074	0.20	0.19	0.10	0.088
$\Delta\xi_{extr}$	-0.11	0.32	-0.36	0.14	0.10

Table 7. DE relaxation parameters of the sample K-RJ-9 in the case of *p-p* recording polarizations for three HG periods ($DE_{initial}$ values are taken from [1]).

	$\Lambda = 0.50$ μm	$\Lambda = 2.0$ μm	$\Lambda = 8.6$ μm
$DE_{initial}$, %	0.21	23	9.5
σ	0.86	0.91	0.15
$\Delta\xi_{extr}$	-0.98	-0.37	0.18

3. Discussion of the observed regularities

Let us analyse the effects of HG period (Λ), recording beam polarizations and of materials glass transition temperature (T_g) on the recording efficiency characterized by $DE_{initial}$ and recording stability characterized by DE deviations σ and $\Delta\xi_{extr}$. Lower deviations correspond to higher HG stability.

3.1. Effect of HG period on the recording efficiency

Arrangement of HG periods (in micrometres) in the decreasing DE order for different polarizations gives the following sequences. W-75, *s-s*: 8.6, 2.0, 0.50; *p-p*: 2.0, 8.6, 0.50; *L-L*: 2.0, 8.6, 0.50; *L-R*: 2.0, 8.6, 0.50. K-RJ-9, *p-p*: 2.0, 8.6, 0.50. As seen, the most efficient HG period depends on recording polarizations. Most often optimal is Λ= 2.0μm.

3.2. Effect of recording polarizations on its efficiency

Arrangement of polarizations in the decreasing DE order gives the following sequences. W-75, Λ = 0.50 μm: *p-p, L-R, L-L, s-s, s-p*; $\Lambda = 2.0$ μm: *L-R, p-p, L-L, s-s, s-p*; $\Lambda = 8.6$ μm: *L-R, p-p, L-L, s-s, s-p*. W-50, $\Lambda = 2.0$ μm: *L-R, p-p, L-Ls-p, s-s*. Amorphous As_2S_3, $\Lambda = 2.0$ μm: *p-p, s-s, L-L, L-R, s-p*. The most efficient recording polarizations depend on the material and HG period. Yet, in most cases they are *L-R* or *p-p*. *L-R* polarizations most often are the most efficient in azobenzene films thus supporting the azochromophore orientation mechanism in the light electrical field.

3.3 Effect of HG period on the recording stability
Arrangement of periods in the increasing σ, $\Delta\xi_{extr}$ order gives the following sequences. W-75, *s-s*: 8.6, 2.0, 0.50; *p-p*: 0.50, 8.6, 2.0; *L-L*: 0.50, 8.6, 2.0; *L-R*: 8.6, 2.0, 0.50. K-RJ-9, *p-p*: 8.6, 2.0, 0.50. As seen, HG are the most stable in the case of 8.6 and 0.50 μm HG periods. As a rule, these HG periods do not coincide with the most efficient HG recording period.

3.4. Effect of recording beam polarizations on its stability
Arrangement of the polarizations according to increasing σ, $\Delta\xi_{extr}$ gives the following sequences. W-75, $\Lambda = 0.50$ μm: *L-L, p-p, s-s, L-R, s-p*; $\Lambda = 2.0$ μm: *s-s, p-p, s-p, L-L, L-R*; $\Lambda = 8.6$ μm: *s-s, L-R, L-L, p-p, s-p*.W-50, $\Lambda = 2.0$ μm: *L-L, p-p, s-p, L-R, s-s*. Amorphous As_2S_3, $\Lambda = 2.0$ μm: *L-L, p-p, L-R, s-s, s-p*. One can see that the recording stability is the highest in the case of *L-L* and s-s polarizations. They do not correspond to the most efficient recording polarizations *L-R* and *p-p*. In fact, we can speak about the anticorrelation of recording efficiency and stability in our experiments.

3.5. Effect of material's glass transition temperature on the HG recording efficiency and stability

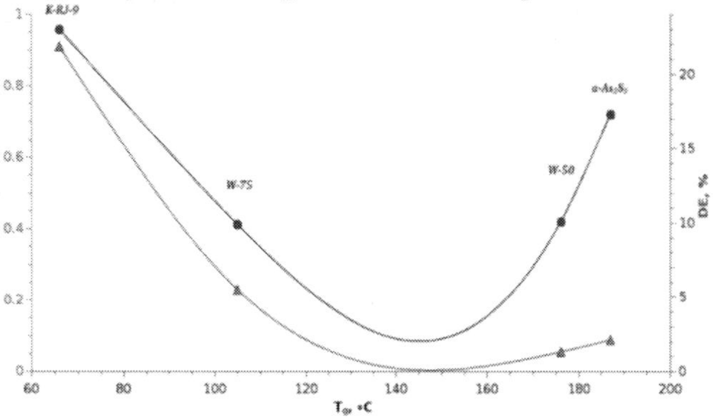

Fig.9. Materials initial (maximum) DE and mean square deviation σ of relative DE versu T_g in the case of *p-p* HG recording at Λ= 2.0μm. The red curve (triangles) corresponds to σ, but the blue one (circles) to DE.

The tendency can be seen in Fig.9 that the lower T_g value of the material corresponds to the lower recording stability. This is an indication to the possible role of the mass transfer in the HG relaxation process because the lower T_g corresponds to the lower viscosity. A similar T_g dependence was recently observed in azobenzene molecular glassy films with respect to the stability of nonlinear optical second-order susceptibility [6].

3.6. Effect of the sample thickness on the recording stability
If one compares the DE deviations σ, $\Delta\xi_{extr}$ in the case of materials with close glass transition temperatures (Table 1) but with strongly different thicknesses such as W-50 and a-As_2S_3 , one can easily see (Tables 2,3) that larger sample thickness facilitates lower stability.

3.7. HG relaxation mechanism
Non- monotonical DE changes after the recording which can be quite large (Figs. 4-8) are not because of the temperature changes in the laboratory. This temperature was monitored during the experiments and changed from 18.5 to 22.6^{0}C. Calculations of thermally induced refractive index changes have shown that they are too small to cause the observed DE changes. Besides, no correlation was observed between the temperature and DE changes. In our opinion, material relaxation is responsible for DE changes. After the HG recording the film is thermodynamically unstable and tends to the thermodynamic equilibrium. The driving force behind the relaxation is the mechanical stress periodically distributed in space as already suggested in [2,4]. It arises during the HG recording due to

the *trans-cis* photoisomerization in azobenzene molecular glassy films [1,3,5], and due to the photo induced As-As bond breaking and As-S bond formation in a-As_2S_3 film [4]. In the case of azobenzene films one reason of arising such stress distribution is that *cis*-isomer occupies larger space than *trans*-isomer [3]. This conclusion is supported by the observed tendency (see Sections 3.3 and 3.6) that larger HG stability corresponds to smaller sample thickness, d, and larger HG period, Λ. It is in accordance with structure perturbation theory [7] by which one can show that the mechanical stress increase due to HG is

$$\Delta\sigma = \text{const}\ d/\Lambda.$$

As the result, the thermostimulated rearrangement of azobenzene chromophores together with their environment takes place resulting in the mass transfer. HG recording stability dependence on T_g supports this conclusion. Near the surface, there is enhanced cooperative segmental mobility and a reduced effective T_g [8]. Therefore, post recording mass transfer is expected to be stronger on the surface than in the volume. Relaxation difference was indeed observed in transmission and reflection modes [1]. Efficient photo-induced mass transfer has also been observed in a-As_2S_3 films [9]. Charge transfer and the interaction of molecular dipoles may also take place

RSE effect observed in this study was much weaker than in fresh a-As_2S_3 films [3]. The reason may be that photoinduced structural changes in fresh a-As_2S_3 films are much stronger than in relaxed a-As_2S_3 films used in this study and the structure changes in azobenzene molecular glassy films due to *trans-cis* photoisomerization. Of course, details of relaxation remain to be established in numerous further studies.

4. Conclusions

Holographic recording efficiency and stability depend on the material, HG period and recording beam polarizations.

Anticorrelation has been observed of HG recording efficiency and stability. Recording was efficient with *p-p* or *L-R* polarizations at 2.0 μm HG period whereas it was stable with *s-s* or *L-L* polarizations at 8.6μm HG period.

An approximate anticorrelation has been observed between recording stability and material's glass transition temperature, T_g.

Relaxational self-enhancement (yet not large) of HG has been found in azobenzene molecular glassy films **W-75** and **K-RJ-9**. In the latter case, also in the reflection mode.

HG relaxation in studied films mainly proceeds by post-recording mass transfer under the influence of mechanical stress.

5. Acknowledgment

The financial support of the Latvian State Research Program on Multifunctional Materials IMIS^2 is greatly acknowledged.

References

[1] Ozols A, Augustovs P, Kokars V, Traskovskis K, and Saharov D 2013 *IOP Conf. Series: Materials Science and Engineering* **49** 012043
[2] Oliveira O N, Jr, Kumar J, Li L, and Tripathy S K in [3], pp. 429-483
[3] Sekkat Z and Knoll W 2002 *Photoreactive Organic Thin Films* (California: Academic Press)
[4] Ozols A, Salminen O and Reinfelde M 1994 *J. Appl. Phys.* **75** 3326
[5] Ozols A, Saharov D, Kokars V, Kampars V, Maleckis A, Mezinskis G, and Pludons A 2010 *J. of Phys: Conf. Series* **249** , doi: 10.1088/1742-6596/249/1/012055

[6] Rutkis M, Traskovskis K, Tokmakovs A and Kokars V 2014 *Joint 12th Russian/CIS/Baltic/Japan Symp. On Ferroelectricity and 9th Int. Conf. Functional Materials and Nanotechnologies,Inst. of Solid State Physics, University of Latvia, September 29-October 2, Riga, 2014, Book of Abstracts,* p.89
[7] Jastrzebski Z D 1987 *The Nature and Properties of Engineering Materials* (New York:Wiley)
[8] Priestley R D, Ellison C J, Broadbelt L J, and Torkelson J M 2005 *Science* **309** 456
[9] Gertners U and Teteris J 2010 *Lithuanian Journ. of Physics* **50** 35.

IOP Conf. Series: Materials Science and Engineering **77** (2015) 012019 doi:10.1088/1757-899X/77/1/012019

Recording of Surface Relief in Azobenzene Containing Low Molecular Weight Organic Glasses

K Klismeta, J Teteris

Institute of Solid State Physics, University of Latvia, 8 Kengaraga Str., LV 1063 Riga, Latvia

E-mail: k.klismeta@gmail.com

Abstract. One of the used methods to induce mass motion on the surface of azopolymer films is holographic recording using laser wavelength corresponding to the absorption band of the azo compound. The sinusoidal light interference pattern on the surface of the sample leads to a sinusoidal surface patterning - a stable surface relief grating (SRG). As confirmed by atomic force microscopy (AFM), these gratings are found to be very large, with depth up to hundreds of nanometres and they diffract light very efficiently. In this work azobenzene containing low molecular weight organic glass K-RJ-8 was experimentally studied. The compound absorbs at the visible spectrum, holographic recording in this range using different polarization states of the recording beams was done. Diffraction efficiency was measured during the recording of the SRG. The relationship between formation of SRG and photoinduced birefringence is discussed.

1. Introduction

One of the properties of azobenzene containing compounds is that under influence of light the azo compound undergoes trans-cis isomerisation [1] and during this process the moieties may align relative to the electric field vector of light. This leads to anisotropy in the material (which can be observed as dichroism and birefringence), thus making these compounds light polarization sensitive [2]. Another phenomenon which strongly depends on the light polarization is photoinduced mass transport [3]. A well known method how to induce mass motion is holographic recording. By using different polarization states of the two recording beams, different interference patterns can be obtained on the surface of the material and a strong effect on the efficiency of the mass transport can be observed [4].

Mass transport phenomenon allows to create a surface relief directly by light, without the post processing of chemical etching. Materials that possess this ability of direct optical patterning may have potential application in various photoactive devices, optical information storage and the production of diffractive optical elements used in telecommunication (e.g. filters, Bragg reflectors) [5].

As opposed to functionalized azo-polymers, low molecular weight organic glasses containing azobenzene have the advantage of a well-defined molecular structure and they exhibit more uniform physical properties. In this work thin films of azobenzene containing low molecular weight organic glass K-RJ-8 were experimentally studied. We attempt to explain distinctive attributes of photoinduced mass transport in azo compounds by studying photoinduced dichroism and birefringence in K-RJ-8.

2. Experimental

The studied compound K-RJ-8 was synthesized in Riga Technical University by Kaspars Traskovskis [6]. The molecular structure of the compound is shown in figure 1, the absorption spectrum is shown in figure 5. The thickness of the thin film used for birefringence measurements was 0.3 μm, for dichroism measurements and for the holographic recording a thicker sample (0.7 μm) was used to prevent the mass transport from reaching the substrate.

Surface relief gratings were recorded using a holographic set-up shown in figure 2.

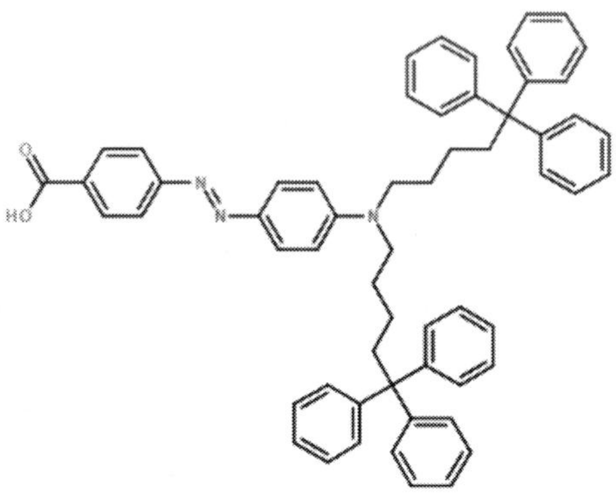

Figure 1. Molecular structure of low molecular weight organic glass K-RJ-8.

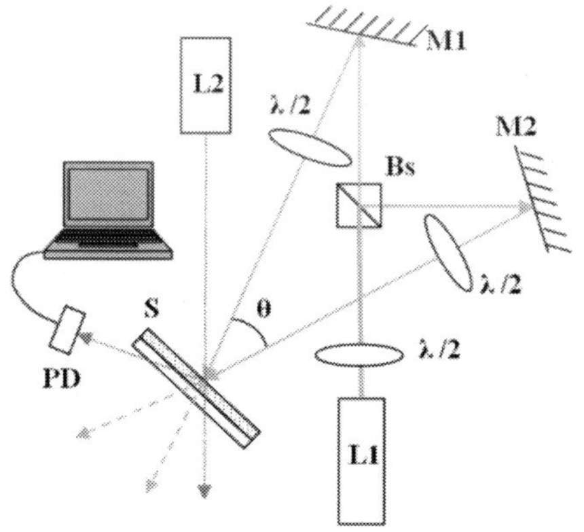

Figure 2. Experimental set-up for holographic recording. L1 – recording laser (491 nm), L2 – reading laser (653 nm), λ/2 – half-wave plate, Bs – beam splitter, M – mirror, S – sample, PD – photodiode.

For holographic recording 491 nm laser (L1) was used. The laser beam was split into two beams with equal intensities by a beam splitter (Bs). Using mirrors (M) the beams were directed to the sample (S) where they interfere. The period of the interference pattern was Λ=1 μm. Half-wave plates (λ/2) were used to set the polarization states of the recording beams. The intensity of the beams was $I_1=I_2=0.18$ W/cm^2. The recording course was monitored by registering the 1st order diffracted beam of 653 nm laser (L2), which is weakly absorbed by the sample, with a photodiode (PD). The diffraction efficiency of the obtained grating was calculated as in (1):

$$\eta = \frac{I_1}{I_0} \tag{1}$$

where I_0 is the intensity of the reading laser and I_1 is the intensity of the 1st order diffracted beam.

Experimental set-up for studying the photoinduced dichroism is shown in figure 3. The polarization of 532 nm pump laser (L) was set by a half-wave plate (λ/2), linearity was improved by Glan-Taylor polarization cube (P). The intensity of the pump beam was I =0.07 W/cm^2. As the probe beam a polarized (A) white light source was used (D). To determine difference of absorption in orthogonal directions, the sample (S) was illuminated with pump beam with polarization parallel or orthogonal to the polarization of the probe beam. The spectra were measured using Ocean Optic HR4000CG spectrometer (PD).

Photoinduced birefringence was measured using experimental set-up shown in figure 4. Birefringence measurements were done using the same wavelength (532 nm) and laser intensity (I =0.07 W/cm^2) as in dichroism experiment.

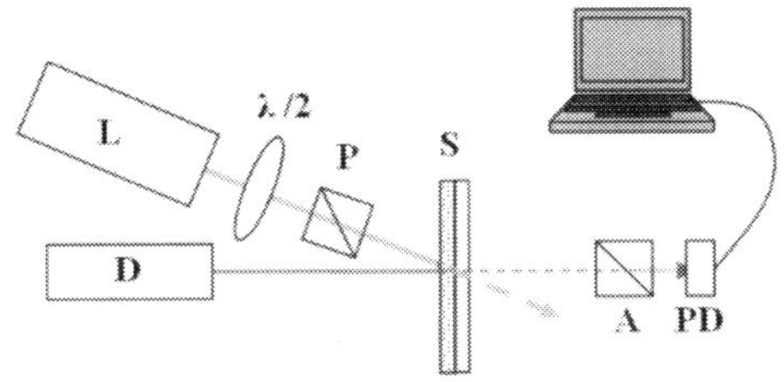

Figure 3. Experimental set-up for absorption spectra observation. L – pump laser (532 nm), D – white light source, λ/2 – half-wave plate, P - polarizer, A – analyzer, S – sample, PD – photodiode.

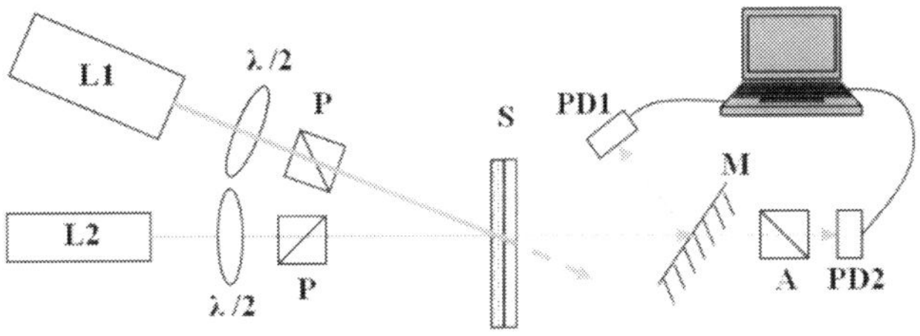

Figure 4. Experimental set-up for photoinduced birefringence measurements. L1 – pump laser (532 nm), L2 - probe laser (532 nm), λ/2 – half-wave plate, P - polarizer, A - analyzer, M – semitransparent mirror, S – sample, PD - photodiode.

Birefringence was induced by 532 nm laser (L1). The polarization of the laser beams was set using half-wave plates (λ/2). The linearity of the polarization was improved by Glan-Taylor polarization cubes (P). Probe beam from another 532 nm laser (L2) passes through the sample (S), a part of it is reflected to a photodiode to take into account the changes of transmittance regarding photobleaching but the rest of it passes through an analyzer (A). If a difference between the refractive indexes in orthogonal directions is present in the sample, the polarization plane of the probe beam shifts and a signal appears on the photodiode behind the analyzer. Birefringence then can be evaluated using (2):

$$\Delta n = \frac{\lambda}{\pi d}\arcsin\left(\sqrt{\frac{I}{I_0}}\right) \tag{2}$$

where I is the probe beam intensity passing through crossed polarizer and analyzer, I_0 – probe beam intensity passing through parallel polarisers (with transmittance changes taken into account), λ - probe wavelength and d is the thickness of the layer of material.

3. Results and discussion

Diffraction efficiency dependence on time for different polarization states of the recording beams is shown in figure 5.

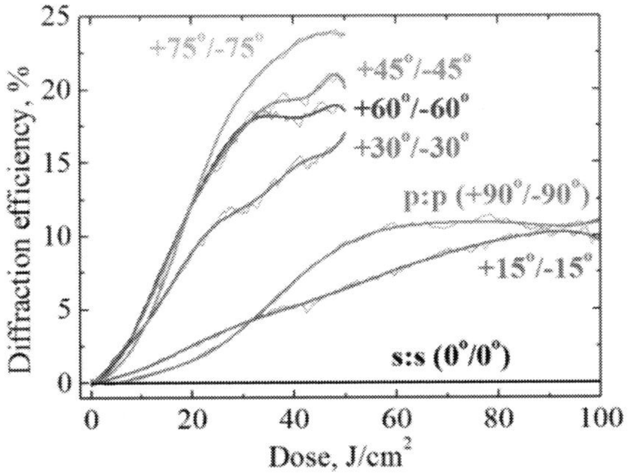

Figure 5. Diffraction efficiency dependence on exposure dose for different polarization states of the recording beams. The intensity of the 491 nm laser beams was I_1=I_2=0.18 W/cm².

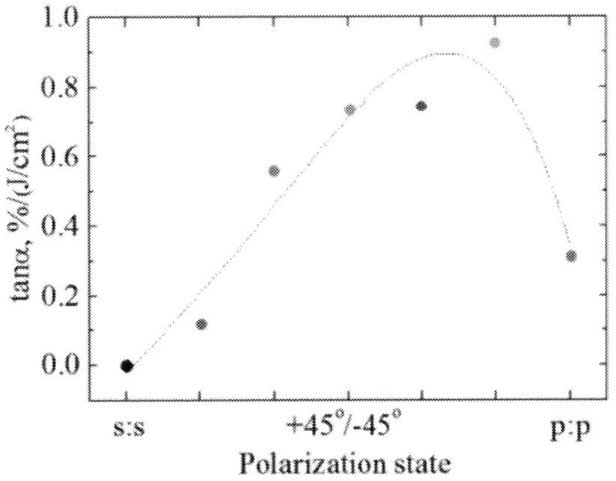

Figure 6. Mass transport characterized by tanα (the slope of the linear part of the diffraction efficiency) for different polarization states of the recording beams.

Diffraction efficiency is directly correlated with the depth of the obtained relief and thus to efficiency of mass transport [7]. Recording was done until saturation was reached which indicates that the surface tension forces have overcome the forces of mass transport. There are three parts of the diffraction efficiency curve - nonlinear beginning, linear part and the saturation. To characterize the sensitivity of the material the slope of the linear part is calculated. The slope (tanα) for different polarization states is shown in table 1. The p-polarization is parallel to the grating vector and s-polarization is orthogonal to the grating vector, respectively.

Table 1. Sensitivity of material for different polarization states of the two recording beams.

Polarization state	Angle between polarization planes	tanα
s:s (0°/0°)	0°	0
+15°/-15°	30°	0.12
+30°/-30°	60°	0.56
+45°/-45°	90°	0.74
+60°/-60°	120°	0.75
+75°/-75°	150°	0.93
p:p (+90°/-90°)	180°	0.31

Very strong dependence of holographic recording on polarization state of the recording beams was observed. The correlation is shown in figure 6, by placing the obtained points according to the angle between polarization planes of the two beams. It can be concluded from these experiments that the p-polarization component of light is required for the mass transport to occur. At the same time the best results could be obtained with the polarization states where the polarizations of both beams are almost orthogonal. This can be explained as follows. Two light beams with the same (in this case p:p) polarizations create a light intensity pattern, leaving some places on the surface of the sample not exposed to light treatment at the interference minima. These regions experience no photosoftening effect and therefore are more rigid and less likely to be affected by mass transport phenomenon. This could be improved by using different polarizations for both beams, when the resulting interference pattern exposes the whole area to light - this pattern has the periodical modulation of crucial p component and the periodical modulation of the s component which provides photosoftening without interfering with the main movement of mass.

Figure 7 shows the absorption spectrum of non-irradiated sample and spectra taken while irradiating the sample with polarized 532 nm laser light. Dichroism in the spectra of the K-RJ-8 thin film shows that there is distinctive anisotropy induced by polarised light. If the sample is not illuminated, the moieties are orientated randomly. Strong absorption of the stable *trans* isomer of the azo compound is observed at 450 nm. During the illumination with polarized 532 nm laser light, a decrease of the absorbance occurs due to photoisomerisation and photoorientation processes [8]. Firstly, a part of the moieties in *trans* form undergo transition to *cis* form by absorbing a photon (increase of the absorption of the *cis* state at 375 nm is observed). Secondly, through this photoisomerisation process the moieties may align relative to the electric field vector of light. That can be observed as a difference between the strength of the absorption in orthogonal directions while the sample is being illuminated - part of the moieties in *trans* form that absorb in the direction orthogonal to pump light polarization accordingly do not absorb in the parallel direction.

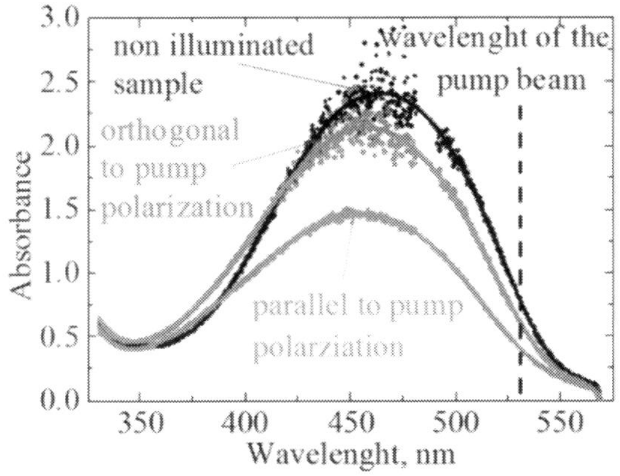

Figure 7. The absorption spectrum of a non illuminated sample and the spectra taken while the sample is being illuminated with linearly polarized 532 nm laser light (intensity I=0.07 W/cm^2) with polarization of the probe beam parallel or orthogonal to one of the pump beam.

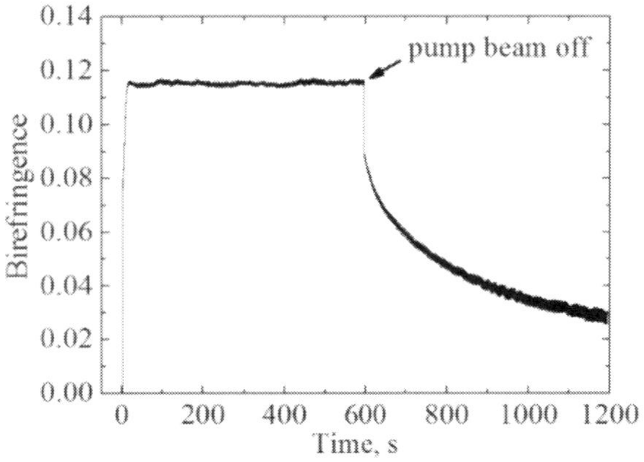

Figure 8. The photoinduced birefringence in time. Birefringence was induced by linearly polarized 532 nm laser (intensity I=0.07 W/cm^2) and measured at the same 532 nm wavelength.

Dichroism in material indicates birefringence and vice versa. Figure 8 shows photoinduced birefringence in K-RJ-8. Birefringence reaches saturation in first seconds and remains constant during the rest of the illumination. After 10 minute time period the pump was switched off and after another time period of 10 minutes the birefringence has dropped significantly. Relaxation of birefringence can be observed contrary to surface relief gratings which are known to be stable [3].

Dichroism and birefringence measurements explains why the holographic recording done with s:s produces no diffracted beam but recording done with p:p or any other polarization state with contains the p component gives far better results. It can be connected to the obvious anisotropy induced by polarized laser light.

4. Conclusions

The studied thin films of compound K-RJ-8 possess very good light sensitivity properties for holographic recording with 491 nm wavelength - large diffraction efficiency (almost 25 %) and recording sensitivity of about 0.9 % cm^2/J were obtained. Very strong dependence of holographic recording on polarization state of the recording beams was observed due to the anisotropy of photoisomersiation and photosoftening processes. Distinctive dichroism and high values of photoinduced birefringence (almost 0.12) were observed in K-RJ-8.

References

[1] Sudesh Kumar G and Neckers D C 1989 *Chem. Rev.* **89** 1915
[2] Todorov T, Nikolova L and Tomova N 1984 *Applied Optics* **23** 4309
[3] Rochon P, Batalla E and Natansohn A 1995 *Appl. Phys. Lett.* **66** 136
[4] Jiang X L, Li L, Kumar J, Kim D Y, Shivshankar V and Tripathy S K 1996 *Appl. Phys. Lett.* **68 (19)** 2618-20
[5] Natansohn A and Rochon P 2002 *Chemical Reviews* **102** 4139
[6] Ozols A, Augustovs P, Kokars V, Traskovskis K and Saharov D 2013 *IOP Conf. Series: Materials Science and Engineering* **49** 012043
[7] Reinfelde M and Teteris J 2011 *Journal of Optoelectronics and Advanced Materials* **13 (11)** 1531-33
[8] Ojanen J and Rantala T T 2009 *The Open Chemical Physics Journal* **2** 37-46

IOP Conf. Series: Materials Science and Engineering **77** (2015) 012020 doi:10.1088/1757-899X/77/1/012020

SbSI Based Photonic Crystal Superlattices: Band Structure and Optics

Sevket Sımsek[1], Husnu Koc[2], Selamı Palaz[3], Oral Oltulu[3], Amırullah M. Mamedov[4,5] and Ekmel Ozbay[4]

[1] Hakkari University, Faculty of Engineering, Department of Material Science and Engineering, 3000, Hakkari, Turkey
[2] Siirt University, Faculty of Science and Letters, Department of Physics, 56000, Siirt, Turkey
[3] Harran University, Faculty of Science and Letters, Department of Physics, 63000, Sanliurfa, Turkey
[4] Bilkent University, Nanotechnology Research Center, 06800, Ankara, Turkey
[5] Baku State University, International Scientific Center, Baku, Azerbaijan

E-mail: mamedov@bilkent.edu.tr

Abstract. In this work, we present an investigation of the optical properties and band structure calculations for the photonic crystal structures (PCs) based on one-dimensional (1D)-photonic crystal. Here we use 1D SbSI based layers in air background. We have theoretically calculated the photonic band structure and optical properties of SbSI based PC superlattices. In our simulation, we employed the finite-difference time domain (FDTD) technique and the plane wave expansion method (PWE), which implies the solution of Maxwell equations with centered finite-difference expressions for the space and time derivatives.

Keywords: Fibonacci Photonic Crystal, SbSI, Supercell

1. Introduction

It is well known that the photonic crystal (PC) based superlattices can play an essential role in the controlling of the optical processes in various devices of optoelectronics [1]. Therefore, great attention is paid to the investigation of the physical properties of PC based superlattices. The PC based superlattices of various types are considered, namely, strictly periodic, disordered, lattices with defects, etc. The structures intermediate between the periodic and disordered structures, or quasi-periodic lattices – the Fibonacci and Thue-Morse superlattices, occupy a special place among the superlattices. On the other hand, one of the topics of interest in the optics of PC is the possibility to tailor emittance/absorptance by changing the distribution of electromagnetic modes. Emittance tailoring by conventional PCs was investigated in [2,3]. One of the structures that may be used in emittance tailoring is quasiperiodic multilayers, like the Fibonacci superlattices [4]. Due to their structural self-similarity, these show regularities in their transmission/reflection spectra. The strong resonances in spectral dependences of fractal multilayers can localize light very effectively [1,5]. In addition, long-range ordered aperiodic photonic structures offer a extensive flexibility for the design of optimized light emitting devices, the theoretical understanding of the complex mechanisms governing optical gaps and mode formation in aperiodic structures becomes increasingly more important. The formation of photonic band gaps and the existence of quasi-localized light states have already been demonstrated for one (1D) and two-dimensional (2D) aperiodic structures based on Fibonacci and the Thue-Morse sequences [1,4]. On the other hand, a one – dimensional Schrödinger equation has been examined with a stepwise quasiperiodic potential whose value is equal to either of the two fixed values sequentially in accordance with the Fibonacci set [1]. For such a potential, an electron was found to have a "critical wave function" with strong spatial fluctuations. This type of wave function correlates with the properties of the energy spectrum. The unusual electron properties of quasiperiodic potentials have also stimulated extensive research of the optical counterparts. However, to the best of our knowledge, a rigorous investigation of the band gaps and optical properties in the more complex types of aperiodic structures has not been reported so far.

In this paper, we investigated the energy spectrum and optical properties in the Fibonacci-type photonic band gap (PBG) structures consisting of ferroelectric material (SbSI) in detail by using the finite-difference time-domain (FDTD) method and the plane wave expansion method (PWE). The choice of the SbSI crystal as the active media for our investigation was associated with its unusual optical and electronic properties. It is well known, that SBSI is the ferroelectric material and its properties are very sensitive to external influences (temperature, electric field, stress and light) [6-8]. Therefore, SbSI based Fibonacci photonic crystal may be used as efficiency material for photonic devices.

2. Computational Details

2.1. Fibonacci Sequences and Model

Quasiperiodic structures are nonperiodic structures that are constructed by a simple deterministic generation rule. In a quasiperiodic system two or more incommensurate periods are superimposed, so that it is neither aperiodic nor a random system and therefore can be considered to intermediate the two [1]. In other words due to a long-range order a quasiperiodic system can form forbidden frequency regions called pseudo band gaps similar to the band gaps of a PC and simultaneously possess localized states as in disordered media [5]. The Fibonacci multilayer structure (well-known quasiperiodic structure) has been studied in the past decade, and recently the resonant states at the band edge of the photonic structure in the Fibonacci sequence have also been studied experimentally [9]. A 1D quasi-periodic Fibonacci sequence is based on a recursive relation, which has the form, $S_{j+1}=\{S_{j-1}, S_j\}$ for $j \gg 1$, with $S_0=\{B\}$, $S_1=\{A\}$, $S_2=\{BA\}$, $S_3=\{ABA\}$, $S_4=\{BAABA\}$ and soon, where S_j is a structure obtained after j iterations of the generation rule. Here, A and B are defined as being two dielectric materials, with different refractive indices (n_A, n_B) and have geometrical layer thickness (d_A, d_B). In place of materials A and B, we used air (A) and strongly anisotropic material - antimony sulfoiodide (SbSI, B) [10]. In Fig. (1) and (2) we schematically show the geometry of conventional PCs and FPCs. 1D photonic crystals are characterized by the refractive index contrast and filling fraction. The refractive index contrast is the ratio of the higher refractive index to the lower refractive index in the multiplexer system. The filling fraction f is the ratio between the thickness of the lower refractive index layer (air) and the period of the PC, i.e., $f=d_1/(d_1+d_2)$. Typical 1D PCs and FPCs are shown Fig.1 and Fig.2

The thickness of the considered layers of air and SbSI $\left(d_{air} = \frac{\lambda}{4n_{air}},\ d_{LN} = \frac{\lambda}{4n_{LN}}\right)$ are $d_{air} = 0.5a$ and $d_{SbSI} = 0.5a$, respectively. The lattice constant is $a = 1\ \mu m$. The filling fraction is set to 0.5. The refractive index contrasts of anisotropic SbSI are taken as following:

n_a=2.87 n_b=3.63 n_c=4.55 T_c = 22 °C (Paraelectric Phase)

n_a=2.87 n_b=3.57 n_c=4.44 $T_c \approx$ 12 °C (Ferroelectric phase) at λ=633 nm.

The refractive index of the background dielectric medium is assumed as air (n_{air}=1.0).

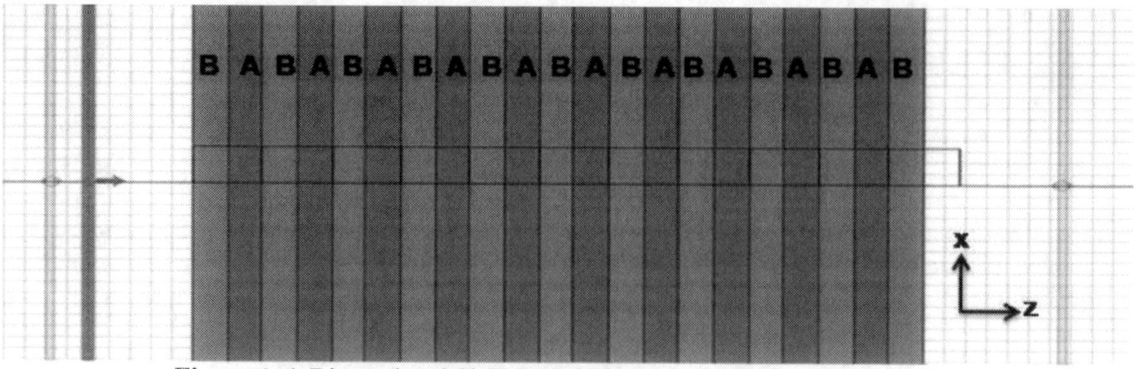

Figure 1. 1-Dimensional SbSI Based Photonic Crystal Structure (PCs).

Figure 2. 1-Dimensional SbSI Based Fibonacci Photonic Crystal Structure (FPCs).

2.2. Finite Difference Time Domain (FDTD) Method and Plane Wave Expansion Method (PWE)

In our calculations, we used the OptiFDTD software package [11]. The OptiFDTD software package is based on the finite-difference time-domain (FDTD) method for transmission spectra and the plane wave expansion method (PWE) for photonic band structure. Equi-frequency contour of SbSI based 1D PCs was calculated with MPB free software [12].

The photonic band structures of the proposed PCs were calculated by solving the Maxwell equations. The Maxwell equation in a transparent, time-invariant, source free, and non-magnetic medium can be written in the following form:

$$\nabla \times \frac{1}{\varepsilon(\mathbf{r})} \nabla \times \mathbf{H}(\mathbf{r}) = \frac{\omega^2}{c^2} \mathbf{H}(\mathbf{r}) \quad (1)$$

Where,

$\varepsilon(\boldsymbol{r})$ is the space dependent dielectric function

c is the speed of light in vacuum.

$\boldsymbol{H}(\boldsymbol{r})$ is the magnetic field vector of frequency ω and time dependence $e^{j\omega t}$.

This equation is sometimes called the Master Equation, and represents a Hermitian eigen-problem, which would not be applicable if the wave equation were derived in terms of the electric field. The Bloch theorem states that, due to infinite periodicity, the magnetic field will take the form:

$$\mathbf{H}(\mathbf{r}) = e^{j\mathbf{kr}} \mathrm{h}_{\mathrm{k}}(\mathbf{r}) \quad (2)$$

Where

$$\mathbf{h}_{\mathrm{k}}(\mathbf{r}) = \mathrm{h}_{\mathrm{k}}(\mathbf{r} + \mathbf{R}) \quad (3)$$

for all combinations of lattice vectors $\boldsymbol{R}$. Thus, Maxwell equation is given in operator form:

$$(\nabla \times j\mathbf{k}) \times \left[\frac{1}{\varepsilon(\mathbf{r})} (\nabla \times j\mathbf{k})\right] \times \mathbf{h}_k = \frac{\omega^2}{c^2} \mathbf{h}_k \quad (4)$$

By solving these equations for the irreducible Brillouin zone, we can obtain the photonic band structure.

FDTD algorithm is one of the most appropriate calculation tools [13]. For solving Maxwell's equations depending on the time, FDTD algorithm divides the space and time in a regular grid. Perfect matched layers (PMLs) can be used in the determination of the boundary conditions [14]. In general, the thickness of the PML layer in overall simulation area is equal to a lattice constant. FDTD solves electric and magnetic fields by rating depending on space and time and deploys that rating in different spatial regions by sliding each field component half of a pixel. This procedure is known as Yee grid discretization. Fields in these grids can be classified as TM and TE polarization. In our calculations, we have used Periodic Boundary Condition (PBC) and (PMLs) at x- and z-directions, respectively.

3. Results and Discussion

3.1. Photonic Band Structure and Transmittance

We calculate the spectral properties in the n-th order (n=10) Fibonacci-type quasiperiodic layered structures consisting of compounds. The photonic band structures of 1D SbSI based PCs have been calculated in high-symmetry directions in the first Brillouin. The band structures with transmittance spectrums for SBSI based photonic crystal are shown in Fig. 3. We can see that there exist five fundamental photonic band gaps (PBG) from Fig.3. The width of the fundamental photonic band gaps are (0.313-0.426) ($\omega a/2\pi c$) for 1th PBG, (0.700-0.788) ($\omega a/2\pi c$) for 2nd PBG, (1.091-1.133) ($\omega a/2\pi c$) for 3rd PBG, (1.424-1.544) ($\omega a/2\pi c$) for 4th PBG and (1.831-1.883) ($\omega a/2\pi c$) for 5th PBG, respectively. When the frequency of an incident electromagnetic wave drops in these PBGs, the electromagnetic wave will be reflected completely by the photonic crystal. It can be seen in Fig.3 that transmittance is zero in these range of frequencies. All pseudogaps exist in the frequencies where the effective refractive index of the structure are positive and the spectral width of the gaps are invariant with the change in the transmittance (see, Tables 1).

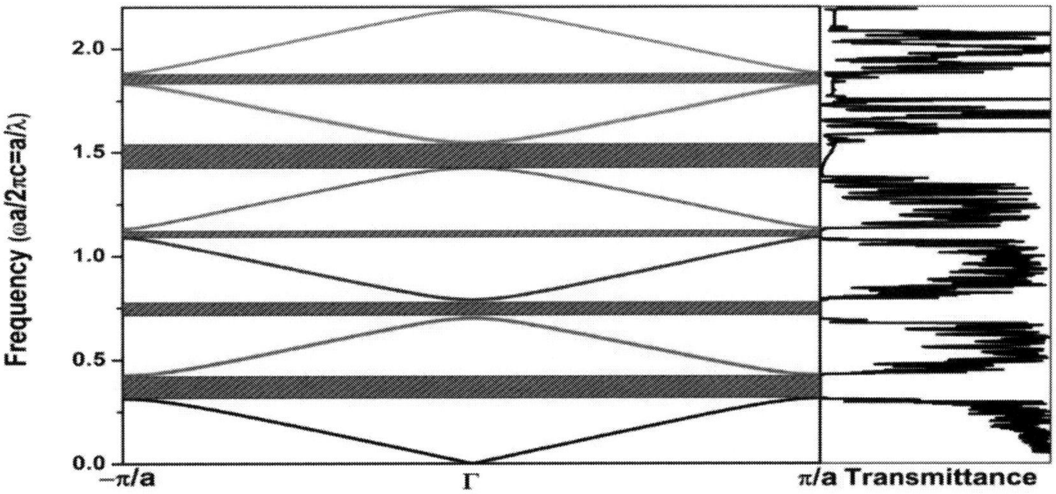

Figure 3. TM Band structure and transmittance spectra of anisotropic SbSI.

Table 1 Variation of full band gap size for TM modes with filling factor for anisotropic SbSI based layers in air background.

Filling Factor	TM1		TM2		TM3		TM4		TM5	
	Band Gap (ωa/2πc)	Gap Size(%)	Band Gap (ωa/2πc)	Gap Size(%)	Band Gap (ωa/2πc)	Gap Size(%)	Band Gap (ωa/2πc)	Gap Size(%)	Band Gap (ωa/2πc)	Gap Size(%)
0.1	(0.424-0.499)	16.165	(0.865-0.992)	13.728	(1.326-1.477)	9.099	(1.806-1.944)	7.350	(2.297-2.391)	4.009
0.2	(0.376-0.493)	26.806	(0.811-0.948)	15.544	(1.295-1.343)	3.596	(1.712-1.792)	4.534	(2.123-2.265)	6.437
0.3	(0.346-0.479)	32.128	(0.797-0.860)	7.628	(1.190-1.289)	8.002	(1.600-1.712)	6.791	(2.047-2.089)	2.041
0.4	(0.326-0.455)	33.007	(0.771-0.793)	2.831	(1.111-1.235)	10.541	(1.542-1.588)	2.940	(1.899-2.012)	5.790
0.5	(0.313-0.426)	30.508	(0.700-0.788)	11.815	(1.091-1.133)	3.814	(1.424-1.544)	8.109	(1.831-1.883)	2.830
0.6	(0.304-0.395)	25.887	(0.650-0.764)	16.020	(1.035-1.088)	5.046	(1.387-1.432)	3.160	(1.708-1.819)	6.296
0.7	(0.299-0.365)	19.961	(0.618-0.722)	15.615	(0.959-1.064)	10.282	(1.318-1.381)	4.618	0	0
0.8	(0.296-0.339)	13.454	(0.599-0.676)	11.969	(0.912-1.010)	10.212	(1.234-1.338)	8.020	(1.564-1.658)	5.822
0.9	(0.295-0.315)	6.645	(0.592-0.630)	6.187	(0.889-0.946)	6.182	(1.189-1.260)	5.748	(1.491-1.574)	5.399

Table 2 Variation of full band gap size for TE modes with filling factor for anisotropic SbSI based layers in air background.

Filling Factor	TE1		TE2		TE3		TE4		TE5	
	Band Gap (ωa/2πc)	Gap Size(%)	Band Gap (ωa/2πc)	Gap Size(%)	Band Gap (ωa/2πc)	Gap Size(%)	Band Gap (ωa/2πc)	Gap Size(%)	Band Gap (ωa/2πc)	Gap Size(%)
0.1	(0.401-0.498)	21.633	(0.829-0.990)	17.704	(1.287-1.466)	13.036	(1.769-1.916)	7.976	(2.263-2.336)	3.156
0.2	(0.346-0.490)	34.383	(0.775-0.925)	17.704	(1.262-1.279)	1.301	(1.622-1.754)	7.810	(2.041-2.196)	7.304
0.3	(0.314-0.471)	39.805	(0.763-0.810)	5.927	(1.109-1.248)	11.815	(1.528-1.619)	5.767	(1.911-2.018)	5.487
0.4	(0.293-0.439)	39.621	(0.708-0.761)	7.180	(1.037-1.161)	11.268	(1.419-1.518)	6.710	(1.790-1.876)	4.653
0.5	(0.280-0.402)	35.689	(0.634-0.748)	16.476	(1.024-1.038)	1.355	(1.311-1.437)	9.144	(1.672-1.776)	6.040
0.6	(0.271-0.366)	29.720	(0.584-0.713)	19.801	(0.935-1.018)	8.539	(1.289-1.301)	0.931	(1.563-1.667)	6.412
0.7	(0.266-0.334)	22.620	(0.552-0.662)	18.221	(0.860-0.979)	12.942	(1.184-1.275)	7.329	(1.517-1.549)	2.075
0.8	(0.263-0.306)	15.114	(0.534-0.612)	13.593	(0.813-0.915)	11.802	(1.102-1.214)	9.646	(1.398-1.507)	7.516
0.9	(0.262-0.282)	7.421	(0.526-0.564)	6.990	(0.791-0.848)	6.934	(1.058-1.129)	6.492	(1.327-1.411)	6.129

IOP Conf. Series: Materials Science and Engineering **77** (2015) 012020 doi:10.1088/1757-899X/77/1/012020

The numerical results of the variation of a full band gap with a changing filling factor from 0.1 to 0.9 are given in Table 1 and Table 2. The variation of band gap sizes (%) as a function of filling factor are shown for TM mode in Fig. 4. It is clear that the size of the gap increases with filling factor for the first band gap. The largest gap size is about 38% when filling factor is as high as 0.4, but it decrease when filling factor continues to increase. On the other hand, the fifth band gap size does not change too much according to filling factor, but it reaches the minimum value when filling factor is 0.7. We also calculated the field distribution of the TM modes in our n-th (7) order Fibonacci sample, following a standard PWE method. The magnitude of the magnetic field (E) at the left interface of the dielectric layer is simply related to H at the right interface of the same layer by using matrix relation [15]. For 1D structures, it is also possible to compute the magnetic field distribution inside the sample (Fig. 5). The insets show the normalized field intensity distribution for the λ=1.55 μm wavelength on the n-th (n=7) interface.

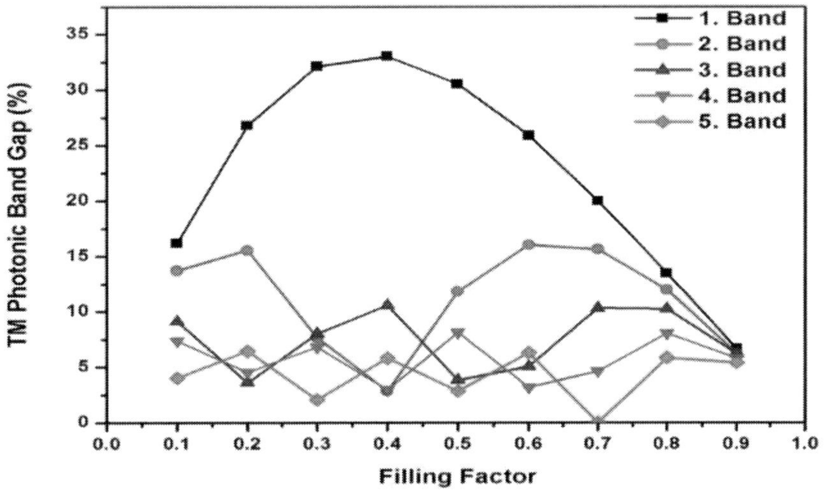

Figure 4. TM Filling factor for anisotropic SbSI.

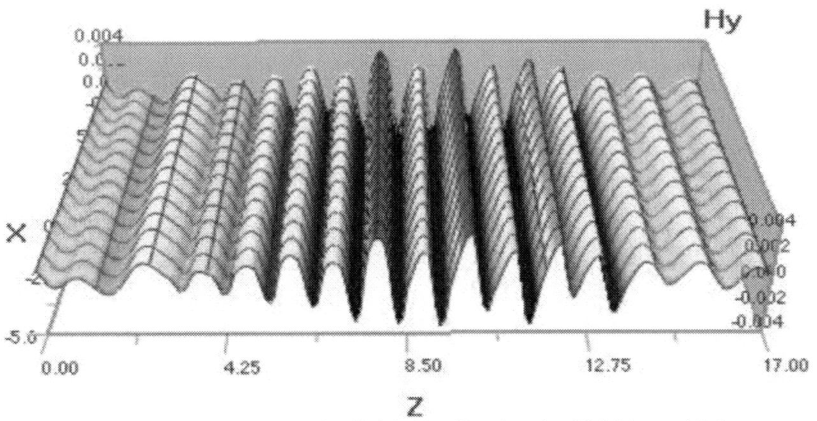

Figure 5. TM Magnetic field distribution in SbSI based PCs.

IOP Conf. Series: Materials Science and Engineering **77** (2015) 012020 doi:10.1088/1757-899X/77/1/012020

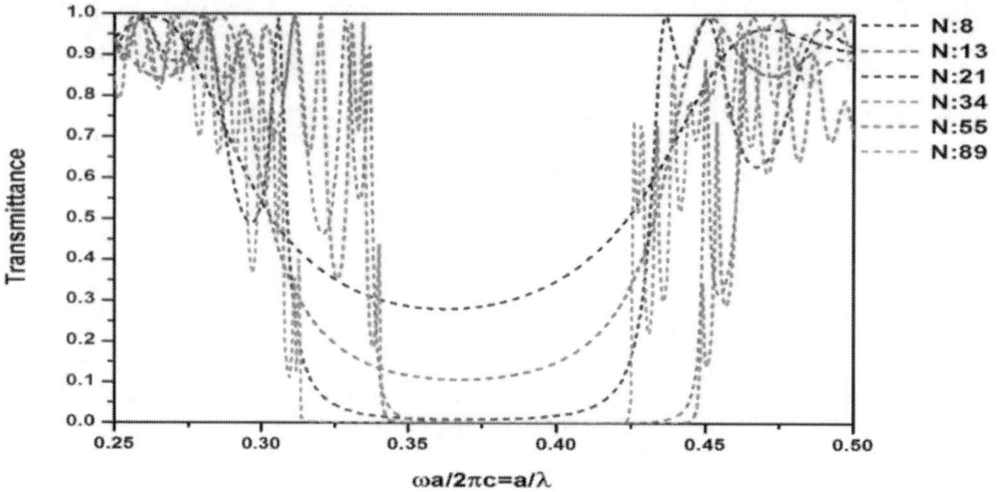

Figure 6. TM Transmittance spectra of 1D SbSI based PCs.

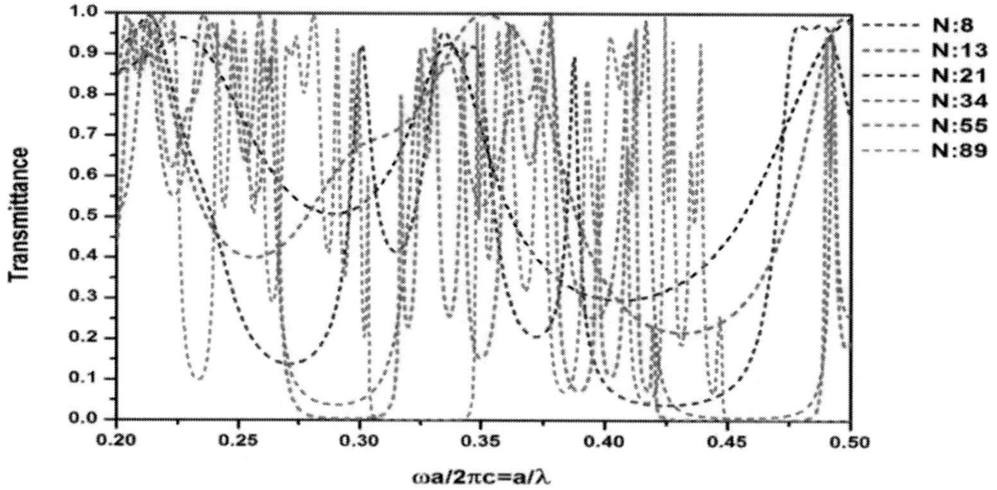

Figure 7. TM Transmittance spectra of 1D SbSI based FPCs.

Figs. 6-7 show the transmittance spectra of SbSI based both normal PCs and FPCs from 8 to 89 layers for the transverse-magnetic (TM)-polarized incident electromagnetic wave. The positions of the minima in the transmission spectrum correlate with the gaps obtained in the calculation. A transmission spectrum of a simple eight-layer (also n=8, 13, 21, 34, 55, 89) 1D SbSI based PC is compared in Fig. 6-7 with an eight-layer (also n=8, 13, 21, 34, 55, 89) 1 D SbSI based FPC. One full period in spectrum is presented which corresponds to the frequency range (0.2-0.5) ($\omega a/2\pi c$). Although there is still a gap in the transmission spectrum of the Fibonacci structure around (0.2-0.5) ($\omega a/2\pi c$) the spectrum modified significantly. Notably, the total number of transmission peaks in both cases equal to the total number of elementary layers in the structure. This is general property of multilayer structures. In addition, the spectral regularities inherent in Fibonacci structures have also been derived in analytical form based on the transfer matrix approach. Fibonacci structures feature a transmission band in the center which first splits into two and then into three subbands (Fig. 7). This central triplet is indicative for the Cantor triadic set and for higher generations definite self-similar and scaling features develop that are inherent in fractals. Different portions of transmission spectra for the same high-order generation of Fibonacci structure do exhibit similar spectral shape which becomes apparent when using the "exciton in quantum dot" lowest state energy expansion [1]. The transmission spectra of Fibonacci structures also show scalability which means that the spectra of different generations have a similar shape when the frequency axis is properly scaled. Notably, the transmission spectra of generic Fibonacci structures exhibit perfect transmission bands. Many among well resolved peaks have a transmission coefficient close to unity.

3.2. Equi-frequency surface

Because, the momentum conservation law is satisfied at the reflection and refraction of waves, it is convenient to analyze the reflection and refraction of a certain wave in the space of wave numbers $\Sigma_{\vec{k}} = \{0;\ k_x, k_y, k_z\}$ by introducing the equi-frequency surface of the wave. This surface is directly described by the dispersion relation of the anisotropic medium at the fixed frequency ω. Then, the group velocity $\vec{V}_g$ of the wave in an anisotropic medium can be found as the frequency gradient in the space of wave vectors [16]. It is known that for an electromagnetic wave propagating in an anisotropic medium with a fixed frequency ω, the equi-frequency surface represents a sphere. In this case, the wave vector $\vec{k}$ and group velocity vector $\vec{V}_g$, which determines the ray direction, are always parallel. However, equi-frequency surface is not spherical for anisotropic media and the vectors $\vec{V}_g$ and $\vec{k}$ are not parallel. By analogy with 3D case, the propagation, reflection, and refraction of the wave in 2D structure can be described in terms of the equi-frequency dependence which can be considered as the section of the dispersion surface $\omega(k_y, k_z)$ in the space of variables $\{\omega, k_y, k_z\}$ by the plane corresponding to constant frequency. It is well known that the analysis of equi-frequency dependences is most efficient in the studies of 2D geometries, especially in solving problems when only orientations of the $\vec{V}_g$ and $\vec{k}$ vector of incident, reflected, and refracted waves are of interest, and are not the amplitudes of the reflected and refracted rays. The equi-frequency dependence has a simple physical meaning for the analysis of 2D geometries: since this dependence describe all the possible waves with the given frequency ω and various wave vectors, the directions of the reflected and the refracted rays can be determined by simply finding the points in equi-frequency dependences of media that satisfy the momentum conservation law at a known orientation of the boundary and a given angle of incidence of the wave. Now, we present some numerical examples for our PC structures. In all of these examples, we exploit symmetry to calculate the equi-frequency surfaces over the irreducible Brillouin Zone of the entire Brillouin Zone. First, we consider the equi-frequency surface of a square lattice of SbSI based dielectric layers in an air medium for the $\vec{H}$–polarization.

Here, the map was discretized using six field points per edge of the unit cell for the first band in Figure 8 (a). The map was discretized using four field points per edge of the unit cell for the second band. The curves shown correspond to equi-frequency surfaces of the lowest order band up to frequencies just below the band gap starting at around 0.11 (2πc/a). For the second band, the curves shown correspond to equi-frequency surfaces of the lowest order band up to frequencies just below the band starting at around 0.16 (2πc/a).

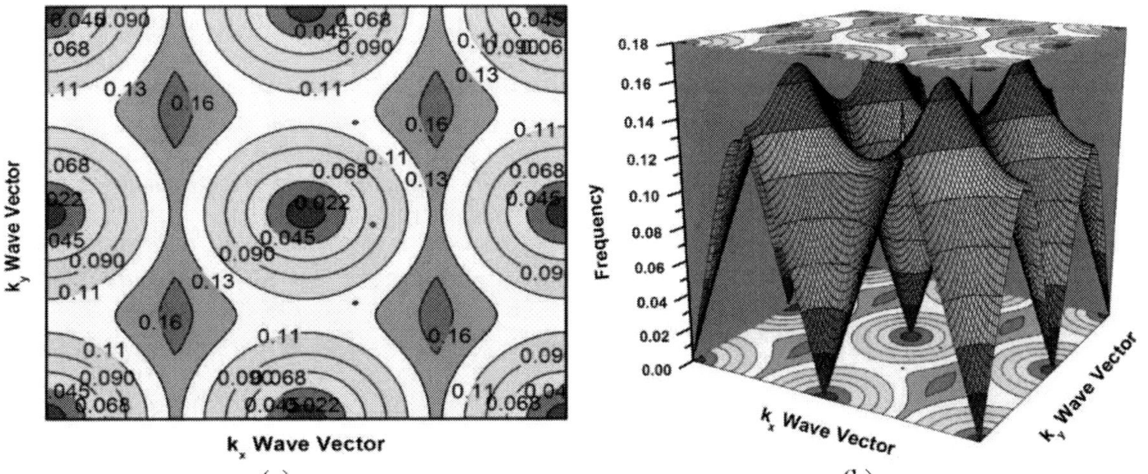

(a) (b)

Figure. 8 Equi-frequency contour of SbSI based 1D PCs for TM1 band in 2D (a) and 3D (b).

3.3. Group velocity

We believe that a pulse of light propagating along the PC let the pulse be finite in dimension along the direction of propagation. Theoretically, such a pulse can be represented as a wave packet formed as a superposition of the modes, but with a different propagation constant [17]. From [17], it follows that the envelope of the wave packet propagates with the velocity $V_g^z = \omega_n(\beta)$, where β is propagation constant. The direct calculation of the derivative of the dispersion relation calculated numerically is not always convenient and can give an error. As shown in [18], the group velocity of the wave packet V_g^z, is equal to the velocity of energy transfer by the mode $n\beta$. Thus, by using the results of [19] the group velocity can always be calculated with more accuracy, irrespective of the number of points in the dispersion curve. Fig. 9 show the results of calculations of the group velocity of the wave packet formed of the differently polarized localized modes in PC with SbSI dielectirc layers in an air background in 1D.

From Fig. 9, it is evident that the components of the group velocity versus the high symmetry direction (0- $\pi/2a$) vary over wide limits. The dependences $V_g^z(\beta)$, which describe the wave packets of localized modes of any polarization and any order, exhibit (generally) a maximum at certain propagation constants. This means that the dispersion of group velocity can be positive, negative, or zero [19]. Therefore, as can be seen in Figure 9, the group velocity of TE-polarized wave packets is almost always higher than that of the TM-polarized packets.

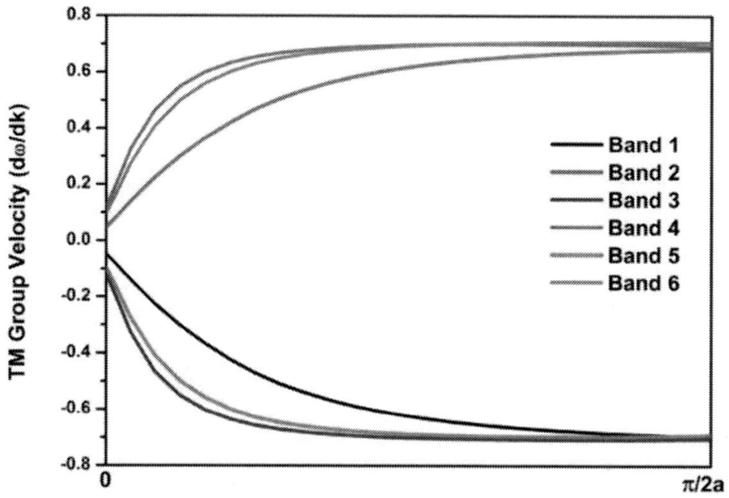

Figure 9. TM Group velocity of anisotropic SbSI.

4. Conclusion

In the present paper the photonic band structures and transmission properties of the 1D SbSI PCs consisting of dielectric layers immersed in air were studied for the first time. We have investigated transmittance spectra of SbSI based both normal PCs and FPCs from 8 to 89 layers. The results show that the number of the repetition period also has a great influence on the average transmittance of the pass band of the both normal PCs and FPCs.

Acknowledgements This work is supported by the projects DPT-HAMIT, DPT-FOTON, and NATO-SET-193 as well as TUBITAK under the project nos., 113E331, 109A015, and 109E301. One of the authors (Ekmel Ozbay) also acknowledges partial support from the Turkish Academy of Sciences.

References

[1] S V Gaponenko 2010 Introduction to Nanophotonics *Cambridge University Press N.Y p.485*

[2] M Maksimovic, Z Jaksic 2006 Emittance and absorptance tailoring by negative refractive index metamaterial-based Cantor multilayers *J. Optics A: Pure & Appl. Opt.* **8** 355-362
[3] A N Poddubny, E L Ivchenko 2010 Photonic quasicrystalline and aperiodic structures *Physica E* **42** 1871 – 1895
[4] A Rostami, S Matloub 2005 Exactly solvable inhomogeneous Fibonacci-class quasi-periodic structures (optical filtering) *Optics Comm.* **247** 247-256
[5] H Rahimi, A Namdar, S R Entezar, H Tajalli 2010 Photonic transmission spectra in one-dimensional Fibonacci multilayer structures containing single-negative metamaterials *Progress In Electromagnetics Research* **102** 15-30
[6] V M Fridkin 1980 Ferroelectric Semiconductors. *New York Consultants Bureau p.314*
[7] H Akkus, A M Mamedov 2007 Linear and Nonlinear Optical Properties of SbSI: First-Principle Calculation, *Ferroelectrics* **352** 148-152
[8] A K Zeinally, A M Mamedov and Sh M Efendiev 1973 A study of SbSI absorption edge, *Ferroelectrics* **6** 119–122
[9] L Dal Negro, C J Oton, Z Gaburro, L Pavesi, P Johnson, A Lagendijk, R Righini, M Colossi, D S Wiersma 2003 Light Transport through the Band-Edge States of Fibonacci Quasicrystals *Phys. Rev. Lett.* 90 055501-055505
[10] D Lusk, I Abdulhalim, F Placido 2001 Omnidirectional reflection from Fibonacci quasi-periodic one-dimensional photonic crystal *Optics Comm.* **198** 273-279
[11] OptiFDTD 10, http://www.optiwave.com/.
[12] S G Johnson, J D Joannopoulos, The MIT Photonic-Bands Package (http://ab-initio.mitedu/mpb/).
[13] A Taflove and S C Hagness 2000 Computational Electrodynamics: The Finite-Difference Time-Domain Method 2nd ed., *MA: Artech House Publishers, Boston*.
[14] J P Berenger 1994 A perfectly matched layer for the absorption of electromagnetic waves *J. Comput. Phys.* **114** 185–200
[15] F L Pedrotti, L S Pedrotti 1987 Introduction to Optics P*retice- Hall Englewood Cliffs NJ*
[16] E H Lock 2008 The properties of isofrequency dependences and the laws of geometrical optics, *Physics-Uspekhi* **51(4)** 375-393
[17] A S Spitsyn and G F Glinskii 2008 Properties of Waveguide Modes in a Photon Crystal Based on Slotted Silicon with a Defect *Semiconductors* **42(10)** 1237-1243
[18] A Yariv, P Yeh 2007 Photonics: Optical Electronics in Modern Communications *Oxford University Press USA*.
[19] A V Kildishev, V M Shalaev 2011 Transformation optics and metamaterials *Physics-Uspekhi* **54(1)** 53-63

IOP Conf. Series: Materials Science and Engineering **77** (2015) 012021 doi:10.1088/1757-899X/77/1/012021

Environment Humidity Effect on the Weight of Carbonized Na-Al-Si Glass Fabrics Recovery after Heating

E Pentjuss, A Lusis, J Gabrusenoks, G Bajars

Institute of Solid State Physics, University of Latvia, 8 Kengaraga Street, Riga, LV-1063, Latvia

E-mail: Evalds.Pentjuss@cfi.lu.lv

Abstract. Na-Al-Si glass fabrics fibres contain Na^+ ions that diffuse to its surface and along with CO_2 and H_2O from atmosphere create here the shell of carbonate hydrates. The heating of fabric leads to weight loss by evolving these substances. In this work the results of weight recovery study at room relative humidity (20% - 50%) and elevated humidity (near 70%) of fabrics after its heating at different temperatures (70^0C - 150^0C) are compared. The experiments shoved the different weight recovery kinetics. The initial exponential stages up to 0.3 h - 0.5 h of the both recoveries are associated with water absorption and differ by its levels. In a case of lower environment humidity the later weight increase are restricted by its value, but at an elevated humidity has a maximum and followed weight increase. The reasons of observed differences are discussed.

1. Introduction

The object of research was industrially produced fabrics of Na-Al-Si glass fibres. It is known that Na^+ ions diffuse to surface of fibres [1] and due to reaction with H_2O and CO_2 from atmosphere form here the carbonated shell. It leads to increase the weight and thickness of fabrics and change its look from bright to mat. Heating of carbonated fabric up to 90-130 ^{0}C and higher temperatures leads to its weight loss [2], associated with evolving of H_2O and CO_2 from samples. Sodium carbonate (Na_2CO_3) is stable up to melting temperature (851 ^{0}C). The known sodium carbonate hydrated forms decompose by heating to sodium carbonate and water. Heptahydrate ($Na_2CO_3{\cdot}7H_2O$, not known in mineral form) begins to decompose over 32 ^{0}C, decahydrate ($Na_2CO_3{\cdot}10H_2O$) over 33.5 ^{0}C, and monohydrate ($Na_2CO_3{\cdot}H_2O$) over 100 ^{0}C. Trona ($Na_3H(CO_3)_2{\cdot}2H_2O$) begins to decompose to Na_2CO_3, water and CO_2 over 56 ^{0}C [2]. Besides, the weight loss by heating moves the shell and fabric out of its equilibrium with environment atmosphere and water and CO_2 absorption tends to restore it. The weight recovery after heating of carbonated fabric during to reaching the equilibrium weight at room conditions was analysed in [3].The experimental weight recovery – time W (t) points for the beginning period up to 0.2 h-0.4 h are well-fitted (R^2 > 0.98, see Table 1 and 2) to regression of

$$W(t) = A_0 - A_1 \exp(-t/t_1) - A_2 \exp(-t/t_2), \qquad (1)$$

where A_0, A_1, A_2 are weight constants and t_1, t_2 are time constants ($t_1 \leq t_2$), lim $W(t)=A_0$ when $t\rightarrow\infty$ and lim $W_1(t)=A_1exp(-t/t_1)=A_1$, and lim $W_2(t)=A_2$, when $t\rightarrow0$. In (1) $A_0=A_1+A_2$. Velocity of weight recovery is

$$v(t)=(A_1exp(-t/t_1))t_1+(A_2exp(-t/t_2))/t_2. \quad (2)$$

In the present study there are compared and analyzed the weight recovery - time characteristic parameters of relations (1, 2) and possible process that leads to equilibrium at room and environment relative humidity (RH) near 70%.

2. Experimental

For every heating temperature there were used one carbonized fabric sample, cut from one lump of fabric. The heating duration of every sample was 1 h. Relative humidity about 70% was limited by saturated salt solution in a glove box. There are used DTG and RTD measurements to analyze the content of carbonated shells in fabrics and dissolved from fabric and evaporated matter in this research.

3. Results and discussion

The shell can be dissolved in water up to 96% of its weight, dissolved by acid industrially. RTD showed that evaporated powder consists of mixture of trona (60%) and sodium carbonate monohydrate. Content of trona in a powder increased during the storing time. The shell of long time (years) stored fabric consists of trona (DTG measurements). The main weight loss after heating of fabrics is under 100 ^{0}C (Figure 1), related with decomposition of trona to sodium carbonate or carbonate monohydrate, water and CO_2. Some weight loss (seen in Figure 1) above 100 ^{0}C could be caused with evolving of residues of lubricator and/or components from glass. In a case of evaporated powder such loss absents. As in the case [3], when after heating the samples are placed in room conditions (humidity and temperature), up to 0.2 h-0.4 h the weight recovery closely follows regression (1), where the first, more fast component (ΔW_1 (t)) is associated in [3] with water absorption on surface of shell and the second one (ΔW_2 (t)) with an additional absorption to released sites of water, diffused into the shell. Some results (lower) tend to associate the first component of the relation (1) rather with absorption of CO_2.

The mean time constants for the first component (t_1) of weight recovery in relation (1) for nine samples (heated to different temperatures) is 0.7 min, and for the second component (t_2) - 20.4 min. At the beginning (t=0) of weight recovery, the mean velocity of recovery is 11.31 mg/h·cm^2 (v_1) and 0.676 mg/h·cm^2 (v_2) accordingly (Table 1). The values of parameters t_1, t_2 are significantly dissipated, that could be associated not only with different treatment temperatures, but with different samples for every of heating temperature also. Amplitude of the second component A_2 in Table 1 dominates at low heating temperatures, but first one A_1 is relative stable. The mean values of amplitudes are comparable (A_1=0.113 mg/cm^2 and A_2=0.17 mg/cm^2). No pronounced correlation was observed between parameters (t_1, A_1, v_1) and heating temperature of samples and humidity of environment, that looks prospective for the stable concentration of CO_2 in atmosphere.

It may be supposed that during mentioned above beginning time interval there are going mainly accumulation of water and CO_2 on the surface and inside the shell. We cannot be able to indicate an increase of thickness of fabric during this period (Figure 3).

The next pronounced increase of weight in Figures 2 and 3 are caused by a notable RH increase of environment along with increase of thickness of samples (Figure 3). As it seen from the weight kinetics in figures, the main part of absorbed weight are installed in sodium carbonate molecules (may be of different kind) and only a small part of weight follows to the fast decrease of RH of environment after 2500 h storing (Figure 2) and 6000 h storing (Figure 3). This part of weight (somewhat more than 12% of absorbed weight in a case of Figure 2) has to be associated with free water in shell and may be in pores of glass and characterizes the new state of water equilibrium of fabric with environment. Later there are observed the weight stabilization (Figures 2 and 3) and some decrease of thickness that indicates to packaging of shell along time (Figure 3).

Table 1. Parameters of regression (1) for time interval up to 0.4 h after heating at different temperatures and velocity components v_1 and v_2 of weight recovery in relation (2) at the recovery beginning (t=0)

T,^{0}C/sample	t_1, h	t_2, h	A_0, mg/cm^2	A_1, mg/cm^2	A_2, mg/cm^2	R^2	$v_1(0)$, mg/h cm^2	$v_2(0)$, mg/h cm^2
74/1a-3	0.01538	0.143	0.3571	0.1148	0.2412	0.9964	7.46	1.69
74/1a-5	0.01844	0.1883	0.4041	0.165	0.2349	0.998	8.95	1.25
78/2n1	0.00693	0.2929	0.3375	0.0835	0.2541	0.9833	12.05	0.87
84/2n2	0.01216	0.8076	0.5457	0.1064	0.4385	0.9976	8.75	0.543
94/2n3	0.00791	0.1418	0.1822	0.0975	0.0843	0.996	12.33	0.59
103/2n4	0.00535	0.165	0.1845	0.0777	0.1068	0.9921	14.52	0.647
114/3a1	0.0195	0.874	0.212	0.111	0.099	0.9821	5.69	0.113
130/3a2	0.0131	0.278	0.1547	0.1119	0.041	0.989	8.54	0.147
150/3a3	0.0063	0.1682	0.1874	0.148	0.0394	0.9967	23.49	0.231

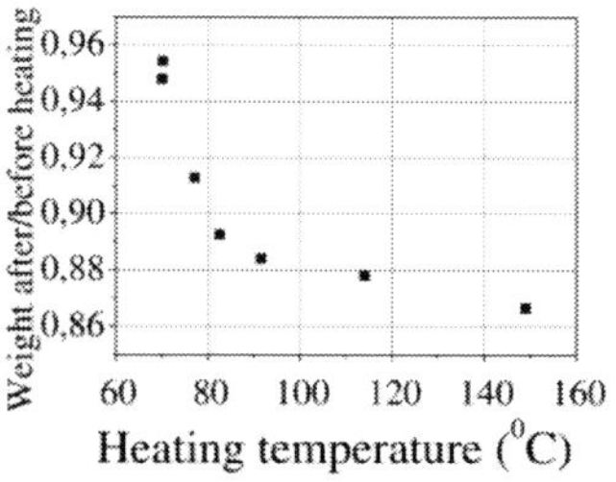

Figure 1. Relative weight decrease of fabric samples, heated at different temperature

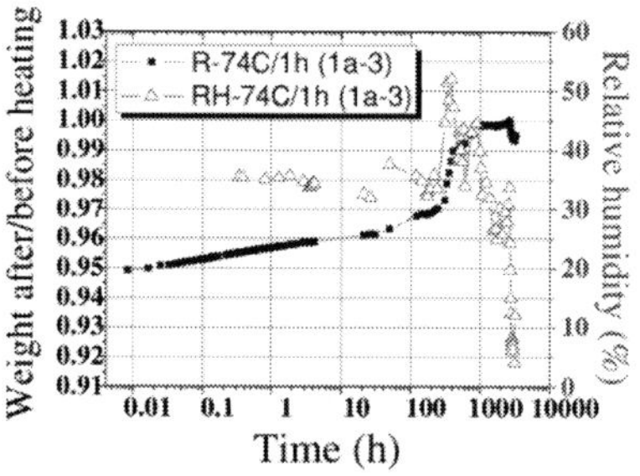

Figure 2. Weight recovery of heated fabric samples 1h at 70 ^{0}C and room RH *vs* time

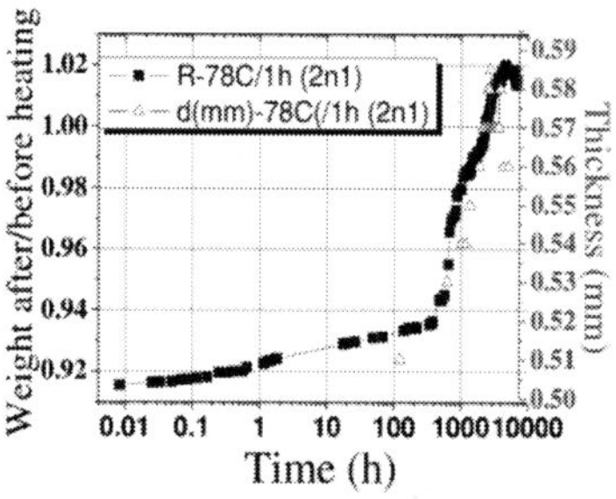

Figure 3. Weight recovery of heated fabric samples 1h at 78 ^{0}C at room conditions and fabric thickness *vs* time

The change of room environment to more stable and elevated humidity level near RH 70 % and room temperature highly elevates the mean values of amplitude A_2 (1.41 mg/cm^2) and velocity v_2 (3.64 mg/h·cm^2) in the beginning time period of weight recovery (Table 2).

Table 2. Parameters of regression (1) for time interval up to 0.3 h after sample heating at different temperatures (during 1 h) and velocity components of weight recovery in relation (2) at the beginning of recovery (t=0)

T,^{0}C/ (sample)	R^2	t_1, h	t_2, h	A_0, mg/ cm^2	A_1, mg/cm^2	A_2, mg/cm^2	$v_1(0)$, mg/h cm^2	$v_2(0)$, mg/h cm^2
70/(1-1)	0.99937	0.0124	0.2358	1.022	0.10994	0.90867	8.89	3.854
72/(1-2)	0.99976	0.0052	0.2635	1.169	0.14231	1.0266	27.26	3.896
76/(1-3)	0.99958	0.0086	0.732	2.4445	0.1379	2.3056	16.07	3.154

The other parameters of second components and all parameters of the first components of relations (1 and 2) were without notable changes, compared to lower humidity case, discussed above. It directly indicates that secondcomponents of relations (1 and 2) characterize the absorption of water. Then symmetrically, the first components of the both relations have to characterize the absorption of CO_2 (it is not a proof). The later weight recovery kinetics (Figures 4, 5, 6) compared to lower humidity environment (Figures 2, 3), have the weight maximum at about 40 h - 50 h by high absorption and desorption of water. The recovered weights overcome its values before heating up to 10 % in its maximum, compared to low humidity cases. As it proceeds from data in literature [2], it could be associated with formation of sodium carbonate heptahydrate, that most commonly occurs and which readily effloresces to form monohydrate by water releasing and evaporation.

The anhydrous sodium carbonate can arose from partly decomposed by heating trona. The heating temperature increases in temperature region above 70 ^{0}C leads to the fast volume increases of anhydrous sodium carbonate and its conversion to sodium carbonate heptahydrate by water absorption from humid atmosphere (Figure 4). It explains the high sensitivity of maximum values in Figures 4, 5, 6 to heating temperatures of samples. When weight minimum is reached, the weight begins the slow and almost linear increase in linear scale along time (Figures 5 and 6 in log scales). The short and fast humidity decrease to its room levels leads to the decrease of weights near its values before heating. It indicates that increase of weights in Figures 5 and 6 could be caused by increase of free water content and hygroscopicity of samples up to at least 6000 h long storing in humid atmosphere. It is seen that equilibrium of fabric with humid environment during this time is not reached.

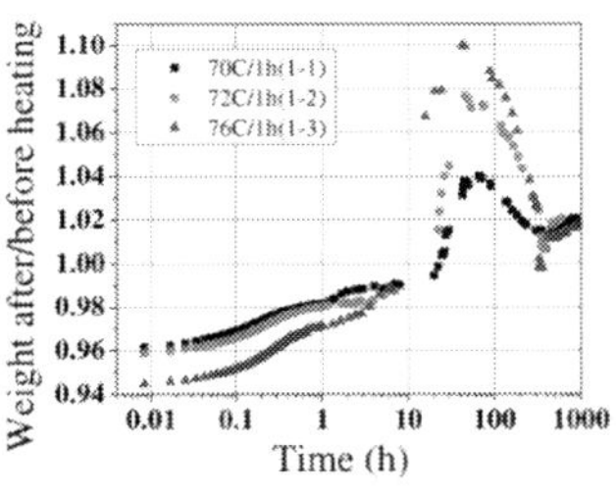

Figure 4. Weight recovery of heated 1h at 70 ^{0}C, 72 ^{0}C, 76 ^{0}C fabric samples at RH 70% and room temperature *vs* time

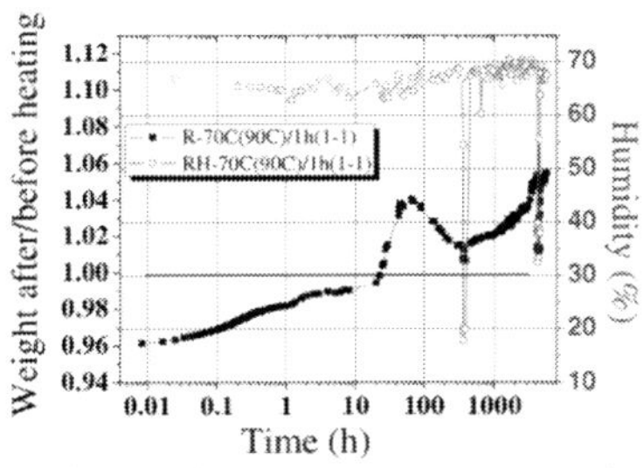

Figure 5. Weight recovery of heated 1h at 70 ^{0}C fabric sample at RH 70% and room temperature *vs* time

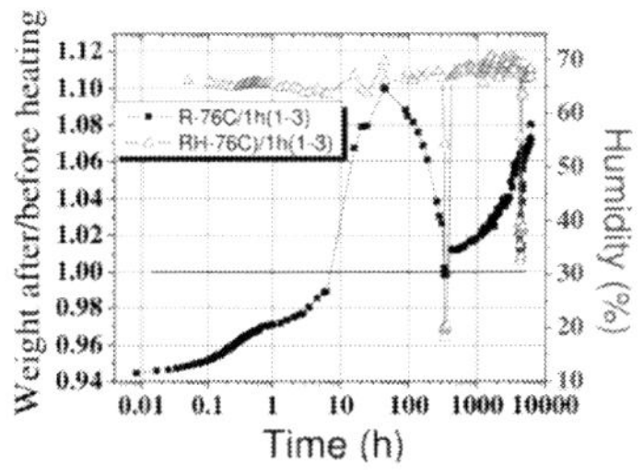

Figure 6. Weight recovery of heated 1h at 76 ^{0}C fabric sample at RH 70% and room temperature, *vs* time

4. Conclusions

At room conditions the main part of recovered weight of carbonated glass fabric samples after heating (70 ^{0}C- 150 ^{0}C / 1h) is associated with water installation in sodium carbonate molecules of carbonated shell and only a small part of recovered weight (above 12%) is associated with recovered free water in shell and may be in pores of glass.

The change of room environment to more stable and elevated humidity level near 70% RH principally changes the kinetics of weight recovery after heating (70 ^{0}C - 76 ^{0}C / 1h) and increases the beginning weight recovery velocities and level, and creates a maximum could be caused with formation of sodium carbonate heptahydrate and its transformation to monohydrate later with water evolving.

References

[1] Veal B W., Lam D J and Karim D P 1980 *Nuclear technology* **51** 136
[2] Kyungmin J C, Timothy C K and Soon-Jai K 2008 *Powder Technology* **184** 58–63
[3] Pentjuss E, Lusis A, Bajars G and Gabrusenoks J 2013 *IOP Conf. Series: Materils Science and Engineering* **49** 012044 doi:10.1088/1757-899X/49/1/012044

IOP Conf. Series: Materials Science and Engineering **77** (2015) 012022 doi:10.1088/1757-899X/77/1/012022

A Comparative Study of Natural Fiber and Glass Fiber Fabrics Properties with Metal or Oxide Coatings

Andrej Lusis[1], Evalds Pentjuss[1], Gunars Bajars[1], Uljana Sidorovicha[2], Guntis Strazds[2]

[1]*Institute of Solid State Physics, University of Latvia, Kengaraga Street 8, Riga, LV-1063, Latvia*

[2]*Riga Technikal University, Rīga, Azenes Street 18, LV-1048, Latvia*

E-mail: lusis@latnet.lv

Abstract. Rapidly growing global demand for technical textiles industries is stimulated to develop new materials based on hybrid materials (yarns, fabrics) made from natural and glass fibres. The influence of moisture on the electrical properties of metal and metal oxide coated bast (flax, hemp) fibre and glass fibre fabrics are studied by electrical impedance spectroscopy and thermogravimetry. The bast fibre and glass fiber fabrics are characterized with electrical sheet resistance. The method for description of electrical sheet resistance of the metal and metal oxide coated technical textile is discussed. The method can be used by designers to estimate the influence of moisture on technical data of new metal coated hybrid technical textile materials and products.

1. Introduction

Rapidly growing global demand for technical textiles (woven and non-woven) for a variety of applications in different industries [1] is stimulated to develop new materials based on hybrid textiles materials (yarns, fabrics). The new materials for the technical textiles are products based on the natural and glass fibre fabrics. They each would assure a certain technical requirements together. The technical textiles fabrics with metal and metal oxide coatings are used, for example, for electromagnetic radiation (EMR) shielding, electromagnetic interference (EMI) and electrostatic discharge (ESD) prevention. The sheet resistance is main technical parameter for EMR, EMI and ESD materials. The fabric applications associated with their electrical properties - resistance and dielectric permeability. The electrical properties are used to characterize coatings and to calculate sheet resistance. The electrical impedance spectroscopy (EIS) is used to measure electrical properties of fabrics. The textiles as porous material have large specific surface area with adsorption capability of moisture from environment. Such textiles are used for production different construction or building materials. In practice are used coated technical textiles with sheet resistance from 10^2 to 10^7 Ohm/sq [2]..

For a comparative analysis of metal and metal oxide coated natural and glass fiber fabrics are used EIS versus influence of moisture. The fibers, yarns and fabrics are porous with high surface area for adsorption water molecules. There are differences on the macro, micro and nano level between constitutions and structures of the natural fiber and glass fiber fabrics. The water absorption is dependent on the moisture in the environment. The moisture has influence on electrical characteristics of coated fabrics. The same time water is active component in some types of electrochemical processes. Fabrics as a porous medium must be characterized before textile fabric applications. The functionalizations with coatings are associated with the distribution of metal or oxide nano particles in a porous medium as well as moisture uptake and corrosion of composite components [3 - 5]).

2. Experimental

A comparative study had been done of bast (flax, hemp) fiber and glass fiber fabrics coated with Al, Ni, Cu, NiO and CuO by DC magnetron sputtering (power 300-580 W, Ar=5,0-6,5 sccm, for oxides Ar=5,0-6,0; O2=1,5 -2,0 sccm).

The bast fiber fabric (BFF) from "Limbažu Tīne, SIA" and E-glass fiber fabrics (GFF) from "Valmiera Glass Fiber, JSC". BFF surface density D_{BFF}= 500 g/m^2 and thickness t_{BFF}=1,4 mm and GFF surface density D_{GFF}=200 g/m^2 and thickness t_{GFF}=0,4 mm.

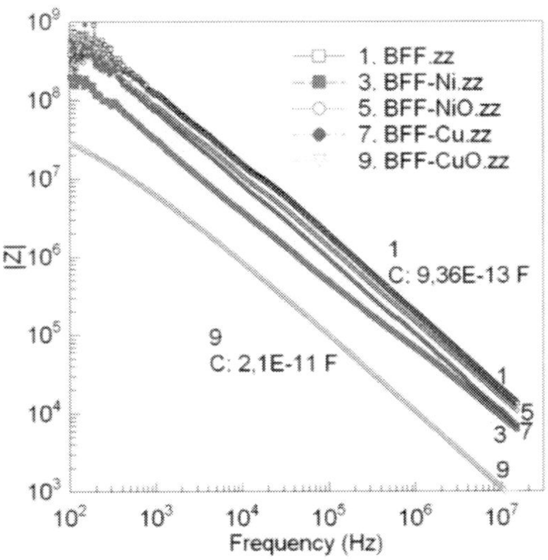

Fig.1. Impedance spectra for dry samples

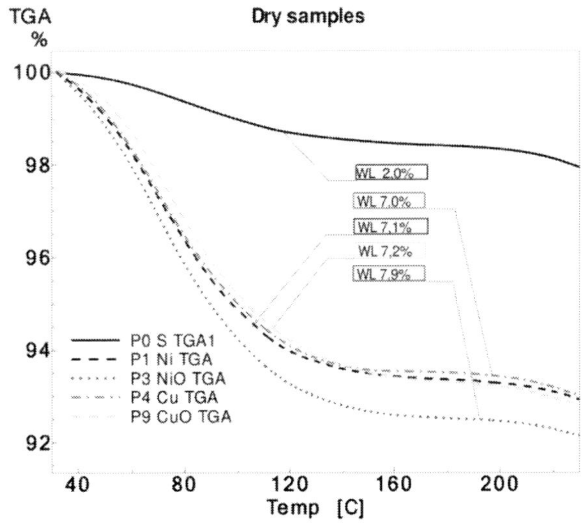

Fig.3. Weight loss for dry samples

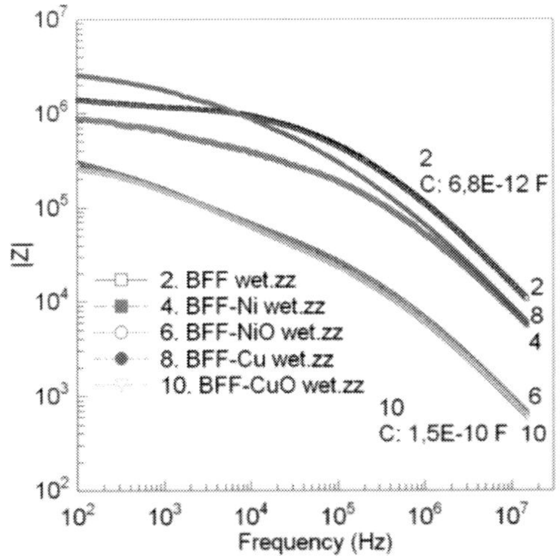

Fig.2. Impedance spectra for wet BFF samples direct from desiccator

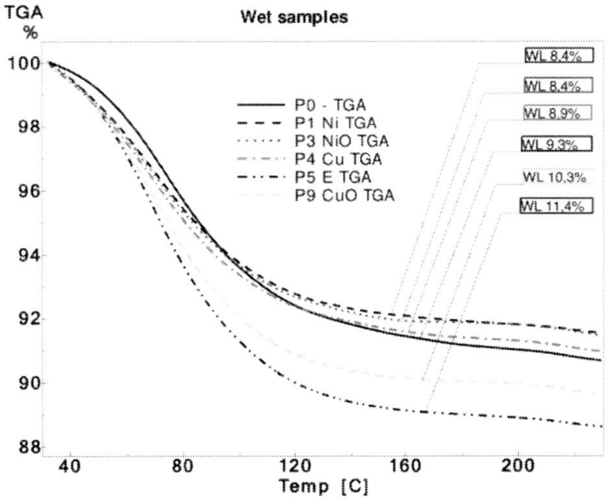

Fig.4. Weight loss (WL) for wet BFF samples direct from desiccator

The bast fiber fabric (BFF) and glass fiber fabrics (GFF) samples before coating had been treated (etched) in argon plasma and after that in the DC magnetron sputtering process had been coated with metal or metal oxide.

Electrical impedance spectroscopy (EIS) and thermogravimetric analyzer (TGA) data are used for characterization of coated fabrics and influence of moisture. For EIS measurements had been done on impedance analyzer HP 4194A (Hewlett-Packard) with HP 16451B sample test fixture (electrode area $S=11{,}34\ cm^2$) and TGA measurements had been done on DTG-60W (Shimadzu). For analyses |Z| spectra was used ZView program (Scribner Associates, Inc).

To characterize the metal coating is a problem in itself. First of all, to functionalize fabrics is to examine the content of moisture and its role on physical properties. The impedance spectroscopy has been used to study moisture in non coated BFF and GFF as well as in coated fabrics. The impedance spectra of such samples are complicate due to heterogeneous and non-homogeneous constitution and dependency from humidity. For humidity studies were prepared two groups of samples: one was to stand at room conditions (dry samples), the other was kept in a desiccator at 100% RH (wet samples). For wet samples measurements had been carried out under moisture free evaporation conditions.

The moisture content has strong influence on impedance modulus |Z| spectra (Fig.1, 2, 5) The moisture has been measured by weight loss with TGA instrument (Fig.3, 4, 6).

The coating as well as the humidity shifts the dry BFF sample |Z| curves to the smaller resistance values (Fig.1 & 2). For wet samples (Fig.2) have a significant impact of moisture on the |Z| values from 100 Hz to 2 MHz's. The |Z| values from 2 MHz to 15 MHz are similar to dry samples. The metal oxide coated BFF have higher water absorption capability then metal coated (Fig.3 & 4).

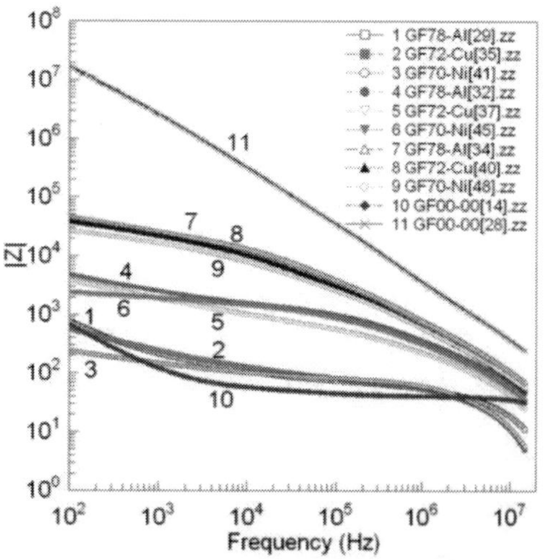

Fig.5. Impedance spectra for wet GFF samples from desiccator measured - 15 min interval.

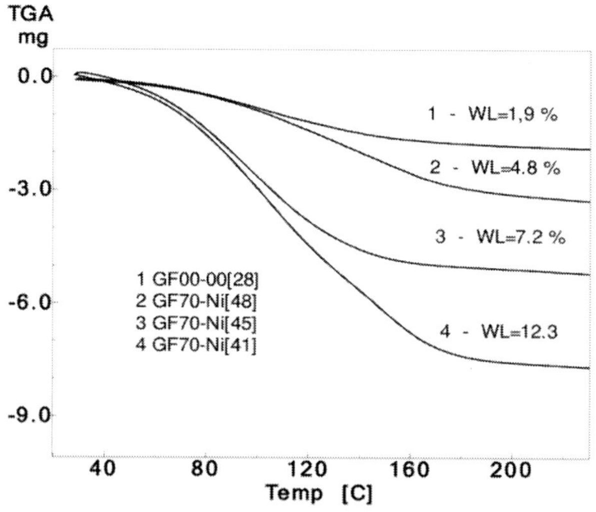

Fig.6. Weight loss (WL) for wet GFF samples from desiccator - measured 15 min interval.

In general |Z| spectra of the GFF and BFF samples have similar behaviour. The coating as well as the humidity shifts the GFF sample |Z| curves to the smaller resistance values (Fig.5) like as BFF |Z| spectra (Fig.1 & 2). For wet samples (Fig.5) have a significant impact of moisture on the |Z| values from 100 Hz to 2 MHz's. There are small differences between |Z| spectra of GFF coated with metal Al, Cu or Ni and grouped 3 groups according TGA data of moisture content (Fig. 6)

The electrical impedance $Z=Z' - jZ''$ of fabrics (BFF and GFF) have complicate charge carrier response for frequencies 100 Hz -15 MHz, i.e., representing their complicate contribution to the resistivity Z'=Real Z. Therefore, the value of sheet resistance in the range of frequencies will vary over a wide range depending on the humidity and will be calculated from the equivalent circuit resistance components. The sheet resistance can be estimated from |Z| spectra by analyses of simple equivalent circuit:

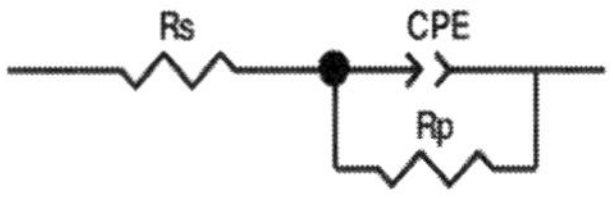

(1)

where Rs, Rp - serial and parallel resistance, CPE – constant phase element. The Rp values had been measured in frequencies region (1-15) MHz. The Rp have been used to calculate BFF and GFF sheet resistance R_{sh} :

$$R_{sh} = Rp*S/t, \qquad (2)$$

where S =11,34 cm2 – electrode area of the sample test fixture, t – fabric thickness t_{BFF} and t_{GFF}. To characterize the moisture influence on fabrics sheet resistance ratio $\Delta R_{sh}/\Delta WL$ or $\Delta R_{sh}/\Delta m$:

$$\Delta R_{sh}/\Delta m = (1/(D*t^2))*(\Delta R_{sh}/\Delta WL) \qquad (3)$$

where Δm - mass of lost moisture from TGA data, D – fabric surface density D_{BFF} and D_{GFF}. The experimental data of BFF and GFF for comparison are summarized for dry and wet fabrics in Table 1

Table 1: Influance moisture on sheet resitance

Sample	Rp, Ohm	WL, %	R_{sh}, Ohm/sq	$\Delta R_{sh}/\Delta m$, Ohm/(sq*g)
1BFF	9,7E+06	2,0	5,6E+11	Reference
11GF00-00[28]	3,9E+06	1,9	2,8E+12	Reference
2BFF wet	3,5E+05	8,3	2,0E+10	
10GF00-00[14]wet	2,3E+01	9,2	1,6E+07	
3BFF-Ni	1,7E+06	7,1	9,8E+10	
5BFF-NiO	5,5E+06	7,9	3,2E+11	
4BFF-Ni wet	1,5E+05	8,4	8,7E+09	1,2E+06
6BFF-NiO wet	2,4E+04	8,4	1,4E+09	4,9E+05
3GF70-Ni[41]wet	6,9E+01	12,3	4,9E+07	Reference
3GF70-Ni[45]vet	1,8E+03	7,2	1,3E+09	1,1E+05
3GF70-Ni[48]wet	1,0E+04	4,8	7,1E+09	1,4E+06

The dry coated BFF and GFF samples have a significant influence of moisture on sheet resistance than for dry non coated fabrics. The wet coated BFF and GFF samples have a similar influence of moisture on sheet resistance for dry coated fabrics.

3. Conclusions

The bast fibre and glass fibre fabrics are characterized with electrical sheet resistance. The method for description of electrical sheet resistance of the metal and metal oxide coated technical textile is developed. The EIS and TGA data are useful for characterization of production hybrid technical fabrics based on natural and glass fibres. The method can be used by designers of a new metal coated hybrid technical textile materials and products to estimate the technical data as well as the influence of moisture on the technical data.

References

[1] Opinion of the European Economic and Social Committee on 'Growth Driver Technical Textiles' Official Journal of the European Union: C 198/14, 10.7.2013. http://www.eesc.europa.eu
[2] Shielding & Conductive Fabrics: http://www.lessemf.com/fabric.html.
[3] Elmer T H 1984 *J. Am. Chem. Soc.* **67** 778
[4] Mišíková L, Liška M and Galusková D 2007 *Ceramics - Silikáty* **51** 131
[5] Mišíková L and Liška M 2008 *Advanced Materials Research* **39-40** 355

IOP Conf. Series: Materials Science and Engineering **77** (2015) 012023 doi:10.1088/1757-899X/77/1/012023

Durable hydrophobic sol-gel finishing for textiles

S Vihodceva[1], **S Kukle**[1] **and J Bitenieks**[2]

[1] Riga Technical University, Institute of Design Technologies, 18 Azenes street, LV-1048, Riga, Latvia

[2] Riga Technical University, Institute of Polymer Materials, 14/24 Azenes street, LV-1048, Riga, Latvia

E-mail: Svetlana.Vihodceva@rtu.lv

Abstract. The surface of cotton textile was modified to create a water-repellent finishing by depositing a modifying coatings using the sol-gel technique. Treated textiles evaluated using scanning electron microscopy, X-Ray powder diffraction (XRD). The wettability of treated fabrics was characterized by water contact angle and drop test. The results showed that the cotton textile treated with 7.5 wt.% zinc acetate dihydrate sol showed excellent hydrophobic properties, water contact angle could reach 145°C without decreasing after 50 hydrothermal treatment cycles.

1. Introduction

Hydrophobic fabrics recently have attracted great attention because of their importance in industrial applications due to their unique characteristics like self-cleaning, anti-contamination, antisticking [1, 2]. Cotton has always been the principal clothing fiber due to its outstanding properties such as softness, comfort, biodegradation, high mechanical stability etc. But the abundant water-absorbing hydroxyl groups on cotton surface make the fiber absorbent and easily stained by the liquids [1, 3, 4]. The hydrophobic properties to cotton textile can be brought by sol-gel technology. Chengyu Pan et. al. [5] were successfully modified cotton textile with $Al(NO_3)_3$ and sodium stearate by a sol-gel process for the hydrophobic properties imparting. Lihui Xu et. al. [1] studied the superhydrophobic properties imparting to the cotton with treatment by the SiO_2 hydrosol and alkylsilanol - hexadecyltrimethoxysilane solution. Montarsolo Alessio et. al. [6] used silica-based sol with hydrophobic additives (N-propyltrimethoxysilane, hexadecyltrimethoxysilane and fluorooctyltriethoxysilane) to prepare hydrophobic coating.

Hence, the first goal of this article is to investigate the action of zinc acetate dihydrate as a hydrophobic finishing agent in the presence of tetraethoxysilane and hydrofluoric acid to cotton textile at a low temperature. The second goal, in the present study are observed one step process to create durable hydrophobic cotton textiles.

[1] Svetlana.Vihodceva@rtu.lv.

2. Experimental

2.1. Materials, chemicals, sol preparation and textile treatment

The commercial plain weave 100% cotton fabric with the thickness 0.40 mm, surface density 187 g/m^2 and porosity 69.80 %. The analytically pure chemicals tetraethoxysilane (TEOS, $C_8H_{20}O_4Si$, Alfa Aesar, Germany), ethanol (C_2H_5OH), hydrofluoric acid (HF, Alfa Aesar, Germany), deionized water and zinc acetate dihydrate $Zn(CH_3COO)_2 \cdot 2H_2O$ (Scharlau, Spain) were used. Sols were prepared by a controlled hydrolysis, by previously described method [7]. Samples were prepared with zinc acetate dihydrate 2.5, 5 and 7.5 wt.% concentration. The fabric samples were dipped into the prepared sol for 10 minutes at room temperature. Subsequently, the samples were dried at 90 °C for 10 minutes with thermal post-treatment at 120 °C for 2 minutes or with combined curing and drying in one step at 90 °C for 30 minutes.

2.2. Modified textiles testing

The resistance of hydrophobic finish to hydrothermal treatment, samples were washed at 40 °C for 30 minutes without steel balls with 5mg/l standard non-phosphate detergent without optical brighteners [8]. After each washing cycle samples were rinsed with distilled water and dried under nominally ambient conditions (i.e. 20 ± 2 °C and a relative humidity of 65 ± 2 %) [9].

The determination of liquid absorbency was made by the drop test in accordance with AATCC Test Method 79 [10]. The water contact angle measurements were made by the drop method by an optical tensiometer Theta Attension (Finland).

Scanning electronic microscope (SEM) measurements were carried out using Tescan Mira (Czech Republic) and Hitachi S-3400N (Japan).

Powders acquired after drying at temperature from 100 to 500 °C were characterized by X-Ray powder diffraction (XRD) recorded for 2θ at a scan rate of 1° min^{-1} using an Ultima+ X-ray diffractometer (Rigaku, Japan) with CuKα radiation.

3. Results and Discussion

3.1. Water absorbency and water contact angle

After the cotton textile treatment by 5 wt. % or 7.5 wt. % zinc acetate dihydrate sol, the water absorbency velocity was decreased from 11 seconds (Fig. 1) to > 30 minutes (Fig. 2). The textile treatment with 2.5 wt.% zinc acetate dihydrate sol doesn't change the water absorbency velocity significantly which increased to approximately 2 minutes. The number of cycles of the hydrothermal treatment (50 cycles) has not affected the water absorbency velocity of the textiles modified by 7.5 wt.% zinc acetate dihydrate sol, the water absorbency velocity remains > 30 minutes; water absorbency velocity of the textile modified by 5 wt. % zinc acetate dihydrate sol was decreased only after 50 cycles of the hydrothermal treatment up to 2 minutes.

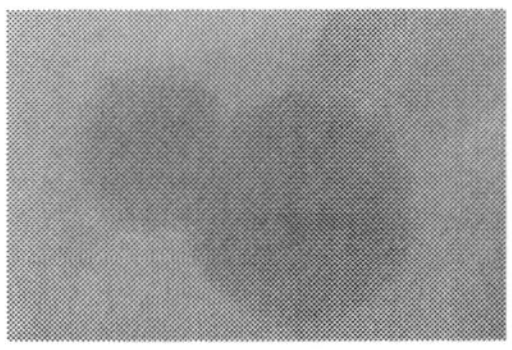

Figure 1. Water absorbency (drop 20 μL) untreated cotton

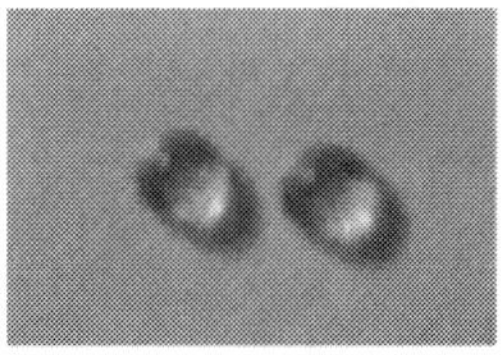

Figure 2. Water absorbency (drop 20 μL) after treatment by 5 wt. % zinc acetate dihydrate sol

For untreated textiles and treated with 2.5 wt.% zinc acetate dihydrate sol the fast water absorbance was observed (Fig. 3). It was found that samples treated with sol within the range of zinc acetate

IOP Conf. Series: Materials Science and Engineering **77** (2015) 012023 doi:10.1088/1757-899X/77/1/012023

dihydrate concentration 5 wt.% - 7.5 wt.% showed hydrophobicity with contact angle not less than 144° before hydrothermal treatment (Fig. 4, Tab. 1). The different decrease of contact angle was observed after hydrothermal treatment. The cotton textile treatment by 5 wt. % zinc acetate dihydrate sol provides the water-repellent properties also after 20 cycles of the hydrothermal treatment. The cotton textile treatment by 7.5 wt. % zinc acetate dihydrate sol provides the water-repellent properties to cotton textiles even after 50 cycles of intensive hydrothermal treatment (Fig. 5).

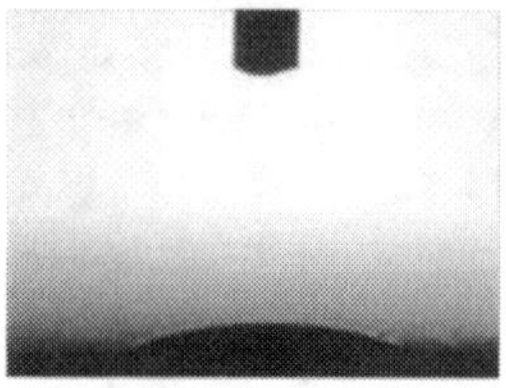

Figure 3. Water contact angle measurements (drop 5 μL) untreated textile

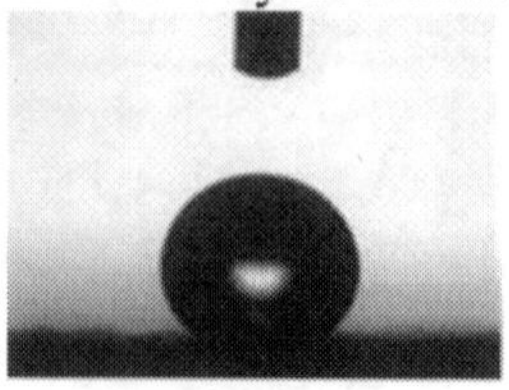

Figure 4. Water contact angle measurements (drop 5 μL) treated by 7.5 wt.% zinc acetate dihydrate sol

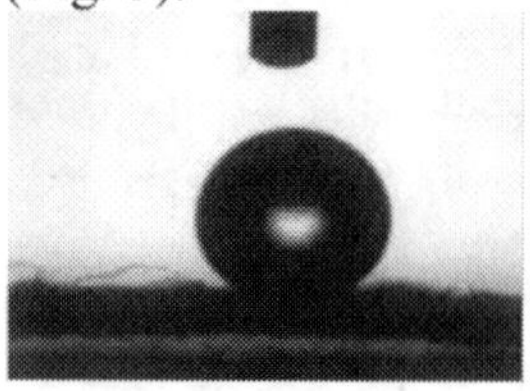

Figure 5. Water contact angle measurements (drop 5 μL) treated by 7.5 wt.% zinc acetate dihydrate sol after 50 hydrothermal treatment cycles

Table 1. Water contact angle of modified textiles and after hydrothermal treatment

Zinc acetate concentration	*Thermal post-treatment temperature*	*Treated*	*1 hydrothermal treatment cycle*	*20 hydrothermal treatment cycles*	*50 hydrothermal treatment cycles*
5 wt.%	90 °C 30 min	148.93	150.04	146.77	104.84
	120 °C 2 min	149.24	144.06	147.69	117.54
7.5 wt.%	90 °C 30 min	144.48	141.60	147.23	151.29
	120 °C 2 min	145.06	139.35	145.43	140.77

3.2. SEM micrographs analysis

The surface of the treated and untreated cotton textile samples was observed and compared using SEM. Untreated cotton textile has grooves and cracks (Fig. 6), in contrast such characteristics completely disappeared on the surface of the treated cotton textile with sol within the range of zinc acetate dihydrate concentration 5 wt. % - 7.5 wt. % (Fig. 8-9); in case of textile treated with 2.5 wt. % zinc acetate dihydrate sol the cracks and grooves are still observed. SEM micrographs (Fig. 7-9) prove that the textiles by zinc acetate dihydrate sol are coated homogenously. Regardless of the zinc acetate concentration (5 wt. % or 7.5 wt. %), there is a thin coating obtained with single particles and particles agglomerates dispersed in the film, some particle agglomerates consolidates and form irregular several micrometer wide and/or long clusters on the textile surface (Fig. 8 and 9).

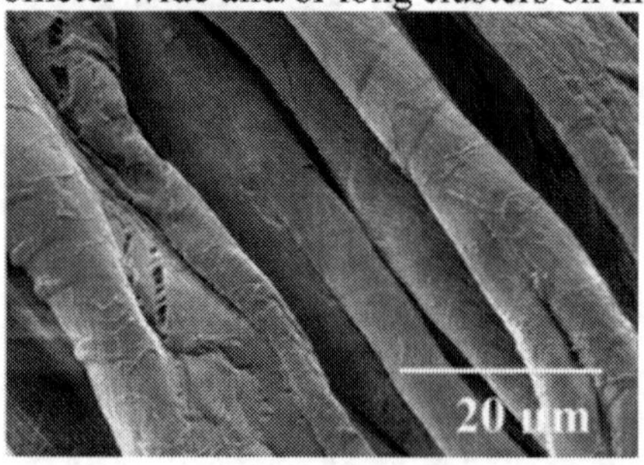

Figure 6. Untreated

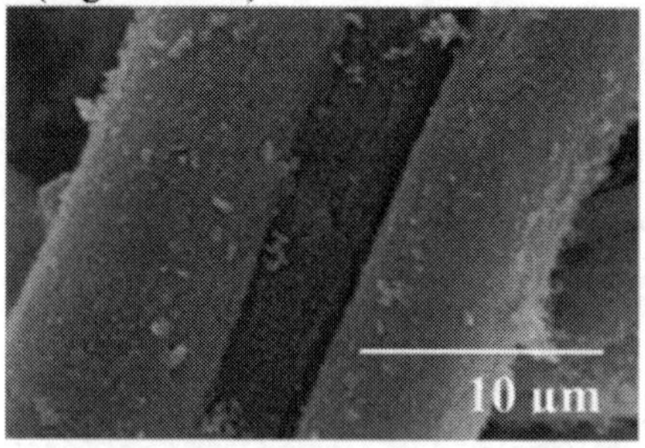

Figure 7. Treated by 2.5 wt.% zinc acetate dihydrate sol

Figure 8. Treated by 5 wt.% zinc acetate dihydrate sol

Figure 9. Treated by 7.5 wt.% zinc acetate dihydrate sol

3.3. XRD

It was concluded that the difference between untreated and treated cotton textile cannot be identified by XRD, that can be explained by the fact that the archived coatings are very thin and treated textile contain a small amount of the modifying chemical elements. XRD analyses of sol powder evince that crystalline phase of the ZnO appears starting from the temperature 300 °C, in the range of the temperature from 300, 400 and 500 °C SiO_2 is also obtained; from the 400 °C the formation of the willemit - $Zn_2(SiO_4)$ begins. At lower temperatures that are appropriate for the cotton textile thermal treatment – from 100 to 200 °C - occurs a partial decomposition of the zinc acetate dihydrate and basic zinc compound formation. The functional coating primarily consist from zinc acetate ($C_4H_6O_4Zn$), silica compounds: $H_2Si_3O_7(H_2O)$ and $H_2Si_{14}O_{29}{\bullet}5.4H_2O$, and probably from insoluble basic zinc compounds or zinc oxide in amorphous phase.

4. Conclusion

The coated cotton textile exhibited excellent hydrophobicity with the water contact angle reached 145 °C after treatment and 150 °C after the 50 hydrothermal treatment cycles, as well high durability of the coating that withstand 50 cycles of the hydrothermal treatment (washing-drying). Zinc acetate dihydrate sol allow to deposit thin, homogenous functional coatings with dispersed single particles and particles agglomerates.

The method is simple and may have some promising applications for cotton materials to generate hydrophobic surfaces.

5. References

[1] Xu L, Zhuang X, Xu B and Cai Z 2011 *Appl. Surf. Sci.* **257** 5491
[2] Wei Q 2009 *Surface modification of textiles* (United Kingdom: CRC Press)
[3] Xu B, Cai Z 2008 *Appl. Surf. Sci.* **254** 5899
[4] Qing Y, Zheng Y, Hu C, Wang Y, He Y, Gong Y and Mo Q 2013 *Appl. Surf. Sci.* **285P** 583
[5] Pan C, Shen L, Shang S and Xing Y 2012 *Appl. Surf. Sci.* **259** (2012) 110
[6] Montarsolo A, Periolatto M, Zerbola M, Mossoti R and Ferrero F 2013 *Tex. Res. J.* **83** 1190.
[7] Vihodceva S and Kukle S 2013 *IOP Conf. Ser.: Mater. Sci. Eng.* **49** 012022
[8] Standard: ISO 105C10-A01 2006 Textiles – Tests for colour fastness – Part A01: General principles of testing
[9] Standard: EN 20139 1992 Textiles – Standard atmospheres for conditioning and testing
[10] Standard: AATCC 79 2010 Textiles – Absorbency of textiles

IOP Conf. Series: Materials Science and Engineering **77** (2015) 012024 doi:10.1088/1757-899X/77/1/012024

Mechanical Pressure Induced Capacitance Changes of Polyisoprene/Nanostructured Carbon Black Composite Samples

K Ozols and M Knite

Riga Technical University, Institute of Technical Physics, Riga, Latvia

E-mail: kozols@ktf.rtu.lv

Abstract. Polyisoprene/nanostructured carbon black (PNCB) composite samples with different amount of carbon black (CB) filler were prepared. Investigations of mechanical pressure induced relative capacitance changes (RCC) depending on frequency (20 Hz – 2 MHz) were conducted. It was found that PNCB samples show pronounced and rather complex RCC effect, which depends on frequency, amount of filler (1 – 10 phr of CB) and pressure (from 0 to 234 kPa). At a certain frequency and a certain filler amount RCC effect changes its sign. Pressure induced capacitance changes at least for low CB filler concentrations are caused by piezopermittivity property of the PNCB composite.

1. Introduction

PNCB composites show pronounced resistance change effect when subjected to mechanical deformation [1]. This effect has been used to develop flexible mechanical pressure and organic solvent vapor sensors [2]. Mechanisms of electroconduction [3] and emergence of dielectric constant [4] for such composites in static regime have been explained. Change of electroconduction under mechanical influence [1,2] is in general rather well understood. Recently it was found that PNCB composites show a piezocapacitance effect as well [5]. To advance the composites for sensor and, possibly, other applications further dielectric spectroscopy studies of PNCB samples influenced by mechanical pressure should be conducted. The current work is dedicated mostly to capacitance change measurements of PNCB composite samples while mechanical pressure is applied.

2. Samples and experiments

2.1. *Development of samples*

Development of PNCB composite samples is as follows. SWR-3L polyisoprene rubber is used as a matrix due to its high elastic properties. Carbon black (high structure Degussa™ Printex™ EX-2 carbon black, the average particle size is 30 nm, surface area 950 m^2/g) is used as electroconductive filler. First, a liquid solution of matrix is obtained by dissolving polyisoprene with curing ingredients in chloroform. Then, colloidal solution of CB in chloroform is prepared whereby CB nanoparticles are dispersed in chloroform using ultrasonication (5 min, specific power 1 W/ml). The CB solution is added to the matrix solution and the obtained mixture is stirred for 24 h. Further, the mixture is poured out into Petri dishes to evaporate chloroform. After evaporation of chloroform the obtained dry, raw PNCB mixture is processed in Two-Roll Mills (LabTech Engineering Company, model "Scientific", cold processing mode used) to disperse CB nanoparticles again. After processing with roll mills a raw

PNCB mixture (Figure 1a) is ready to be used for sample creation during vulcanization. A piece of a raw PNCB mixture is put into a press form together with brass electrodes (thickness of electrodes is 0.05 mm).

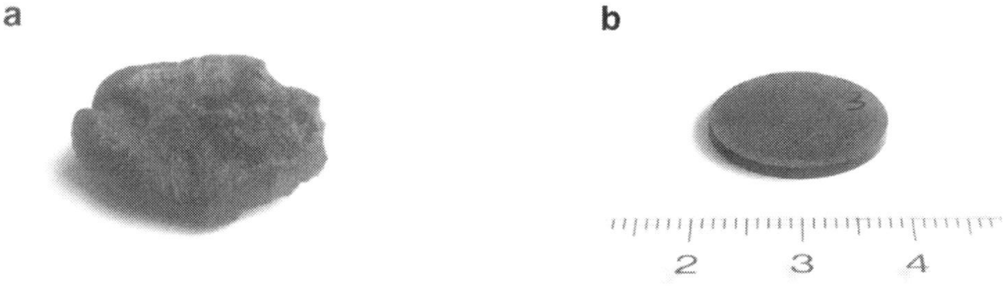

Figure 1. Images: a) PNCB composite mixture after roll mixing and before vulcanization, b) finished PNCB composite sample with brass electrodes.

The press form is then put into a hot press (Rondol™, curing temperature 150^{0}C, pressure 3 MPa) for 15 minutes. As an additional result of vulcanization of the PNCB mixture, the electrodes are attached tightly to the PNCB composite. After cooling the press form finished PNCB composite sample is removed (Figure 1b). The thickness of the whole sample is 1.1 mm, (1 mm is for the composite), the diameter of the sample is 18 mm. Samples with different CB concentrations were created, from 0.5 phr CB (0.5 mass parts of CB per 100 parts of rubber) to 10 phr of CB. Testing of samples was conducted no less than 24 h after samples were created.

2.2. Measurements

Dielectric spectroscopy measurements were conducted using an Agilent E4980A Precision LCR meter (frequency range 20 Hz - 2 MHz). The mechanical pressure was applied using weights with different mass. During measurements the weights were isolated from the samples with a teflon spacer.

3. Results and Discussion

3.1. Measurement results

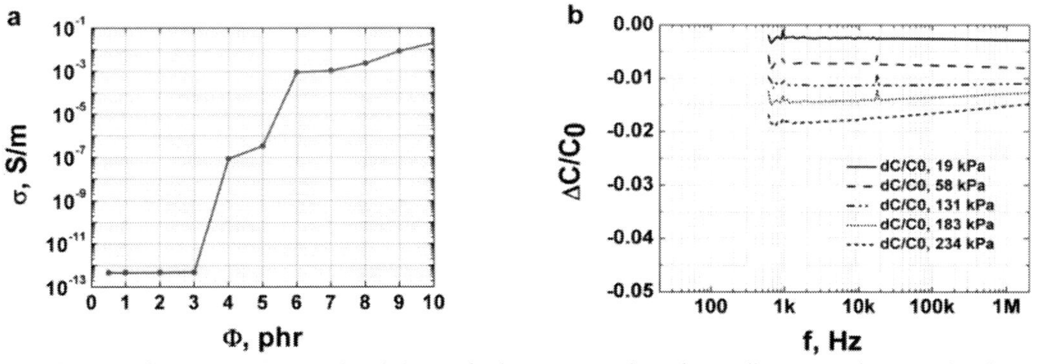

Figure 2. Graphs: a) DC conductivity of the composite depending on the conductive filler concentration, b) The RCC effect depending on pressure for sample containing 3 phr of CB.

According to Knite et al [1] in the case of DC conductivity, as mechanical pressure is applied, PNCB samples show the largest conductivity change when the concentration of the electroconductive filler fits to a region of percolation transition. For samples used in this work percolation transition would correspond to CB concentrations approximately from 3 to 6 phr (Figure 2a). RCC (i.e. $\Delta C/C_0$) effect depending on frequency and applied different values of mechanical pressure for PNCB sample containing 3 phr of CB is shown in Figure 2b. As it can be seen on the graph the maximum RCC effect for the given sample is small - around 2%. The effect is negative - the capacitance is decreasing at all applied pressures and at all measured frequencies. Measurement results for the lowest frequencies are not shown, because the capacitance of the sample is too small (~12 pF at low frequency) for precise low frequency readings on the LCR meter. The DC conductivity of the sample containing 3 phr of CB is just on the limit of percolation transition and therefore in the case of piezoconductivity the effect caused by mechanical pressure also would be small. The largest RCC effect (20% at maximum pressure) is found for the next sample, containing 4 phr of CB (Figure 3a). Conductivity of this sample (10^{-7} S/m) resides at the lower part of percolation transition. The sign of the RCC effect is negative, too, but in this case large frequency dependence is observed. At 2 MHz maximum RCC effect is only 4%.

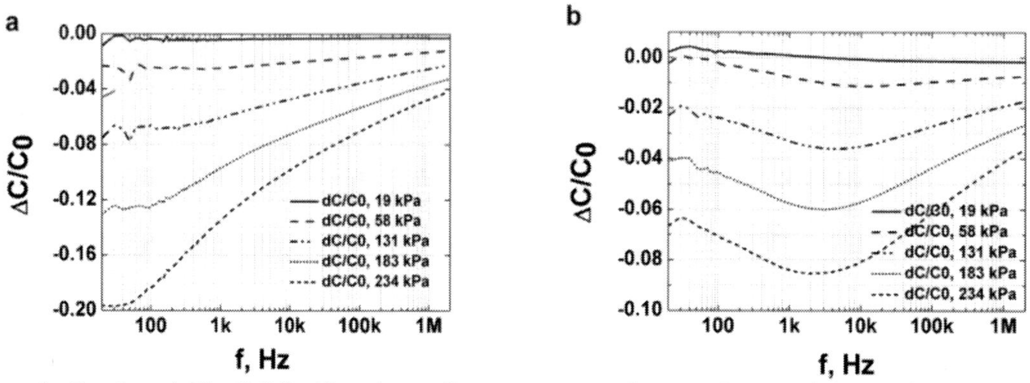

Figure 3. Graphs: a) The RCC effect depending on pressure for sample containing 4 phr of CB, b) The RCC effect depending on pressure for sample containing 6 phr of CB.

Further increase of CB leads to decrease of RCC effect while at the same time it shifts the RCC maximum frequency to higher values (see Figures 3a and 3b). Adding more CB to the matrix (7 phr of CB) introduces a change in sign of RCC effect at different frequencies (Figure 4a).

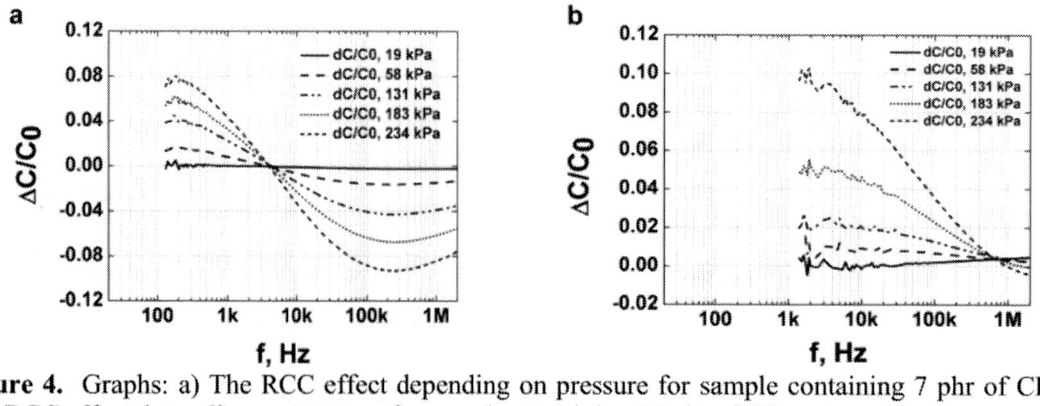

Figure 4. Graphs: a) The RCC effect depending on pressure for sample containing 7 phr of CB, b) The RCC effect depending on pressure for sample containing 10 phr of CB.

At frequency around 200 Hz compression of the sample increases the capacitance of the sample up to around 8% (at maximum pressure), at 4 kHz RCC effect equals zero for all pressures applied, but at higher frequencies, negative RCC effect is observed with a maximum effect around 300 kHz (max RCC around 9% at 234 kPa). Measurement results for the lowest frequencies are not shown, because dielectric loss in the composite becomes too large (tan δ = 600 at 100 Hz) for precise capacitance readings on the LCR meter. Further CB amount increase in PNCB sample (10 phr of CB, Figure 4b) shifts a zero-change frequency up to around 650 kHz. Well expressed positive RCC effect can be observed (around 10% at 234 kPa, at 2 kHz). On the contrary, maximum negative RCC effect is just around 1%.

3.2. Proposed mechanism of capacitance change

The observed RCC effect is rather challenging to explain since the effect changes its value and sign depending on the CB concentration and frequency. The electroconductive channels in the PNCB composites are formed between CB aggregates by means of tunneling barriers [1,3], where tunneling currents can flow. A network of these electroconductive channels are responsible for the piezoresistance effect [1,2]. Conductive nanoparticle aggregates form microcapacitors [4]. Conductivity percolation at least for polymer/CNT composites is tightly bound to permittivity percolation [6]. Based on these facts, we conclude that the same network of electroconductive channels, which is responsible for piezoresistance effect in the PNCB composite, does form a network of capacitive channels (microcapacitor chains), which mostly determine the overall capacitance of the sample and are responsible for the RCC effect. It should be noted that piezoconductance effect (figures not shown in this paper) for the same samples, same frequency range and for all the concentrations of CB mentioned here is always negative. That means that the compression of PNCB sample, at least for CB concentrations up to 10 phr and pressures up to 234 kPa, always leads to interruptions in conductive and, simultaneously, capacitive channels. Since the underlying physics of conductance and capacitance are different, that could explain why the RCC effect is not always negative. Currently this phenomenon is not entirely clear to us.

Since in our case PNCB samples represent capacitors with an elastic dielectric layer between electrodes and these samples show a decrease of capacitance while being compressed (this is true at least for PNCB samples with low CB filler amount) then it confirms, that piezocapacitance effect of the samples is based on piezopermittivity property of the PNCB composite.

4. Conclusions

The measurements conducted showed, that both negative and positive RCC effects in PNCB composite samples exist. It has been found that inversion of sign of the RCC effect depends on the CB filler concentration and partially on the AC frequency. The explanation for observed effects is based on the concept, that PNCB composite represents an electrical circuit consisting of a large amount of microscale resistive and capacitive elements, which not only form a network of electroconductive channels but simultaneously form a network of capacitive channels. Compression of PNCB samples leads to a change of parameters of resistive and capacitive elements in the composite. Thus, the overall capacitance (and conductance) of the sample is changed. The observed effects simultaneously confirm the piezopermittive property of the PNCB composite (at least for the cases with low filler concentrations). In PNCB and similar composites, which could be compressed or stretched, RCC effects together with conductivity change effects may help to explain the structure of composites and provide information about the movement of nanoparticles during mechanical influence.

5. Acknowledgements

This work is financially supported by the Council of Science of Latvia and by Latvian National Research Program in Materials Science.

References

[1] M. Knite, V. Teteris, A. Ķiploka, J. Kaupuzs, Polyisoprene-carbon black nanocomposites as tensile strain and pressure sensor materials, Sensors and Actuators A 110 (2004), p142–149

[2] M. Knite, K. Ozols, J. Zavickis, V. Tupureina, I. Klemenoks, and R. Orlovs, Elastomer–Carbon

Nanotube Composites as Prospective Multifunctional Sensing Materials, JNN, Vol. 8, 1-6, 2008

[3] I. Balberg, A comprehensive picture of the electrical phenomena in carbon black–polymer composites, Carbon 40 (2002) 139-143.

[4] P. Pötschke, S.M. Dudkin, I. Alig, Dielectric spectroscopy on melt processed polycarbonate—multiwalled carbon nanotube composites, Polymer 44 (2003) 5023–5030

[5] M.Knite, J.Zavickis, G.Sakale, K.Ozols & A.Linarts, Advanced smart polymer/nanographite composites for environmental pollution control, in book Green design, Materials and Manufactoring Processes Bartolo et al. (eds) 2013, Taylor & Francis Group, London, ISBN 978-1-138-00046-9, 587-592

[6] Brian P. Grady, Carbon Nanotube-Polymer Composites: Manufacture, Properties, and Applications, John Wiley & Sons, Inc, 2011, ISBN: 9781118084380, p268–269

IOP Conf. Series: Materials Science and Engineering **77** (2015) 012025 doi:10.1088/1757-899X/77/1/012025

Template-based synthesis of nickel oxide

N Mironova-Ulmane[1], **A Kuzmin**[1], **I Sildos**[2]

[1] Institute of Solid State Physics, University of Latvia, LV-1063 Riga, Latvia

[2] Institute of Physics, University of Tartu, 51014 Tartu, Estonia

E-mail: nina@cfi.lu.lv

Abstract. Nanocrystalline NiO has been produced using a facile template-based synthesis from nickel nitrate solutions using cellulose as a template. Thus obtained oxides were studied by scanning electron microscopy, x-ray diffraction, Raman scattering spectroscopy, photoluminescence spectroscopy and confocal spectromicroscopy. The filamentary/coral morphology of the samples has been evidenced and is built up of agglomerated nanocrystallites with a size in the range of about 26-36 nm. The presence of two-magnon contribution in Raman scattering spectra suggests the existence of antiferromagnetic ordering at room temperature. Finally, the observed near-infrared photoluminescence band at 850 nm has been tentatively attributed to the defect-perturbed Ni^{2+} states at the surface.

1. Introduction

Synthesis of materials using various template techniques becomes popular during last years due to simplicity and often low costs [1, 2, 3]. Such approach involves fabrication of the desired material within the pores or channels of a template that can be retained or removed thereafter. Several template materials have been utilized up to now to produce nanosized and mesoporous nickel oxide: eggshell membranes [4] or pinewood [5] as biotemplates, inorganic macroporous skeletons built up of close-packed polystyrene (PS) spheres [6], soft template octylamine [7], filamentous carbon [8], polyacrylonitrile [9], anodic aluminium oxide [10], cellulose [11] and agarose gel [12].

In this study we report on a facile template-based synthesis of nickel oxide from nickel nitrate solutions using cellulose as a template. We have found that thus obtained NiO samples retain the morphology of the template, which can be, however, influenced by the solution concentration. In particular, we will focus on the lattice dynamics, antiferromagnetic ordering and near-infrared photoluminescence of cellulose-templated NiO in comparison with micro- and nanocrystalline NiO produced by other methods [13, 14, 15].

2. Experimental

Cellulose-templated nickel oxide powder samples were produced by thermal decomposition of nickel nitrate hexahydrate salt at high temperature. First, aqueous solutions of nickel nitrate hexahydrate $Ni(NO_3)_2{\cdot}6H_2O$ were prepared at two concentrations (0.01 M and 0.05 M) to check the influence of the solution concentration on the amount (thickness) of nickel oxide layer surrounding the template framework. Next, the Whatman cellulose filter paper (Grade No. 2, Whatman 1002-110), used as a template, was impregnated with the solution, air dried at room temperature and heated for 1 hour at 650°C in tubular furnace to decompose nitrate and

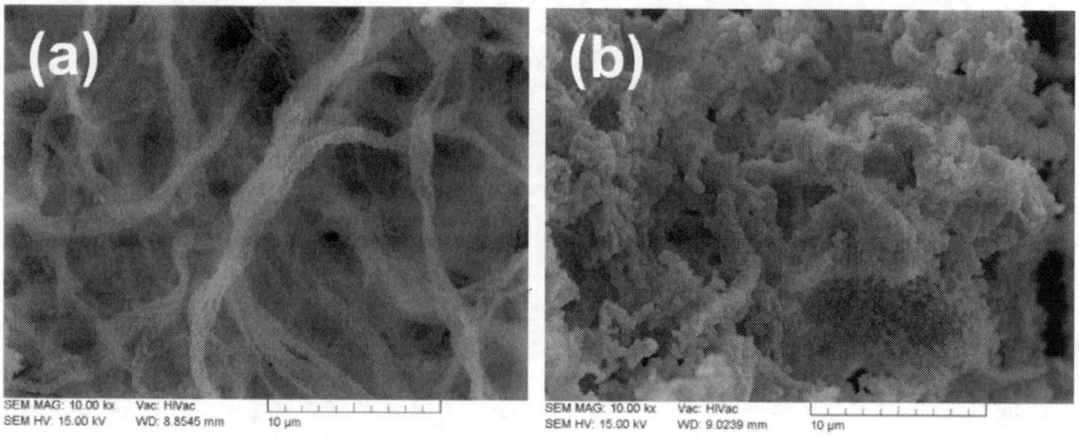

Figure 1. SEM images of cellulose-templated NiO: (a) 0.01 M and (b) 0.05 M.

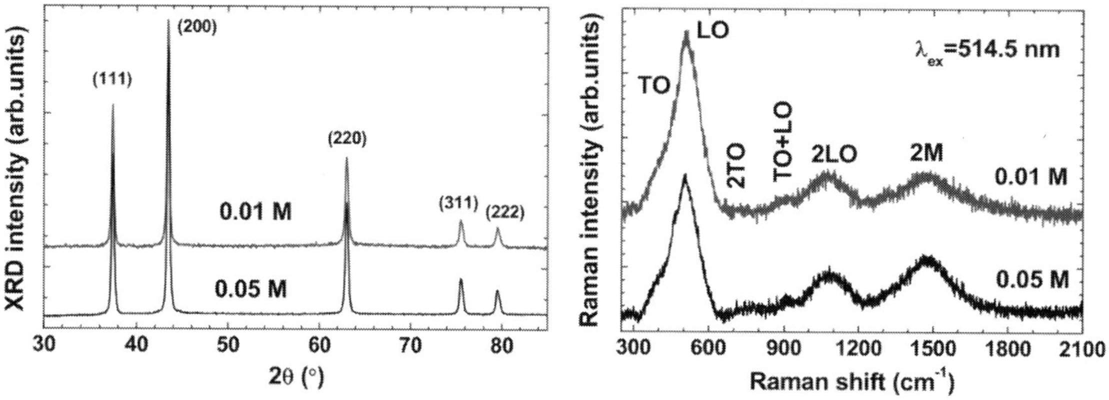

Figure 2. Room temperature x-ray diffraction patterns (left panel) and Raman scattering spectra (right panel) of cellulose-templated NiO.

to burn completely cellulose template. The final product, cellulose-templated NiO, was collected and used in further analysis. It has black color, which indicates the presence of nickel vacancies [16, 17].

The morphology of two samples was studied by scanning electron microscopy (SEM) using TESCAN MIRA microscope with secondary electron detector, whereas their phase composition was detected by x-ray diffraction using Bruker D2 PHASER θ-θ diffractometer equipped with copper anode x-ray tube.

Raman scattering spectra were measured in back-scattering geometry at room temperature (20°C) through 50× microscope objective using a Renishaw inVia micro-Raman spectrometer equipped with an argon laser (514.5 nm, max cw power P_{ex}=10 mW). The spectral signal was dispersed by 2400 grooves/mm diffraction grating onto a Peltier-cooled CCD detector.

Confocal microscopy and optical photoluminescence spectroscopy were performed using a confocal microscope with spectrometer Nanofinder-S (SOLAR TII). Spectra were collected in back-scattering geometry at room temperature (20°C). Diode-pumped solid-state (DPSS) Nd:YAG laser (532 nm, max cw power P_{ex}=150 mW) was used as the excitation source, and

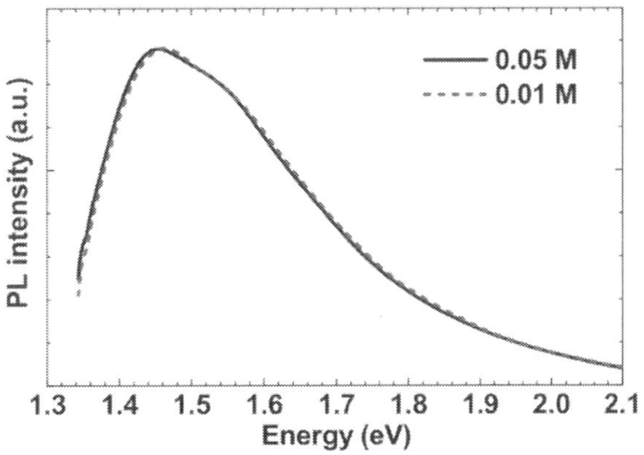

Figure 3. Photoluminescence spectra of cellulose-templated NiO at room temperature. The excitation laser wavelength was $\lambda_{ex} = 532$ nm.

the spectra were dispersed by 150 grooves/mm diffraction grating mounted in the 520 mm focal length monochromator. The elastic laser light component was eliminated by the edge filter (Semrock LP03-532RE). Peltier-cooled back-thinned CCD camera (ProScan HS-101H) was used as a detector in photoluminescence experiments, whereas the Hamamatsu R928 photomultiplier tube (PMT) was employed in confocal-spectral imaging experiments. To avoid possible sample damage, the laser power at the sample was controlled by a variable neutral-density filter with an optical density (OD) in the range OD=0-3.

3. Results and discussion

SEM images of cellulose-templated NiO are shown in figure 1. The grains with a size below 100 nm are observed in the sample produced from 0.01 M nitrate solution. They form filamentary structure surrounding cellulose framework, which was fired during annealing at 650°C. An increase of the nitrate solution concentration up to 0.05 M results in an agglomeration of grains into larger particles with a size of up to 500 nm, and the morphology becomes coral-like.

X-ray diffraction patterns of cellulose-templated NiO are shown in figure 2 (left panel). Five Bragg peaks (111), (200), (220), (311) and (222), due to the fcc NiO phase (JCPDS, No. 04-0835), are well visible. Their positions correspond to the average lattice parameter $a_0 = 4.174 \pm 0.003$ Å, which is in good agreement with the reported data for bulk NiO [18, 19]. The average size of nanocrystallites was calculated from the line broadening of the XRD patterns using the Scherrer's formula, and is equal to about 26–36 nm. This result suggests that the grains, observed by SEM in figure 1 for both samples, are composed of many crystallites.

The Raman scattering spectra of cellulose-templated NiO are shown in figure 2 (right panel). The temperature-dependent [20, 21, 22, 23] and pressure-dependent [24] Raman scattering studies of bulk NiO allow one to identify contributions from one-phonon (TO and LO modes), two-phonon (2TO, TO+LO, and 2LO modes) and two-magnon (2M) excitations. While in face-centered cubic (fcc) structure the first-order Raman scattering is forbidden, the intensity of one-phonon scattering (the band at $\sim$500 cm^{-1}) is expected to increase significantly in defect-rich NiO [20]. In single-crystal, microcrystalline and nanocrystalline (down to 23 nm) NiO the intensity of the one-phonon band is usually smaller [14, 22, 25] or comparable [15, 20] to that of the two-phonon band at $\sim$1100 cm^{-1}. However, a significant gain of the one-phonon band intensity was observed by us for smaller NiO nanoparticles with a size of 13 nm [15]. Thus, an increase of surface/bulk ratio and of defect concentration is responsible for the large intensity of

IOP Conf. Series: Materials Science and Engineering **77** (2015) 012025 doi:10.1088/1757-899X/77/1/012025

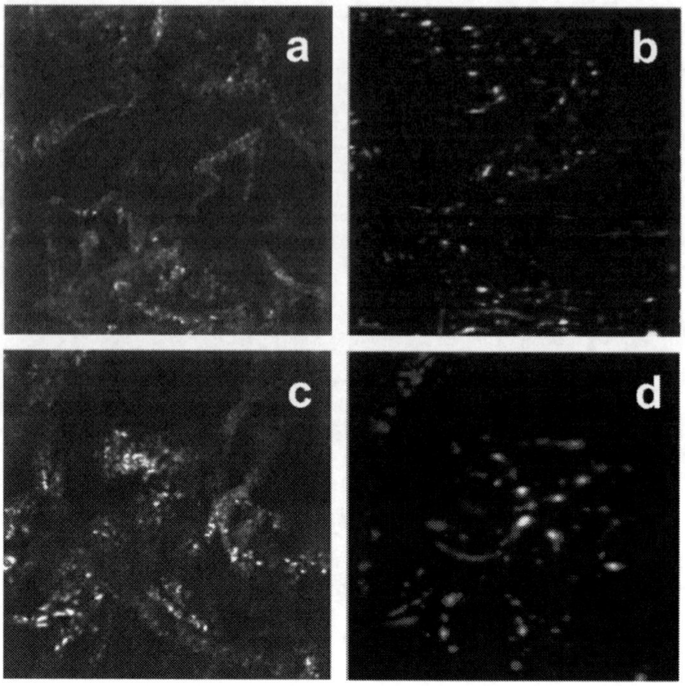

Figure 4. Confocal (a,c) and spectral (b,d) images of cellulose-templated NiO (0.01 M (a,b) and 0.05 M (c,d)). The image sizes are 98 μm×95 μm. The spectral images were acquired at 850 nm.

the one-phonon band at ~500 cm^{-1}. Finally, the presence of the two-magnon contribution (the band at ~1500 cm^{-1}), having intensity comparable to the two-phonon 2LO band at ~1100 cm^{-1}, indicates an existence of antiferromagnetic order at room temperature.

Luminescence of bulk NiO has been rarely studied in the past [13, 26, 27]. Two broad photoluminescence (PL) bands have been detected in [13] at 1.46 eV (850 nm) and 2.4 eV (~513 nm) in green colored NiO and $Ni_cMg_{1-c}O$ (c=0.95, 0.98, 0.99) solid solutions under the low-energy laser excitation. At the same time, two broad emission bands, violet at 3.2 eV and green at 2.6 eV, have been observed under vacuum ultraviolet (VUV) and soft x-ray (XUV) excitations in [27].

Room temperature PL spectra of cellulose-templated NiO, excited by 532 nm laser, are shown in figure 3. The broad near-infrared band peaks at ~1.46 eV (850 nm) and is identical in both samples. The band is similar to that observed by us previously in green colored NiO and $Ni_cMg_{1-c}O$ solid solutions [13]. Since the black color of cellulose-templated NiO is due to the large concentration of nickel vacancies both in the bulk and at the surface [28], we tentatively attribute the origin of the PL band to the defect-perturbed Ni^{2+} states at the surface.

The intensity of near-infrared PL band at ~1.46 eV (850 nm) was further used for mapping of a distribution of photoluminescence centers in cellulose-templated NiO samples. The confocal and spectral images were simultaneously acquired at 532 nm and 850 nm wavelengths, respectively, and are shown in figure 4. The image scanning was performed in the XY directions using two high-speed galvanometer-based mirror scanners, and the optical signals were detected by two PMTs. The confocal images give information on the reflectivity of the sample surface at 532 nm and evidence clearly the filamentary/coral structure of two samples. The spectral images show non-homogeneous spacial distribution of photoluminescence centers: the PL intensity is larger at bright points. Note that the location of bright regions in confocal images differs from that with stronger PL intensity in spectral images, supporting their different origin. The heterogeneous distribution of the PL signal can be explained by particular excitation conditions at the sample points leading to PL nonradiative quenching, for example, due to local temperature variation upon laser heating.

4. Conclusions

Nanocrystaline NiO was synthesized from 0.01 M and 0.05 M nickel nitrate solutions by template-based method using cellulose as a template. Thus obtained NiO powders possess filamentary morphology of the template, have crystallites size in the range of about 26-36 nm and are antiferromagnetically ordered at room temperature. The large amount of nickel vacancies is responsible for the black color of the samples. Finally, the near-infrared photoluminescence band was observed at 1.46 eV (850 nm) in both samples at room temperature under 532 nm laser excitation and is tentatively attributed to the defect-perturbed Ni^{2+} states located at the surface. The distribution of the photoluminescence centers is non-homogeneous as is evidenced by confocal spectromicroscopy.

Acknowledgments

This work was supported by Latvian Science Council Grant no. 187/2012.

References

[1] Huczko A 2000 *Appl. Phys. A* **70** 365
[2] Wade T L and Wegrowe J E 2005 *Eur. Phys. J. Appl. Phys.* **29** 3
[3] Jones M R, Osberg K D, Macfarlane R J, Langille M R and Mirkin C A 2011 *Chem. Rev.* **111** 3736
[4] Rath M K, Choi B H, Ji M J and Lee K T 2014 *Ceram. Int.* **40** 3295
[5] Liu Z, Fan T and Zhang D 2006 *J. Am. Ceram. Soc.* **89** 662
[6] Yan H, Blanford C F, Holland B T, Parent M, Smyrl W H and Stein A 1999 *Adv. Mater.* **11** 1003
[7] Shi J and Wu E 2013 *Micropor. Mesopor. Mater.* **168** 188
[8] Ermakova M A, Ermakov D Y, Lebedev M Y, Rudina N A and Kuvshinov G G 2000 *Catal. Lett.* **70** 83
[9] Qiu Y, Yu J, Zhou X, Tan C and Yin J 2008 *Nanoscale Res. Lett.* **9** 173
[10] Needham S A, Wang G X and Liu H K 2006 *J. Power Sources* **159** 254
[11] Duma V, Popp K, Kung M, Zhou H, Nguyen S, Ohyama S, Kung H and Marshall C 2004 *Chem. Eng. J.* **99** 227
[12] Kim G P, Park S, Nam I, Park J and Yi J 2013 *J. Power Sources* **237** 172
[13] Kuzmin A, Mironova-Ulmane N and Ronchin S 2003 *Proc. SPIE* **5122** 61
[14] Cazzanelli E, Kuzmin A, Mariotto G and Mironova-Ulmane N 2003 *J. Phys.: Condens. Matter* **15** 2045
[15] Mironova-Ulmane N, Kuzmin A, Grabis J, Sildos I, Voronin V I, Berger I F and Kazantsev V A 2011 *Solid State Phenom.* **168-169** 341
[16] Sato H, Minami T, Takata S and Yamada T 1993 *Thin Solid Films* **236** 27
[17] Kohmoto O, Nakagawa H, Isagawa Y and Chayahara A 2001 *J. Magn. Magn. Mater.* **226230** 1629
[18] Slack G A 1960 *J. Appl. Phys.* **31** 1571
[19] Kuzmin A and Mironova N 1998 *J. Phys.: Condens. Matter* **10** 7937
[20] Dietz R E, Parisot G I and Meixner A E 1971 *Phys. Rev. B* **4** 2302
[21] Lockwood D, Cottam M and Baskey J 1992 *J. Magn. Magn. Mater.* **104-107** 1053
[22] Pressl M, Mayer M, Knoll P, Lo S, Hohenester U and Holzinger-Schweiger E 1996 *J. Raman Spectrosc.* **27** 343
[23] Grimsditch M, McNeil L E and Lockwood D J 1998 *Phys. Rev. B* **58** 14462
[24] Massey M J, Chen N H, Allen J W and Merlin R 1990 *Phys. Rev. B* **42** 8776
[25] Mironova-Ulmane N, Kuzmin A, Steins I, Grabis J, Sildos I and Pärs M 2007 *J. Phys.: Conf. Ser.* **93** 012039
[26] Mochizuki S and Saito T 2009 *Physica B* **404** 4850
[27] Sokolov V I, Pustovarov V A, Churmanov V N, Ivanov V Y, Gruzdev N B, Sokolov P S, Baranov A N and Moskvin A S 2012 *Phys. Rev. B* **86** 115128
[28] Ferrari A M, Pisani C, Cinquini F, Giordano L and Pacchioni G 2007 *J. Chem. Phys.* **127** 174711

IOP Conf. Series: Materials Science and Engineering **77** (2015) 012026 doi:10.1088/1757-899X/77/1/012026

Production of glycolic acid from glycerol using novel fine-disperse platinum catalysts

E Sproge[1], S Chornaja[1], K Dubencovs[1], V Kampars[1], L Kulikova[2], V Serga[2] and D Karashanova[3]
[1]Institute of Applied Chemistry, Riga Technical University, Latvia
[2]Institute of Inorganic Chemistry, Riga Technical University, Latvia
[3]Institute of Optical Materialsand Technologies, Bulgarian Academy of Sciences, Bulgaria

E-mail: elina.sproge@rtu.lv

Abstract. Using extractive-pyrolytic method fine-disperse Pt containing composites were synthesized and tested in catalytic glycerol oxidation. Catalyst activity and selectivity to glycolic acid was determined oxidizing glycerol in mild conditions. It was concluded that only iron containing platinum catalysts were selective to glycolic acid. Selectivity to glycolic acid reached 53–60% with glycerol conversion 12–56%.

1. Introduction

Selective heterogeneous oxidation of glycerol is one of the most frequently glycerol transformation methods studied today. Fine-disperse composites consisting of noble metal and its support are found their application as effective catalysts in the glycerol oxidation. In the glycerol oxidation number of valuable products can be obtained and one of these products is glycolic acid [1]. Glycolic acid has a broad application field – it is used in textile, food, cosmetic, pharmaceutical and plastics industries [2]. Conventionally glycolic acid is produced from petroleum resources using different hazardous and complicated methods [2, 3]. More eco-friendly glycolic acid preparation methods are enzymatic processes [4]; the main drawback of these methods is low concentration of glycolic acid formed [5]. Selective liquid-phase oxidation of glycerol also is environmentally friendly and it could be used for glycolic acid production. Until now the best selectivity to glycolic acid (52–70% with glycerol conversion of 30–35%) was reached only using supported gold catalysts, besides supported on carbon nanofibers [6, 7].

In this work using extractive-pyrolytic method, we have synthesized several novel fine-disperse platinum catalysts, which contains iron. Their activity and selectivity was studied during liquid-phase glycerol oxidation and compared with already published catalysts [8].

2. Experimental

2.1. Catalyst preparation

Two different types of catalysts were studied – catalysts supported on carriers (Al_2O_3, Y_2O_3, Lu_2O_3, C, Fe_3O_4), which are published in [8], and novel 71wt%Pt+29wt%α-Fe_2O_3 and Pt/Fe composites. For catalyst preparation platinum containing organic extract was used as precursor. The method for the production of metal-containing organic extract is described in [9, 10]. Amounts of extracts were calculated based on each metal's loading in the composite. For the preparation of supported catalysts, the carriers were impregnated with organic extracts and then the mixture was dried at room temperature and calcinated at 300 °C in air for 5 min. For the preparation of Pt/Fe composite a fine-disperse iron powder was added to Pt-containing organic extract and the mixture was calcinated at 600 °C in air for 30 min. Composite with Fe:Pt molar ratio 0.98:0.02 was obtained ($Fe_{0.98}Pt_{0.02}$). To produce 71wt%Pt+29wt%α-Fe_2O_3 composite an equimolar mixture of iron caproate in caproic acid and tri-n-octylammonium hexachloroplatinate in toluene was used as precursor, which was heated from room temperature up to 500 °C and annealed for 5 min in air.

After preparation catalysts were characterized by X-ray diffraction (XRD) using diffractometer D-8 Advance (Bruker AXS) with CuK_α radiation (λ = 1.5418 Å). The average size of Pt crystallites was calculated by line broadening peak (111) using the Scherrer equation. Specific surface area (S.S.A.) of the powders was measured using HROM-3 chromatograph by BET method at temperature of liquid nitrogen. TEM measurements were carried out by means of transmission electron microscope JEOL JEM 2100 at 200 kV accelerating voltage and at magnifications 40 000–200 000x. Histograms of particle size distribution were obtained by counting onto the micrographs, and the mean particle diameter (d_{Pt}) was calculated using the formula: $d_{Pt} = \sum(d_i \cdot n_i)\cdot\left(\sum n_i\right)^{-1}$, where n_i was the number of particles specific diameter.

2.2. *Catalyst testing*

Novel platinum catalyst activity and selectivity in the glycerol oxidation were tested in thermostatted slurry bubble reactor (50 ml capacity) equipped with gas supply system. Glycerol oxidation process parameters were as follows: c_o(glycerol) = 0.3 M, c_o(NaOH) = 0.7–1.5 M, n(glycerol)/n(Pt) =211–300 mol/mol, 60 °C, P(O_2)= 1atm, reaction time 4–7 h. Analysis of the reaction mixture was performed by high-performance chromatograph UFLC Shimadzu Nexera.

3. Results

3.1. *Catalyst characterisation*

In Figure 1 typical TEM microphotograph of synthesized 4.8wt%Pt/Fe_3O_4 catalyst is presented. The corresponding size distribution histogram of Pt nanoparticles in the 4.8wt%Pt/Fe_3O_4 catalyst is shown in Figure 2. Electronic microscopy revealed that average size of platinum nanoparticles in this catalyst is 2.4 nm (Table 1). Other catalyst characteristic parameters are shown in Table 1.

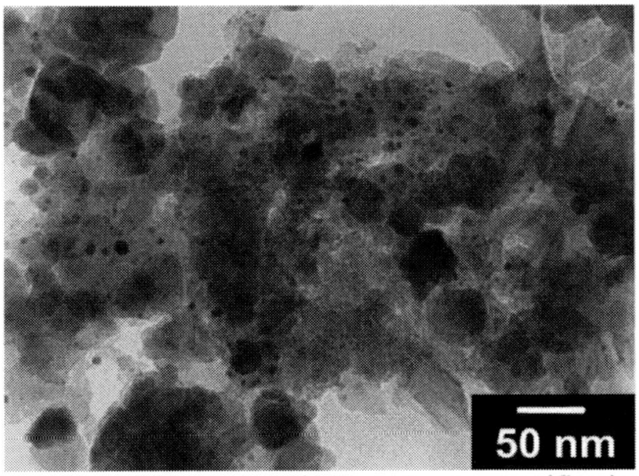

Figure 1. TEM microphotograph at magnification 100 000 x of the 4.8wt%Pt/Fe_3O_4 catalyst.

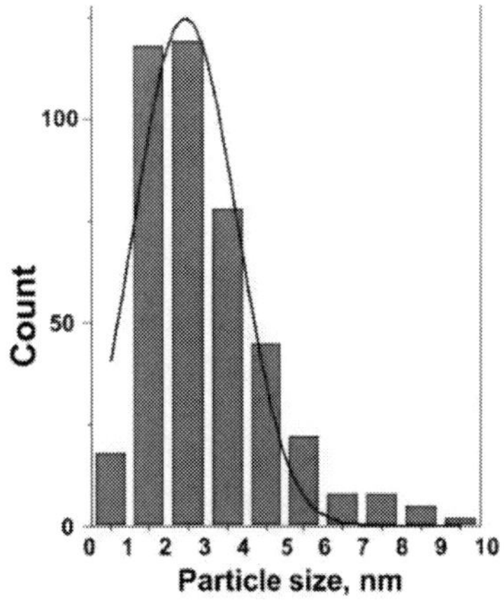

Figure 2. The size distribution histogram of Pt nanoparticles in the 4.8wt%Pt/Fe_3O_4 catalyst.

3.2. Glycerol oxidation

Prepared platinum catalysts were tested in glycerol oxidation in alkaline water solutions. Method is mild and environmentally friendly, because solvent is water, oxidant is molecular oxygen, experiments are carried out in atmospheric pressure and low temperature (60 °C), but the only by-product of the reaction is also water. The results of research are represented in Table 1. As it can be seen from Table 1, all catalysts were active – glycerol conversion was above 20%. Mostly platinum catalysts were selective to glyceric acid whose selectivity varied from 37 to 55%, except catalysts containing Fe. In [8] glycolic acid was firstly observed as the main product of reaction oxidizing glycerol over Pt/Fe_3O_4 catalysts. From Table 1 it can be seen that also novel 71wt%Pt+29wt%α-Fe_2O_3 and Pt/Fe composites had catalytic activity and these catalysts were selective to glycolic acid, unfortunately their activity wasn't as good as for catalysts selective to glyceric acid. Selectivity by glycolic for all Fe containing platinum catalysts was similar (53–60%). Selectivity of glycolic acid didn't depend on glycerol conversion and oxidation time. With increase in glycerol conversion it stayed practically constant. Similar results were reached in [6] – oxidizing glycerol over supported gold catalysts for 1–7 h, selectivity of glycolic acid negligible decreased (for 2%) when glycerol conversion increased from 37 to 100%.

Table 1. Glycerol oxidation over platinum catalysts.

Catalyst	d_{Pt} (nm) by XRD (or TEM*)	S.S.A. of catalyst (m^2/g)	Oxidation time, h	Glycerol conv., mol%	Selectivity, mol% GLYA	TART	LACT	GLYC	OXAL
4.8wt%Pt/Al_2O_3	20	44	7	73	53	14	19	10	2
4.8wt%Pt/Y_2O_3	7	-	7	70	50	8	30	8	2
4.8wt%Pt/Lu_2O_3	23	14	7	65	55	11	20	11	2
4.8wt%Pt/C	20	161	7	68	37	18	30	9	4
2.4wt%Pt/Fe_3O_4	-	34	7	31	37	2	0	60	1
4.8wt%Pt/Fe_3O_4	2.4*	30	7	56	36	2	8	53	1
71wt%Pt+29wt%α-Fe_2O_3[a]	11-35	-	4	31	31	2	10	54	3
			7	47	30	2	9	55	3
Pt/Fe[b]	-	-	4	37	31	2	5	60	3
			7	48	30	2	4	60	2
			4[c]	12	31	3	2	55	8
			7[c]	20	33	2	3	57	5

Reaction conditions: c_0(glycerol) = 0.3 M, c_0(NaOH) = 1.5 M, [c]–0.7 M, n(glycerol)/n(Pt) = 300 mol/mol ([a]–211 mol/mol), P(O_2) = 1atm, 60 °C. Catalyst calcination temperature: 300 °C ([a]–500 °C, [b]–600 °C); catalyst calcination time: 5 min ([b]–30 min). GLYA – glyceric acid, TART – tartronic acid, LACT – lactic acid, GLYC – glycolic acid, OXAL – oxalic acid

To determine influence of NaOH/glycerol molar ratio on Pt/Fe catalyst activity and selectivity, oxidation tests with two different initial NaOH concentrations were carried out when other oxidation parameters were kept constant (Table 1). Initial NaOH concentration 1.5 M corresponds to the NaOH/glycerol molar ratio of 5, while 0.7 M corresponds to the NaOH/glycerol molar ratio of 2.3. As it is known formation of glycolic acid is strongly dependent on amount of H_2O_2 which appears in reaction mixture. There is a perception that H_2O_2 is responsible for C–C bond cleavage, thus leading to the formation of glycolic acid [1, 6, 11, 12]. In [6] measures of H_2O_2 amount showed that in the mixture with greater NaOH concentration, H_2O_2 formed less and selectivity to glycolic acid decreased respectively from 37 to 21%. Glycerol conversion wasn't affected. From Table 1 it can be seen that

IOP Conf. Series: Materials Science and Engineering **77** (2015) 012026 doi:10.1088/1757-899X/77/1/012026

decrease of initial NaOH concentration 2 times led to decrease in glycerol conversion 2 times. Selectivity to glycolic acid reduced, contrary to the results of [6], but the decrease of selectivity was small (3–5%).

4. Conclusions

Novel fine-disperse Fe containing platinum catalysts were selective to glycolic acid oxidizing glycerol in alkaline water solutions. Better results were reached when initial NaOH concentration was greater. The best results (selectivity by glycolic acid 60%, with glycerol conversion 48%) was achieved using Pt/Fe catalyst over following oxidation parameters: c_o(glycerol) = 0.3 M, c_o(NaOH) = 1.5 M, n(glycerol)/n(Pt) = 300 mol/mol, $P(O_2)$ = 1atm, 60 °C. Catalyst calcination temperature: 600 °C; catalyst calcination time: 30 min.

References

[1] Liang D, Gao J, Sun H, Chen P, Hou Z and Zheng X 2011 *Appl. Catal.*, B **106** 423
[2] Hinago H, Nagahara H and Aoki T. Pat. EP 2361900 A2 (31.08.2011)
[3] Zhang Z, Xin L, Qi J, Wang Z and Li W 2012 *Green Chem.* **14** 2150
[4] Datta D and Kumar S 2012 *IJCEA* **3** 141
[5] DiCosimo R, Payne M S, Panova A, Thompson J and O'Keefe D P. Pat. US7198927 B2 (04.03.2007)
[6] Gil S, Marchena M, Sanchez-Silva L, Romero A, Sanchez P and Valverde J L 2011 *Chem. Eng. J.* **178** 423
[7] Gil S, Munoz L, Sanchez-Silva L, Romero A and Valverde J L 2011 *Chem. Eng. J.* **172** 418
[8] Sproge E, Chornaja S, Dubencovs K, Zhizhkun S, Kampars V, Serga V, Kulikova L and Palcevskis E. 2013 *J. Serb. Chem. Soc.* **78** 1359
[9] Cornaja S, Kulikova L, Serga V, Kampars V, Dubencovs K, Zizkuna S and Muravjova O. Pat. EP 2332899 B1 (20.03.2013)
[10] Palcevskis E, Kulikova L, Serga V, Cvetkov A, Cornaja S, Sproge E and Dubencovs K 2012 *J. Serb. Chem. Soc.* **77** 1799
[11] Ketchie W C, Murayama M and Davis R J 2007 *Top. Catal.* **44** 307
[12] Ketchie W C, Fang Y-L, Wong M S, Murayama M and Davis R J 2007 *J. Catal.* **250** 94

IOP Conf. Series: Materials Science and Engineering **77** (2015) 012027 doi:10.1088/1757-899X/77/1/012027

Importance of FTIR Spectra Deconvolution for the Analysis of Amorphous Calcium Phosphates

Agnese Brangule[1,2], Karlis Agris Gross[1]
[1]Institute of Biomaterials and Biomechanics, Riga Technical University, P.Valdena 3, LV-1048,
[2]Riga, Latvia Riga Stradiņš University, Dzirciema 16, LV-1007, Riga, Latvia.

Keywords: amorphous calcium phosphate, bone, spectroscopy, FTIR-DRIFT, deconvolution

Abstract.
This work will consider Fourier transform infrared spectroscopy – diffuse reflectance infrared reflection (FTIR-DRIFT) for collecting the spectra and deconvolution to identify changes in bonding as a means of more powerful detection. Spectra were recorded from amorphous calcium phosphate synthesized by wet precipitation, and from bone. FTIR-DRIFT was used to study the chemical environments of PO_4, CO_3 and amide. Deconvolution of spectra separated overlapping bands in the υ_4PO_4, υ_2CO_3, υ_3CO_3 and amide region allowing a more detailed analysis of changes at the atomic level. Amorphous calcium phosphate dried at 80 °C, despite showing an X-ray diffraction amorphous structure, displayed carbonate in positions resembling a carbonated hydroxyapatite. Additional peaks were designated as A1 type, A2 type or B type. Deconvolution allowed the separation of CO_3 positions in bone from amide peaks. FTIR-DRIFT spectrometry in combination with deconvolution offers an advanced tool for qualitative and quantitative determination of CO_3, PO_4 and HPO_4 and shows promise to measure the degree of order.

Introduction

Amorphous calcium phosphate (ACP) plays an important role in the formation of biomaterials [1]. The diversity of possible arrangements within the apatite structure requires sensitive tools to detect changes from different synthesis conditions, especially for amorphous structures, that are even more difficult to characterize. Fourier transform infrared (FTIR) spectroscopy offers a method for monitoring changes in the chemical groups. This work will show how FTIR-DRIFT spectroscopy provides a simple, non-destructive technique to obtain yet unexplored information about amorphous calcium phosphates.

The aim of this study was to apply curve fitting deconvolution for investigating overlapping bands in the υ_4PO_4, υ_2CO_3, υ_3CO_3 band region in synthesized amorphous calcium phosphates and υ_3CO_3 and amide band in natural human bone. Deconvolution is based on J.K. Kaupinens et. al. methodology introduced in the 1980's [2] – a mathematical procedure for resolving overlapped peaks in a complex FTIR spectrum, also referred to as "resolution enhancement" without changes to the experimental spectral resolution [3].

The vibration band υ_4PO_4 was chosen in this study due to the υ_4PO_4 562 cm^{-1}, 575 cm^{-1} and 603 cm^{-1} peak overlap with the HPO_4 apatitic peak at 550 cm^{-1} [4]. Chemical determination of HPO_4^{2-} ions by colorimetry cannot be distinguished from PO_4^{3-}. Another method by Gee and Dietz [5] that condenses HPO_4^{2-} ions into pyrophosphates $P_2O_4^{4-}$ cannot be used for powders containing carbonate CO_3^{2-} ions. Carbonate ions interfere with $P_2O_4^{4-}$ ions and partially prevent $P_2O_4^{4-}$ ion formation [6]. Consequently, this spectral method is thus the only viable technique for determining the HPO_4^{2-} content in carbonated ACP.

IOP Conf. Series: Materials Science and Engineering **77** (2015) 012027 doi:10.1088/1757-899X/77/1/012027

Carbonate ion concentration can be determined by heating ACP to 1200 °C to liberate CO_2 for collection in an absorption cell. Carbonates can then be determined by calorimetric titration [7]. This analysis method only gives information on the quantity of carbonates. Deconvolution of υ_2CO_3 and υ_3CO_3 bands in FTIR spectra gives information about the apatite structure. Le Geros observed and interpreted IR peak splitting for the CO_3 to represent A and B type carbonate substitution models – representing OH and PO_4 replacement respectively [8]. Some researchers also reported A1 type, A2 type and B type substitution doublets [9-11] or A-B type apatites, and "nonapatitic" or "labile" CO_3^{2-}, PO_4^{3-} and HPO_4^{2-} [12,13].

Human bone consists mainly of collagen fibers and inorganic compounds, which can be approximated as carbonate containing hydroxylapatite [14,15]. Deconvolution of FTIR spectra will be shown to give information about overlapped amide I, amide II and amide III peaks in collagen, and inorganic υ_3CO_3 bands in the 900 – 1900 cm^{-1} region [15].

Materials and Methods

Materials

Three different carbonate containing ACPs were synthesized at room temperature and dried for FTIR analysis. The first two powders were prepared by mixing a solution containing $Ca(NO_3)_2$ and 30 % ammonia with another solution consisting of $(NH_4)_2HPO_4$ and $(NH_4)_2CO_3$. Drying of these first two wet powders was conducted in air at 20 °C for 72h, or in a convection oven at 80 °C for 24h. The third ACP was a Zn enriched ACP was prepared by mixing the solution ($Ca(NO_3)_2$, $Zn(NO_3)_2$ and 30 % ammonia) with $(NH_4)_2HPO_4$ and $(NH_4)_2CO_3$ solution followed by freeze-drying at -50 °C for 48 h.
Bone was sourced from human vertebra - stored at room temperature in air for 40 years.

Methods

Fourier transform infrared spectroscopy – *diffuse reflectance infrared reflection.* The functional group changes were determined by FTIR–DRIFT (PerkinElmer Spectrum One,) at 450–4000 cm^{-1}, at a resolution of 4 cm^{-1}, with an average of 8 scans, sampled by diamond sampling sticks. A special preparation method was not required for synthesized powders. Powder for bone analysis was obtained from the inside of bone vertebra.

Table 1. Domains, assignments and literature for nanocrystalline apatite

Domain, Assignments	IR (cm^{-1})	Reference
HPO_4 apatitic	551	12
υ_4PO_4	562, 575 and 603	12
PO_4 non-apatitic	617	12
non-apatitic	866	12
υ_2CO_3 type B	871	12
υ_2CO_3 type A	880	12
υ_2CO_3 type B	1460-1470	12
υ_2CO_3 type B doublet	~1455 and ~1410	11
non-apatitic	1500	12
type A	1540	12
υ_2CO_3 type A1 doublet	~1540 and ~1455	11
υ_2CO_3 type A2 doublet	~1565 and ~1505	11
B + non-apatitic	1420	12
Amide I	~1660 – 1690	15
Amide II	~1500 – 1600	15
Amide III	~1242	15

X-ray powder diffraction (XRD). Powders were characterized by X-ray diffraction (Bruker D8 ADVANCE diffractometer). Diffraction patterns were recorded from 5° to 60° using Cu K radiation (λ = 1.54 Å generated at 40 mA and 40 kV) at a step size of 0.2°. Analysis. The FTIR spectra were viewed and smoothed with freeware software Specwin32. Baseline correction and curve-fitting analysis was performed using MagicPlotStudent software with wavenumbers from literature as a guide for υ4PO4, υ2CO3, υ3CO3, amide I, amide II, amide III groups (Table 1). Deconvolution involved both Lorentzian and Gaussian curve fitting.

Results and Discussions

All synthesized ACPs displayed an X-ray amorphous structure - a broad low intensity peak centred at 30°, Fig 1. All three patterns appear similar.
FTIR-DRIFT spectra of three ACPs are shown in Fig 2. Phosphate bands (υ_1, υ_2 and υ_4PO_4) and carbonate bands (υ_2, υ_3CO_3) were detected. By using information from the literature (Table 1) we can infer that PO_4 bands overlap HPO_4 bands. Only small shifts in the IR spectra suggest a change in the chemical group placement.

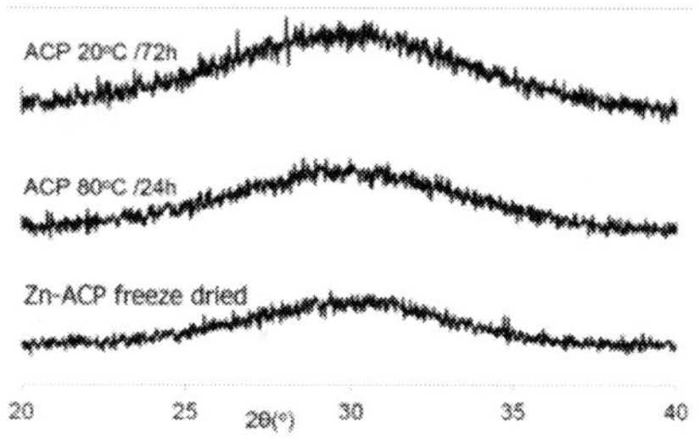

Fig 1. XRD patterns of air dried ACP, 80 °C dried ACP, freeze-dried Zn-ACP.

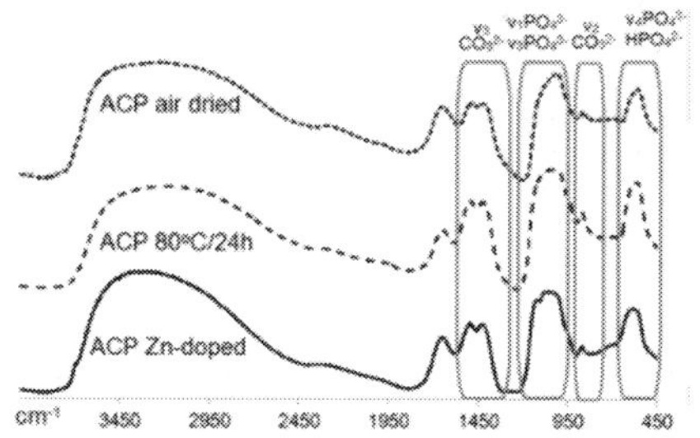

Fig 2. FTIR-DRIFT spectra of air dried ACP, 80 °C dried ACP, and freeze-dried Zn-ACP

Deconvolution of the 500-700 cm^{-1} region of all three FTIR spectra revealed the presence of HPO_4^{2-} (550 cm^{-1}) only in Zn-doped ACP (Fig 3).

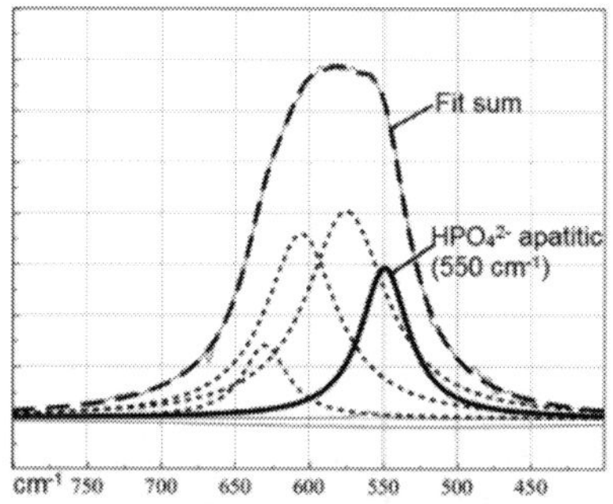

Fig 3. Deconvolution of Zn doped ACP showing $\upsilon_4PO_4^{3-}$ and HPO_4^{2-} band (450—700 cm^{-1})

Spectra obtained in the υ_2CO_3 band region (800 – 920 cm^{-1}) show a considerable change in peak shape and position. A shoulder at ~880 cm^{-1} was observed for the 80 °C dried powder, that was not present for the air dried powder. Using peak assignments from the literature, we can infer the presence of an A type substituted carbonate group. More detailed information on the structure of apatites can be obtained after deconvolution of the spectra. Deconvolution of the 800 – 920 cm^{-1} region displays only non-apatitic carbonate in air-

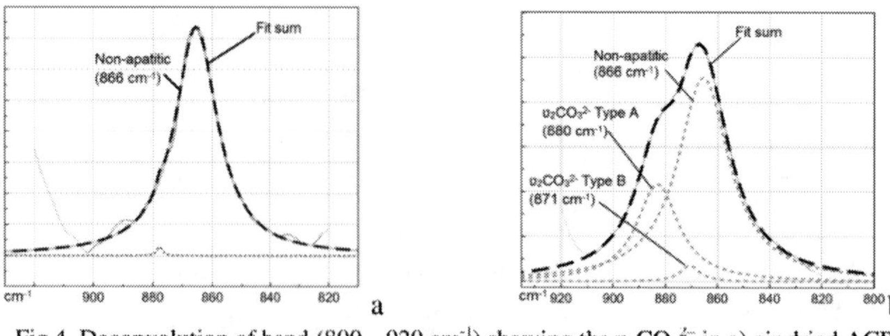

Fig 4. Deconvolution of band (800—920 cm^{-1}) showing the $\upsilon_2CO_3^{2-}$ in a) air dried ACP and b) 80°C dried ACP

dried ACP but additional A-type and B-type carbonate for 80 °C dried powders (Fig 4a and b). The carbonate type can be changed by drying at different conditions [16]. Air-dried powder mainly contained non-apatitic CO_3^{2-}, but drying at 80 °C produced more A-type and B-type carbonates as shown by peaks at 880 cm^{-1} and 871 cm^{-1}, respectively.

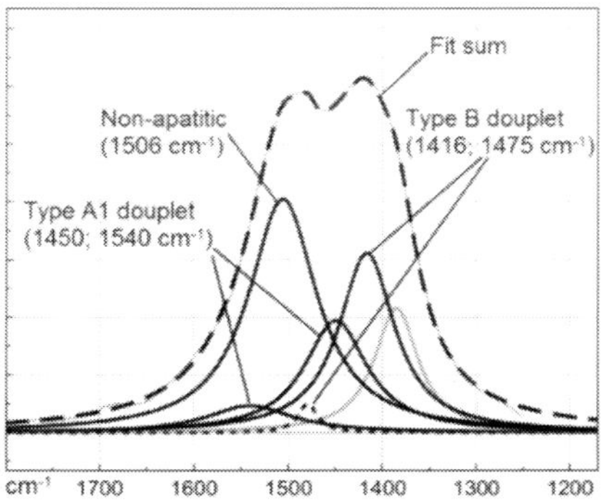

Fig 5. Deconvolution of $\upsilon_3CO_3^{2-}$ band (1200—1800 cm^{-1}) for 80 °C dried ACP

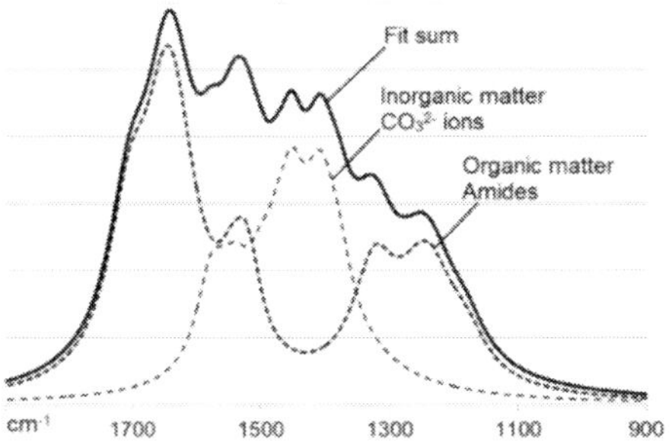

Fig 6. Deconvolution of human bone showing convoluted organic amide and inorganic carbonate bands (900 —1900 cm^{-1})

Additional information on carbonates powders can be sourced from the υ_3CO_3 (1400 - 1600 cm^{-1}) region. Higher crystallinity in relation to carbonates is interpreted by more carbonate occupying the A-type or B-type positions, characteristic of crystalline apatite. Poorly crystalline powder displayed an A 2 carbonate peak (at 1506 cm^{-1}) overlapping a non-apatitic CO_3^{2-} (at ~1500 cm^{-1}), Fig 5. The CO_3^{2-} is not easily determined, but the different CO_3^{2-} positions becomes clear from the numerous peak positions.

The most benefit from deconvoluting FTIR spectra is gained for human bone, Fig 6. The 1000 – 1900 cm^{-1} region has an overlap of amide and CH_2 peaks with carbonate peaks. Deconvolution of the FTIR spectra show convoluted organic amide bands, Fig 7. and inorganic carbonate υ_3CO_3 bands, Fig 8. Deconvolution of the organic band shows that Amide I, Amide II and Amide III bands dominate in this region. Deconvolution of the inorganic υ_3CO_3 bands shows a type A1 doublet, a type A2 doublet and a type-B doublet in this region. Two further peaks were added so that the fit sum peak corresponds to the recorded peak. A type-A2 substituted carbonate can be identified from the broad shoulder at ~1565 cm^{-1}.

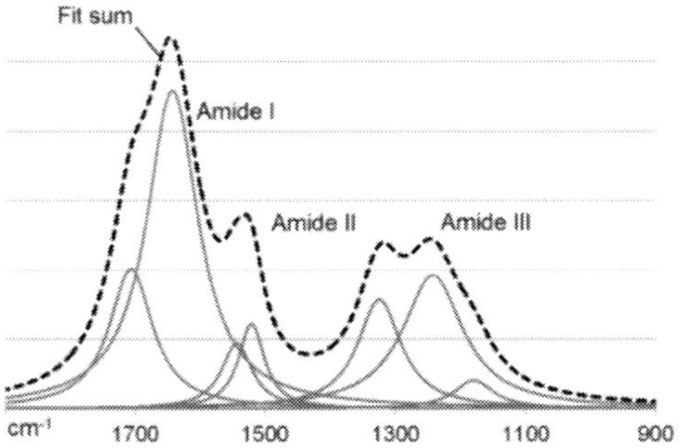

Fig 7. Deconvolution of human bone showing deconvoluted and convoluted amide band

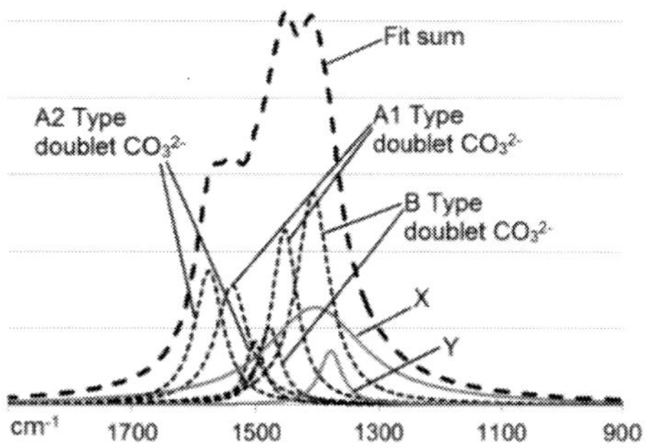

Fig 8. Deconvolution of human bone showing deconvoluted and convoluted carbonate band

Conclusions

Deconvolution of FTIR spectra offers a more detailed qualitative and quantitative analysis of carbonate ("non-apatitic" vs "apatitic") and orthophosphates (HPO_4^{2-} and PO_4^{3-}). The carbonate band (1400 - 1600 cm^{-1}) for the air dried powder shows non-apatitic positions, but drying at 80 °C showed peaks that coincided with crystalline carbonated apatite.

FTIR-DRIFT provides a sensitive, non-destructive tool for the analysis of bone, and with deconvolution can distinguish the inorganic carbonate bands from the organic amide bands.

Deconvolution of ACP spectra provides a new ability that may be used to track the transition of the amorphous phase to a crystalline apatite in bones as well as providing a tool to establish the influence of processing and drying conditions on the state of the amorphous calcium phosphate. The change in the bonding marks the first change from an amorphous to a crystalline state and so this tool will be able to bridge the gap in understanding between the amorphous state and nanocrystalline states.

Acknowledgements. A special thanks to M. Pilmane from the Riga Stradins University of Latvia, for providing the vertebra bone for FTIR analysis.

References

[1] LeGeros R Z In Brown P W 1994 Constantz B (Eds.) Hydroxyapatite and Related Materials 3-28
[2] Kauppinen J K *et al* 1981 *Appl. Spectrosc.* **35** 271–76
[3] Barth A Harris P 2009 Infrared spectroscopy – Past and Present. In: Biological and Biomedical Infrared Spectroscopy 1–52
[4] Rey C 1990 *et al Calcif. Tissue Int.* **46** 384-94
[5] Gee A Dietz V R 1953 *Ann. Chem.* **25** 1320-1324
[6] Elliott J C 1994 Structure and Chemistry of the Apatites and Other Calcium Phosphates
[7] Markovic M *et al* 2004 *J. Res. Natl. Inst. Stand. Technol.* **109** 553-568
[8] LeGeros R Z *et al* 1969 **25** 5-7
[9] Elliott J C *et al* 2002 *Adv. X-Ray Anal.* **45** 172-181
[10] Wopenka B Pasteris J D 2005 *Materials Science and Engineering*: *C* **25** 131–143
[11] Fleet M E *et al* 2004 *Am. Mineral.* **89** 1422-1432
[12] Eichert D *et al* 2007 J.B. Kendall (Eds.), Nova Science Publishers 93-145
[13] Rey C *et al* 1989 *Calcif. Tissue Int.* **21** 267-273
[14] Farre B *et al* 2014 *J. Afr. Earth Sci.* **92** 1-13
[15] Figueiredo M M *et al* 2013 *J. Struct Biol* **181** 207–222 doi:10.1016/j.jsb.2012.12.005.
[16] Brangule A Gross K A 2015 *Key Eng. Mat.* **631** 99-103, doi:10.4028/www.scientific.net/KEM.631.99

RCBJSF–2014–FM&NT IOP Publishing
IOP Conf. Series: Materials Science and Engineering **77** (2015) 012028 doi:10.1088/1757-899X/77/1/012028

Determination of methemoglobin in human blood after ionising radiation by EPR

M. Polakovs[1], N. Mironova-Ulmane[1], A. Pavlenko[1], A. Aboltins[2]

[1]Institute of Solid State Physics, University of Latvia, Riga, Latvia
[2]Nuclear Medicine Department, P.Stradins Clinical University Hospital, Riga, Latvia
e-mail: maksims.polakovs@gmail.com

Abstract. In the present work presents results of investigations of radiation influence on blood of patients examined by radio-isotopes diagnosis (Tc^{99m}), blood of Chernobyl clean-up workers and human blood irradiated by LINAC using Electron Paramagnetic Resonance (EPR). The EPR spectroscopy reveals information on electronic states of transition metal ions, particularly Fe^{3+} in different spin states. It is shown that EPR spectra of blood of patients before examination has signal from metal-protein transferrin (g=4.3) and after administration of radio-isotope proves signal of Fe^{3+} (methemaglobin) in the high spin state (g=6.0). The EPR spectra of Chernobyl liquidator display number of signals including low and high state of ion Fe^{3+} (g = 2.0 and g= 6.0), and transferrin (g=4.3). The EPR spectra of irradiated human blood by LINAC (linear accelerator) have only signal Fe^{3+} (methemaglobin) in low-spin state with g = 2.0.

1. Introduction

Paramagnetic centres in human blood include primarily the molecular complexes containing iron Fe^{3+} (transferrin, methemoglobin) or copper Cu^{2+} ions (ceruloplasmin) and free radicals [1]. EPR spectroscopic studies of iron and copper proteins have been published by Taiwo [2]. The unique ability of in vivo EPR to measure clinically significant exposures to ionizing radiation has been demonstrated by group of scientists [3]. Studies [4-10] were dedicated to explore radiation damage to blood and hemoglobin. In these studies, various physical techniques have been used to determine structural variations of hemoglobin after irradiation: EPR spectroscopy [4, 10], Mossbauer [9] and IR spectroscopy [10], absorption spectroscopy [11]. The new data on structural changes of oxyhemoglobin under irradiation have been investigated by Mössbauer spectroscopy to estimate the probability of protein degradation products resulting from the effect of external factors on the oxyhemoglobin and scheme of oxyhemoglobin radiolysis process proposed [9]. Confocal micro-Raman and FT-IR spectroscopies have been used for detection of radiation influence of hemoglobin of patients examined by radio-isotopes diagnosis (Tc^{99m}). After irradiation some tiny changes of the Raman scattering bands which connected with out-of-plane porphyrine bending vibrations have been observed. Additionally, several scattering bands attributed latter to methemoglobin have been detected [11]. It was assumed that radiation exposure of blood leads to transition from hemoglobin (Fe^{2+}) to methemoglobin (Fe^{3+}) with a delocalization of iron from porphyrine ring plane [11]. In the present work results of EPR studies of radiation influence on blood of patients examined by radio-isotopes

IOP Conf. Series: Materials Science and Engineering **77** (2015) 012028 doi:10.1088/1757-899X/77/1/012028

diagnosis (Tc^{99m}), of Chernobyl clean-up workers and irradiated human blood by LINAC are presented.

2. Materials and Methods

Three different ways of blood irradiation and follow-up studies have been performed. Part of the blood samples has been received from the P. Stradins Clinical University Hospital of Latvia, where patients underwent bone diagnostics by Tc^{99m} radioisotope. Venous blood has been donated by consenting patients before and after radioisotope administartion and collected freely in air in the glass tubes containing anticoagulant (heparin or EDTA). Administrated activity of Tc^{99m} to the patient had 600-1000 MBq range individually prescribed. Blood without any anticoagulant was also tested and confirmed no difference to EPR spectra with respect to EDTA or heparin in blood. Activity of injected isotope Tc^{99m} to the patient has been measured by dose calibrator "Curiementor 4" Physikalisch-Technische-Werkstätten (PTW). The calibrator has been proved against the Physikalisch Technische Bundesanstalt (PTB) traceable standard with uncertainty less than 5%.
The other part of blood samples have been collected from Chernobyl liquidators undergoing medical examination at the Centre of Occupational and Radiological Medicine of P.Stradins Clinical Hospital of Latvia. Some liquidators have documented individual dose in the dose register, some having individual dose reconstructed based on beta activity of the teeth [12].
Few blood samples have been voluntarily donated by people from our laboratory group. These samples have been irradiated on LINAC with distinctive doses 0.5 Gy, 1.0 Gy, 2.0 Gy, 5.0 Gy, 10.0 Gy of 6 MV x-rays (photon beam). The absorbed dose values are traceable to BIPM (Bureau International des Poids et Mesures).
The EPR spectra of blood have been measured on a BRUKER EMX-6/1 spectrometer equipped with an Aspect 2000 data system. The spectra have been recorded at microwave power 6.2 mW, applying magnetic field modulation of 100 kHz and amplitude 1 mT. The g-factors of EPR signals have been determined by reference to the external magnetic field value measured by a BRUKER ER 035 Gaussmeter and of the microwave frequency measured by a SYSTRON DONNER 6235A frequency counter. EPR spectra of blood have been measured at temperatures 60 - 200 K using Dewar mounted to the X-Band waveguide type cavity, by fully remote operated control system via the acquisition software WinEPR of the EMX spectrometer. The Digital Temperature Control System, ER 4141VT makes use of gaseous nitrogen as coolant. Additionally, the EPR spectra of fresh blood have been measured at room temperature using the "AquaX" cell. The EPR signal intensities in blood have been measured against standard EPR signals with known spin number using the MgO (Cr^{3+}) standard crystal placed in the resonant cavity.

3. Results and Discussion

The ion of iron in the blood commonly occurs as ferrous Fe^{2+} with electronic configuration Ar ($3d^6$) in one of two different spin states (see Fig.1). When ion of iron Fe^{2+} in porphyrin ring of hemoglobin binds the oxygen it changes the spin state. Actually ion of iron Fe^{2+} is in high spin state (S=2) in dioxyhemoglobin (venous blood) and low spin state (S=0) in oxyhemoglobin (arterial blood) (see Fig.1).
Ionizing radiation change valence of iron from Fe^{2+} to Fe^{3+}. Ions of Fe^{3+} have electronic configuration Ar ($3d^5$) in the high S=5/2 and low S=1/2 spin states (see Fig.2). Human blood contains number of non-heme ions of iron e.g. blood plasma glycoprotein transferrin, which unable to transport oxygen but used as iron depot in the body.

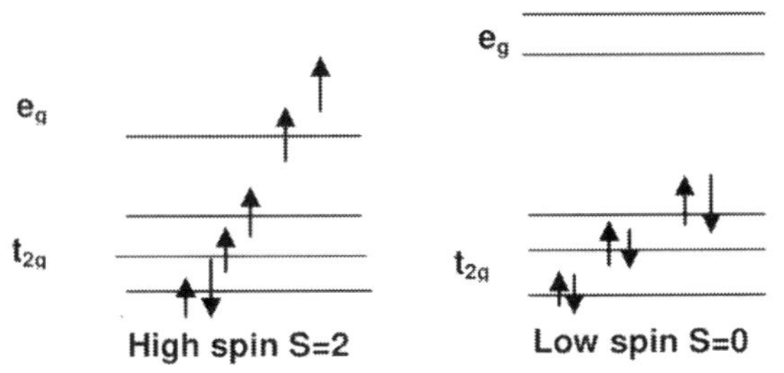

Figure 1. Electomic configuration of Fe^{2+} (ferrous) [Ar]$3d^6$ in low and high spin states in hemoglobin.

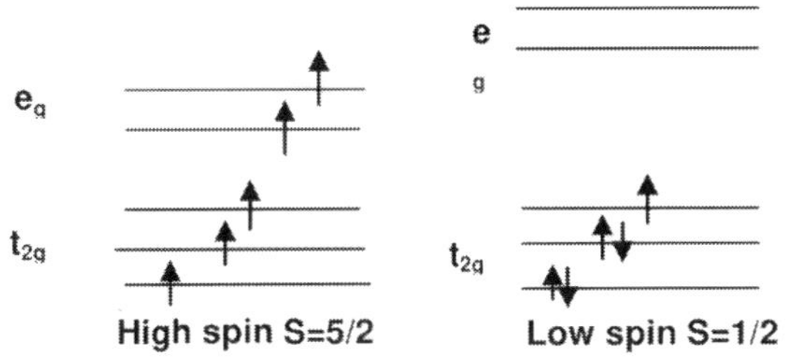

Figure 2. Electomic configuration Fe^{3+} low and high spin state Fe^{3+} (ferric) [Ar]$3d^5$.

The EPR measurements have been done before and after Tc^{99m} administration for bone diagnosis (see figure 3.). Taking into account the activity (600- 1000 MBq) and type of nuclide injected (Tc^{99m}) one could estimate the dose absorbed by the blood [13]. The assessed blood absorbed dose was in the range 50 – 100 mGy. The EPR spectra before Tc^{99m} diagnosis have signal with g-factor 4.3 related to Fe^{3+} ion in high spin state (transferrin). Our investigations showed that not all samples of human blood of none-irradiated blood have EPR signal with g-factor 4.3. Elevated amount of transferrin could be attributed to the individuals having iron deficiency in their organisms. The EPR measurements of samples after Tc^{99m} diagnosis showed none difference in EPR signal with g-factor 4.3 related to transfferin, but demonstrates very high rise in signal with g-factor 6.0 related to Fe^{3+} ion in high spin state (S=5/2, originated from dioxyhemoglobin).

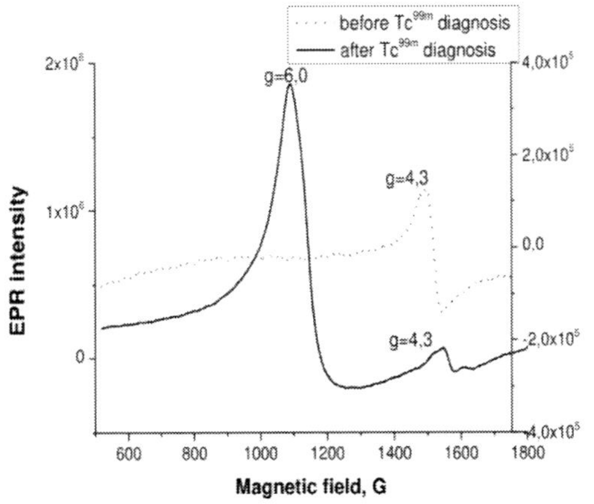

Figure 3. EPR spectra of human blood before and after Tc^{99m} diagnosis.

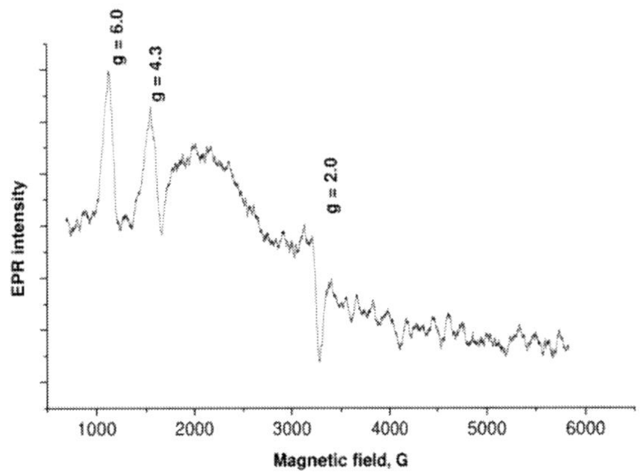

Figure 4. EPR spectrum of blood of Chernobyl clean up worker. (Documented dose in register 0.211 Gy. $^{90}Sr/^{90}Y$ activity measured on teeth in 1997 - 80 Bq/g).

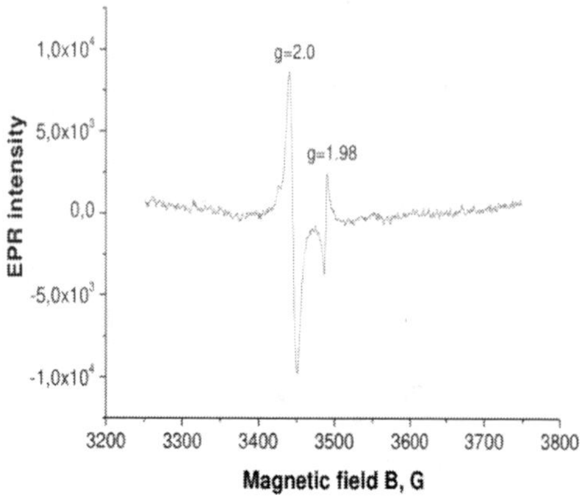

Figure 5. EPR spectrum of human blood after radiation 1 Gy by LINAC.

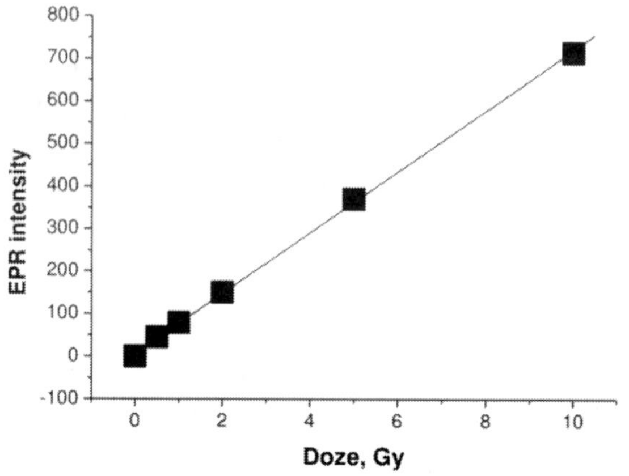

Figure 6. Dependence of EPR signal (g=2.00) intensity of irradiated blood by LINAC.

The EPR spectra for blood samples from Chernobyl clean-up workers have been obtained as well (some of the spectra see on Fig. 4). The blood of Chernobyl clean-up workers with increased level of methemoglobin have been selected for further studies. It is necessary to point out, that obtained EPR spectra have several signals with g-factors 2.0 (methemoglobin and free radical), 4.3 (transfferin), 6.0 (methemoglobin). Overview of results of the measurements let us to assume that some of Chernobyl clean-up workers have methemoglobin content (Fe^{3+} in high spin state) higher than average population. The case of individual presented on Figure 4 suffered both on iron deficiency (high amount of transferrin) and at the same time on anaemia (high amount of methemoglobin). The supposition raised during the studies was that ion Fe^{2+} in hemoglobin is oxidized to the Fe^{3+} in heme by internal radiation exposure. The internal irradiation is resulted from $^{90}Sr/^{90}Y$ and probably other radionuclides incorporated in tooth and further calcified tissues during clean-up activities in Chernobyl. Correlation between EPR signal of methemoglobin and measured activity of the $^{90}Sr/^{90}Y$ in tooth of Chernobyl clean-up workers has been observed. The excess of methemoglobin is reported to the other Chernobyl clean-up workers having high documented doses in dose register [12, 13].

Part of human blood samples have been irradiated by LINAC with 6 MV x-rays (photon beam). Following doses have been delivered to the samples 0.5 Gy, 1.0 Gy, 2.0 Gy 5.0 Gy, 10.0 Gy. The EPR spectra of irradiated samples have only one notable signal with g-factor 2.0 (see Fig. 5), free radical or delocalized electron. The EPR signal with g-factor 2.0 attributed to Fe^{3+} ion with electronic configuration at low spin state (S=1/2, methemoglobin), which was originated from oxyhemoglobin. This fact could be explained that human venous blood has been collected in the sample tubes containing air for future irradiation by LINAC. During sample collection and transport venous methemoglobin bounded the oxygen in the tube, becoming an oxyhemoglobin, which under irradiation turned again to methemoglobin. Additional irradiation of blood samples indicated good linear dependency of EPR signal intensity from the dose received by the blood (see Fig. 6).

4. Conclusion

The samples of human blood before and after three different ways of irradiation have been studied. Blood samples donated by patient involved in Tc^{99m} bone diagnostic have notable EPR signal due to Fe^{3+} ion at high spin states in methemoglobin. Very weak EPR signal of Fe^{3+} ion at low spin states in methemoglobin of patient diagnosed using Tc^{99m} is subject for further investigation. EPR signal at g=4.3 attributed to blood plasma glycoprotein transferrin has been detected as well.
The EPR spectra of Chernobyl clean-up workers have numerous signals related to Fe^{3+} ion in high and low spin states, transferrin and unpaired electron or free radical. The observed correlation between level of methemoglobin in blood of Chernobyl clean-up workers, documented individual dose and/or activity of $^{90}Sr/^{90}Y$ content in teeth provided certain physical background for explanation of diseases prevalence among Chernobyl liquidators compare to the general population [12,14].
The EPR spectra of irradiated blood on LINAC with different doses of 6 MV megavoltage x-rays have signal of Fe^{3+} ion at the low-spin state in methemoglobin. Linear dependence between EPR signal intensity and blood expose has been observed.
Comparing results on EPR spectra intensity of Fe^{3+} ion before and after irradiation one can conclude the different types of radiation could lead to the same effect, enhanced level of methemoglobin after exposure. Ionizing radiation can cause change of hemoglobin to methemoglobin or in other words ion Fe^{2+} heme in hemoglobin is oxidized to the Fe^{3+} in heme by radiation. Iron ion changes the valence due to influence of ionizing radiation, but retains spin state configuration.

Acknowledgments

This work was supported by the Latvian State Research Program.

References

[1] Kubiak , Krzyminiewski R and Dobosz B 2013 *Current topic in Biophysics*, **36**, 7-13
[2] Taiwo F 2003 *Spectroscopy* **17** 53–63
[3] Swartz H M, Khan N, Buckey J,Comi R,Gould L,Grinberg O, Hartford A, Hopf A, Hou H, Hug E, Iwasaki A, Lesniewski P and Salikhov I 2004 NMR Biomed.**17**:335–351
[4] Pulatova M K, Sharygin V L, Shlyakova T G, Sipyagina A E and Wasserman A M 2009 *Biophysics* **54** 323
[5] Dreval V I and Sichevskaia L V 2000 *Biofizika* **45**(6), 1086–1088
[6] Geoffrey P J 1998 *Radiat. Phys. Chem.* 53 (**5**), 511–523
[7] Puchala M, Szweda-Lewandowska Z and Kiefer J 2004 J. *Radiat. Res.* (Tokyo) 45 (**2**) 275–279
[8] Holman H-Y N, Goth-Goldstein R, Blakely E A, Bjomstad K, Martin M C and McKinney W R *2000 Proceedings of SPIE*, 57-63
[9] Oshtrakh M I *Nucl. Instr. and Meth. in Phys.* Res. B 185 129–135
[10] Ahmed M M and Maha A A 2007 *Radiation Physics and Chemistry* 76 1600–1605
[11] Polakovs M, Mironova-Ulmane N, Kurjane N, Reinholds E, Grube M, 2008 *Proc. SPIE* 714214:1-8
[12] Mironova-Ulmane N, Pavlenko A and Zvagule T 2001 *Radiat. Prot. Dosim.* 96:237-240
[13] International Commission on Radiological Protection ICRP report 53. London, UK: ICRP; 1988:215
[14] Mironova-Ulmane N, Pavlenko A, Eglite M 2005 *in book. Recent Advances in Multidisciplinary Applied Physics*, pp.9-19, Elsevier (ISBN 0 08 0444 696-5)

IOP Conf. Series: Materials Science and Engineering **77** (2015) 012029 doi:10.1088/1757-899X/77/1/012029

ODS steel raw material local structure analysis using X-ray absorption spectroscopy

A Cintins[1], A Anspoks[1], J Purans[1], A Kuzmin[1], J Timoshenko[1], P Vladimirov[2], T Gräning[2], and J Hoffmann[2]

[1] Institute of Solid State Physics, University of Latvia, LV-1063 Riga, Latvia
[2] Institute for Applied Materials-Applied Materials Physics, Karlsruhe Institute of Technology, Eggenstein-Leopoldshafen, PO Box 3640, 76021 Karlsruhe, Germany

E-mail: `cintinsarturs@gmail.com`

Abstract. Oxide dispersion strengthened (ODS) steels are promising materials for fusion power reactors, concentrated solar power plants, jet engines, chemical reactors as well as for hydrogen production from thermolysis of water. In this study we used X-ray absorption spectroscopy at the Fe and Cr K-edges as a tool to get insight into the local structure of ferritic and austenitic ODS steels around Fe and Cr atoms and its transformation during mechanical alloying process. Using the analysis of X-ray absorption near edge structure (XANES) and extended X-ray absorption fine structure (EXAFS) we found that for austenitic samples a transformation of ferritic steel to austenitic steel is detectable after 10 hours of milling and proceeds till 40 hours of milling; only small amount of α-phase remains after 80 hours of milling. We found that the Cr K-edge EXAFS can be used to observe distortions inside the material and to get an impression on the formation of chromium clusters. In-situ EXAFS experiments offer a reliable method to investigate the ferritic to austenitic transformation.

1. Introduction

Oxide dispersion strengthened (ODS) steels have exceptional thermal conductivity and low thermal expansion while demonstrating high-temperature creep, corrosion and irradiation resistance. Therefore, they have a great potential for an application as structural materials for concentrated solar power plants, jet engines, chemical reactors as well as for hydrogen production from thermolysis of water. Previous developments were focused mainly on nanostructured ODS ferritic (Fe-Cr alloys) steels with very promising applications in, e.g., fusion power reactors up to about 550°C [1].

Recently the use of the austenitic ODS steels for high temperature fusion applications and concentrated solar power plants has been suggested by the group from Karlsruhe Institute of Technology (KIT) in [2]. It is expected that the dispersion of nanoscaled oxide particles with still unknown (x)Y_2O_3-(1-x)TiO_2 structure will possibly suppress the gas bubble growth and related void swelling under neutron irradiation [3, 4]. In addition ODS particles could reduce irradiation hardening and increase high temperature resistance up to 800°C. Moreover, austenitic steels are non-magnetic and do not suffer from a ductile-brittle-transition, which are typical drawbacks of ferritic/martensitic steels [5].

ODS steels have been studied by different characterization techniques, including transmission electron microscopy (TEM), atom probe tomography (APT) and small angle neutron scattering

Table 1. Nominal chemical compositions of the ferritic (F) and austenitic (A) samples with different milling time.

Sample	Milling time (h)	Fe	Cr	Ni	W	Ti	Y
F0h	0	85.7	13	-	1.0	0.3	-
F40h	40	78.45	20	-	0.9	0.25	0.4
A40h	40	69.8	16	13	0.8	0.1	0.3

[6]. The use of both TEM and APT is strongly restricted to a limited number of particles or by a very small probed region as well as by the challenging requirements for sample preparation. In contrast, synchrotron radiation X-ray absorption spectroscopy (XAS) provides a practical way to study elements of matrix and minority, giving unique and direct information on their chemical and physical state in the sample [7, 8].

The aim of this work was to study powders of ferritic and austenitic ODS steels produced by a mechanical alloying process with X-ray absorption spectroscopy, which gives insight into local structure around Fe and Cr atoms. We have studied the dependence of the local structure on the time of mechanical alloying as well as on the thermal treatment.

2. Experimental and data analysis

In our study we used two groups of samples: ferritic and austenitic ODS steels, which are further labeled with F and A, respectively.

All samples were prepared using a mechanical alloying process in a water-cooled attritor ball mill (Zoz CM02) in argon atmosphere using Zr_2O_3 as a grinding media. A ball-to-powder ratio (BPR) of 10:1 was chosen. The milling time was in the range of 0 to 40 hours for ferritic steels and from 0 to 80 hours for austenitic steels. After milling (i.e. 10, 40 or 80 hours), the powder was taken out of the container and stored in an Argon-filled glove-box. All powders were packed in argon environment and thus transported to the synchrotron for XAS measurements.

A prealloy with a chemical composition of FeCr13W1Ti0.3 (F-prealloy, F0h) was used to produce a ferritic ODS steel. Hence, Fe_3Y powder was added to this powder and mechanically alloyed for 40 hours to produce the F samples (e.g. F40h). To produce the A-samples, a two-step process is required. In the first step Y_2O_3 with particle size about 1 μm was added to F-prealloy instead of Fe_3Y and milled for 40 hours (A-prealloy, A0h). Afterwards pure nickel powder was added and a subsequent milling treatment was carried out to produce A samples (e.g. A40h for sample mechanically alloyed for 40 hours). The initial particle size of the the A0h is important for the ferrit-austenite transition. With a bigger particle size, the probability of been hit by the grinding medium and consequential mechanical alloying with nickel is increased. Thus assumingly leads to a faster transition of bigger ferrite particles into austenite. For this reason the particle size of ferrite particles used for mechanical alloying with nickel should be uniform and smaller than 150 μm. The nominal chemical composition of the samples given in Table 1 does not change during the mechanically alloying process, e.g. the composition of the samples A0h, A10h, A20h, A40h, and A80h is the same. Please note that the titanium was not added during mechanical alloying, it was present already in the F-prealloy.

The Fe and Cr K-edges x-ray absorption spectra were measured in transmission mode at the ELETTRA storage ring (Trieste, Italy) XAFS bending-magnet beamline [9] in the temperature range from 20°C to 1000°C. The x-ray radiation was monochromatized by a 40% detuned Si(111) double-crystal monochromator, and the beam intensity was measured using two ionization chambers filled with argon and krypton gases. The L'Aquila-Camerino vacuum furnace [9] was used for high temperature measurements. The samples for high temperature measurements

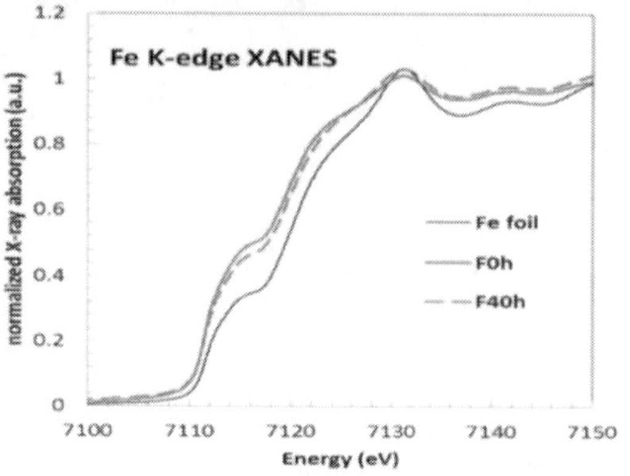

Figure 1. The Fe K-edge XANES spectra of ferritic samples and iron foil.

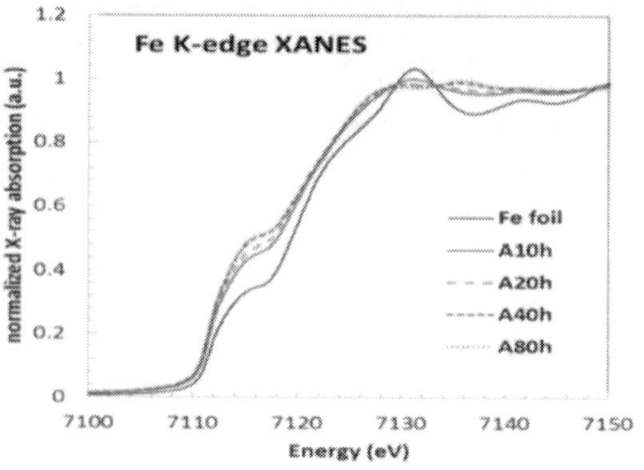

Figure 2. The Fe K-edge XANES spectra of austenitic samples and iron foil.

were prepared by pressing the proper amount of the ODS powder well mixed with boron nitride powder into a pellet. The proper amount of the ODS powder was deposited on the Millipore filter for room temperature measurements. In both cases we have paid special attention to sample homogeneity and thickness to optimum conditions at the Fe or Cr K-edge.

The X-ray absorption near edge structure (XANES) and extended X-ray absorption fine structure (EXAFS) $\chi(k)$ were extracted and analyzed following the conventional procedure [7, 8] using the EDA software package [10]. Theoretical backscattering amplitude and phase-shifts functions, employed in the best-fit EXAFS analysis, were calculated using ab-initio FEFF8 code [11].

3. Results and discussion

3.1. X-ray absorption near edge structure (XANES)

The Fe K-edge XANES spectra are shown in Figs. 1 and 2 and clearly reveal the phase of the matrix. Austenitic steels undergo a phase transition after 40h of milling, which is best seen in the XANES spectra in the region of around 7131 eV, where single white line peak characteristic for α phase splits into two peaks (at about 7129 and 7135 eV) characteristic for γ phase. There are no significant changes in the spectra of the austenitic samples with milling times of 40 hours and 80 hours.

At the same time there is no phase transition in the F-steels after 40 hours of milling as seen in Fig. 1. All the features before and after the absorption edge correspond to the reference Fe K-edge XANES spectrum of iron foil.

3.2. Extended X-ray absorption fine structure (EXAFS)

We found that the Fe and Cr K-edges EXAFS spectra of austenitic steel samples show significant dependence on the treatment procedure, therefore we intensified our efforts in studying in detail their behaviour.

The quality of the Fe K-edge EXAFS spectra (Fig. 3) is very good. For the A-series of samples a noticeable change with an increase of the mechanical alloying time can be observed in the EXAFS spectra and respectively in their Fourier transform (Fig. 3). As one can notice, a reduction of the signal amplitude occurs with an increasing of milling time, which is caused by the increase of the disorder in the steel matrix. The second effect is that there are significant structural changes appearing after 40 hours of milling, which are best observed in the regions of 4.5-5.5 Å^{-1} in k-space and of 3-6 Å in R-space.

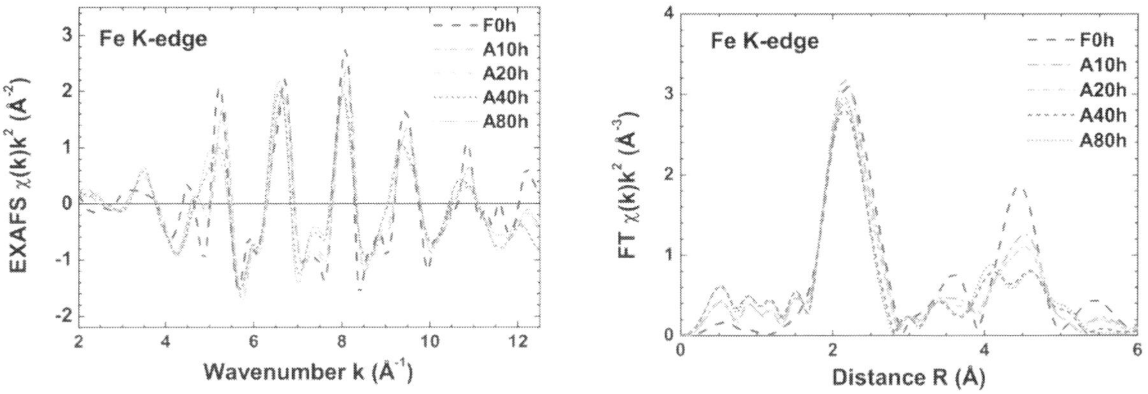

Figure 3. The Fe K-edge EXAFS spectra $\chi(k)k^2$ and their Fourier transforms (phase uncorrected) of prealloy (F0) and austenitic samples with mechanical alloying time from 10 to 80 hours.

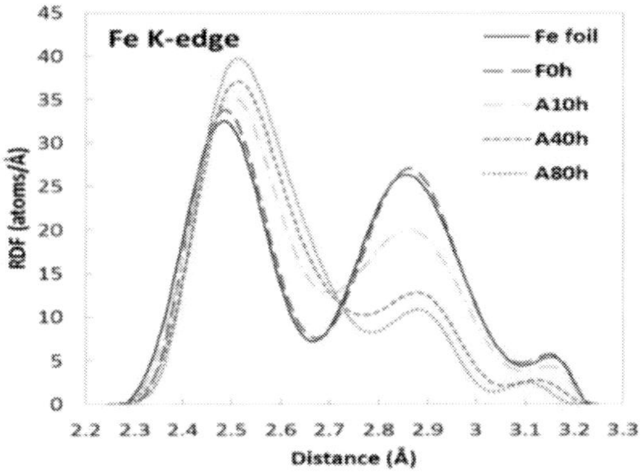

Figure 4. RDFs Fe-M (M=Fe,Cr,Ni) reconstructed from the Fe K-edge EXAFS spectra for prealloy F0 and austenitic samples.

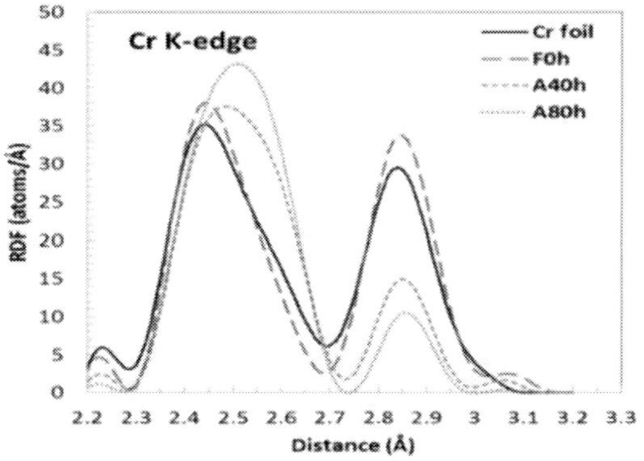

Figure 5. RDFs Cr-M (M=Fe,Cr,Ni) reconstructed from the Cr K-edge EXAFS spectra for prealloy F0 and austenitic samples.

To study these phenomena in more details we have reconstructed the radial distribution functions (RDFs) for the first coordination shell of iron from the EXAFS spectra using advanced approach described in [10]. The advantages of this method are that it is model free and does not require any assumption on the shape of the RDF. Note that this method allows one to analyse the coordination shells where only single-scattering contributions are present. In our case the method can be used for the analysis of peaks in FTs up to about 3 Å in Fig. 3. This R-space range corresponds to 14 atoms in two nearest coordination shells (at about 2.49 Å and 2.87 Å), which appear as a single peak in the Fourier transform, for the pure α-phase and to 12 atoms in the first coordination shell for the pure γ-phase [12].

As one can see in Fig. 4, the reconstructed RDFs confirm findings of the XANES analysis. The second peak at about 2.85 Å decreases strongly for longer milling times. At the same time it does not disappear completely, indicating that a small amount of α-phase still remains even after 80 hours of milling, while the majority of the material is transformed into austenitic phase.

Similar results have been obtained from the analysis of the Cr K-edge EXAFS spectra (Fig. 5). Chromium atoms follow the transformation of the matrix, but their local environment is more

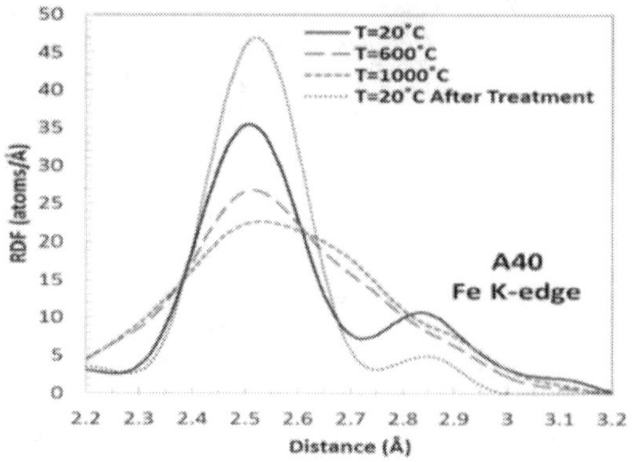

Figure 6. Reconstructed RDF Fe-M (M=Fe,Cr,Ni) for the A40h sample at three temperatures upon heating and after thermal treatment.

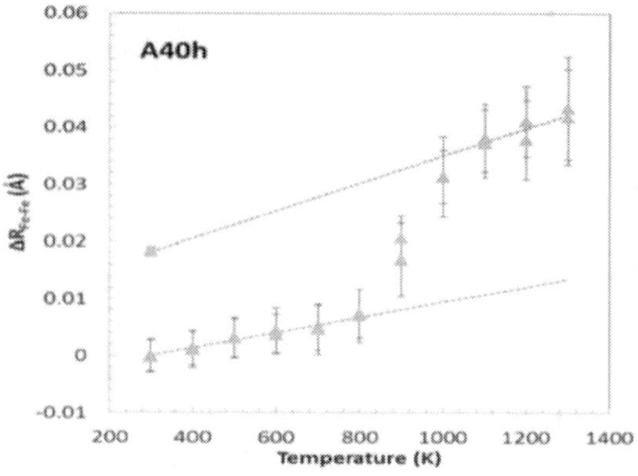

Figure 7. Temperature dependence of the first coordination shell radius of iron in the A40h sample obtained from the conventional EXAFS analysis.

disordered, as is indicated by the broader peaks in RDFs. Since the metallic radii of Fe and Cr are close, this can be an indication that chromium can be distributed in different locations of the material, e.g., forming Cr clusters at the grain boundaries or around oxide nanoparticles.

The reconstructed RDFs for the sample A40h (austenitic sample milled for 40 hours), which was heated from 20°C to 1000°C and then cooled back down to 20°C, are shown in Fig. 6. As expected, the RDF peaks become broader during heating due to an increase of thermal disorder. One can observe that A40h sample does not return to the original structure after the complete heat treatment: the final RDF looks more pronounced, and the amount of the α-phase indicated by the second peak in the RDF at about 2.85 Å is much smaller than before treatment, which is a clear indication that the material becomes more ordered, and the γ-phase becomes dominant. This has also been observed by [13, 14].

The conventional analysis, based on the best fit of the first coordination shell EXAFS signals within the single-scattering approximation, revealed more details about the phase transformations (Fig. 7). As one can notice, structural transformation starts at 600°C and is fully finished above 800°C. This transformation is irreversible, because when the samples are cooled back to 20°C the structure does not return into the starting state, but stays in the same phase as at the high temperature.

4. Conclusions

The nanostructure of ferritic and austenitic ODS steels was studied at the Fe and Cr K-edges X-ray absorption spectroscopy as a function of milling time. Also in-situ EXAFS measurements were performed upon thermal treatment of austenitic ODS samples up to 1000°C to get insights about the local structure around Fe and Cr atoms and their transformation in the presence of ODS precipitates.

We found that the Cr K-edge EXAFS can be used to observe distortions inside the material and to get an impression on the formation of chromium clusters.

Before XAS investigation the necessary milling time was uncertain and needed to be determined. The authors sought for the shortest milling time to be economically efficient and the longest time to receive a homogenous distribution of all elements. Also the amorphization process described by [15] should be investigated. After the milling process the mechanically alloyed powder is fabricated via extrusion or hot isostatic pressing to a near-net-shape geometry.

For austenitic samples a transformation of ferritic steel powders to the fcc (austenitic) structure after 10 hours of milling was detected, and its amount increased upon milling up to 40 hours. Between 40 and 80 hours only a small shift of the Fe K-edge was recognizable in the XANES spectra, indicating that the ferritic/austenitic ratio is moving towards an equilibrium ratio during milling. A small amount of α-phase was detected even after a milling time of 80 hours and for a chemical composition which favors the formation of a complete austenitic phase.

To get deeper insights to subsequent fabrication process and the optimal temperature we performed *in-situ* experiments. A40h was chosen, due to its complete mechanical alloying, which is revealed on the basics of no peak shift compared to A80h in Figs. 2, 4 and 5. The results shown in Figs. 6 and 7 show a complete ferrite to austenite transition after heating up to 900 K. These heating experiments and their results simulate the fabrication process which follows the mechanical alloying and reveal a uniform austenitic steel in a near-net-shape geometry after 40 h of milling time.

So *in-situ* EXAFS experiments offer a reliable method to investigate the ferritic to austenitic transformation in consideration of pinning effects [4, 16, 17], due to ODS precipitates.

Acknowledgments

The experiment at ELETTRA synchrotron facility has received funding from the European Community's Seventh Framework Programme (FP7/2007-2013) CALIPSO under grant agreement No. 312284. This work, supported by the European Community, was carried out within the framework of the European Fusion Development Agreement. The views and opinions expressed herein do not necessarily reflect those of the European Commission.

References

[1] Baluc N, Boutard J L, Dudarev S L, Rieth M, Brito Correia J, Fournier B, Henry J, Legendre F, Leguey T, Lewandowska M, Lindau R, Marquis E, Muoz A, Radiguet B, Oksiuta Z 2011 *J. Nucl. Mater.* **417** 149
[2] Reiser J, Rieth M, Möslang A, Greuner H, Armstrong D E J, Denk T, Gräning T, Hering W, Hoffmann A, Hoffmann J, Leiste H, Mrotzek T, Pippan R, Schulmeyer W, Weingärtner T, and Zabernig A 2014 *Adv. Eng. Mater.* DOI: 10.1002/adem.201400204
[3] Odette G R, Alinger M J, Wirth B D 2008 *Annu. Rev. Mater. Res.* **38** 471503
[4] Hirata A, Fujita T, Wen Y R, Schneibel J H, and Liu C T 2011 *Nature Mater.* **10** 922
[5] Le Flem M, Gentzbittel JM, Widen P 2013 *J. Nucl. Mater.* **442** S325
[6] Klimiankou M, Lindau R, Moslang A 2003 *J. Cryst. Growth* **249** 381387
[7] Rehr J J and Albers R C 2000 *Rev. Mod. Phys.* **72** 621
[8] Aksenov V L, Kovalchuk M V, Kuzmin A, Purans J, and Tyutyunnikov S I 2006 *Crystallogr. Rep.* **51** 908
[9] Di Cicco A, Aquilanti G, Minicucci M, Principi E, Novello N, Cognigni A and Olivi L 2009 *J. Phys.: Conf. Series* **190** 012043
[10] Kuzmin A 1995 *Physica B* **208-209** 175
[11] Ankudinov A L, Ravel B, Rehr J J and Conradson S D 1998 *Phys. Rev. B* **58** 7565
[12] Basinski Z S, Hume-Rothery W, and Sutton A L 1955 *Proc. R. Soc. Lond. A* **229** 459
[13] Enayati M H, and Bafandeh M R 2008 *Journal of Alloys and Compounds* **454** 228
[14] Yingli X, and Zhangjian Z 2013 *Journal of Physics: Conference Series* **419** 012052
[15] Suryanarayana C, Klassen T, Ivanov E 2011 *Journal of Materials Science* **46** 6301
[16] Oka H, Watanabe M, Hashimoto N, Ohnuki S, Yamashita S, Ohtsuka S 2013 *J. Nucl. Mater.* **442** 164
[17] Kim J H, Byun T S, Hoelzer D T, Kim S-W, Lee B H 2013 *Mater. Sci. Eng. A* **559** 101

IOP Conf. Series: Materials Science and Engineering **77** (2015) 012030 doi:10.1088/1757-899X/77/1/012030

EXAFS study of the local structure of crystalline and nanocrystalline Y_2O_3 using evolutionary algorithm method

I. Jonane, J. Timoshenko, A. Kuzmin

Institute of Solid State Physics, University of Latvia, Kengaraga street 8, LV-1063 Riga, Latvia

E-mail: ingajonaane@gmail.com

Abstract. Temperature-dependent local structure and lattice dynamics of yttria (Y_2O_3) were studied by X-ray absorption spectroscopy. Novel method, which combines the reverse Monte Carlo and evolutionary algorithm techniques, was applied for the analysis of extended X-ray absorption fine structure at the Y K-edge. This approach allowed us to reconstruct 3D atomic structure models of crystalline and nanocrystalline Y_2O_3.

1. Introduction

Yttria (Y_2O_3) is a material with a number of interesting technological applications, for instance, in the production of various ceramic materials and optical elements for rare-earth ion doped lasers [1]. Recently it was realized that addition of Y_2O_3 nanoparticles may significantly improve the high-temperature creep properties of steels [2, 3]. Such oxide dispersion-strengthened (ODS) steels are considered now as a promising structural materials for concentrated solar power plants and jet engines, and, in particular, for fusion and fission nuclear reactors [4].

Crystalline cubic Y_2O_3 has a body-centered cubic structure (space group $Ia3$) with the unit cell consisting of 80 atoms. Yttrium ions occupy two non-equivalent Wyckoff positions with different local environment [5] and different local symmetry: Y1 8(b) site with centrosymmetric C_{3i} symmetry and Y2 24(d) site with noncentrosymmetric C_2 symmetry [6]. First-principles calculations of Y_2O_3 electronic properties [1] reveal slightly different Mulliken charges for two yttrium sites: $q(\text{Y1}) = +2.08$ and $q(\text{Y2}) = +2.10$.

Extended X-ray absorption fine structure (EXAFS) spectroscopy may be used to probe the local atomic structure of the material [7]. In this study we analyzed temperature-dependencies of the Y K-edge EXAFS for crystalline and nanocrystalline Y_2O_3 using a novel reverse Monte Carlo and evolutionary algorithm (RMC/EA) approach [8]. It allows us to study not only the first, but also further coordination shells around absorbing atom (yttrium), and to reconstruct a 3D-structure model of the material. We followed the evolution of this model upon temperature increase from 20 up to 1000°C with the aim to analyze lattice dynamics of Y_2O_3 and, in particular, to investigate possible differences in the local environment of Y1 and Y2 ions. The results, obtained for crystalline Y_2O_3, are then compared with the ones, obtained for nanocrystalline Y_2O_3 with the particle size of about 50 nm.

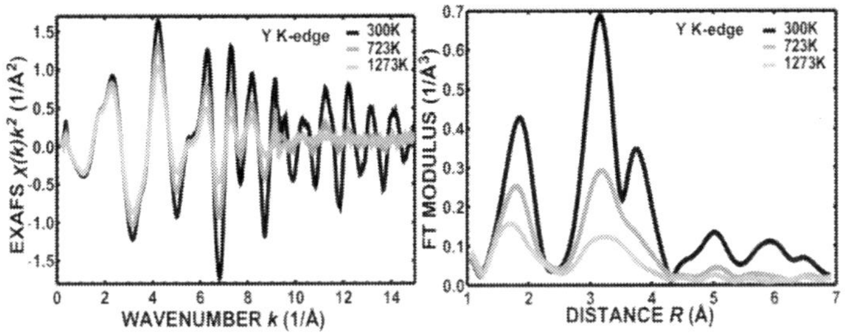

Figure 1. Temperature dependent Y K-edge EXAFS spectra $\chi(k)k^2$ and their Fourier transforms for crystalline Y_2O_3 at three selected temperatures.

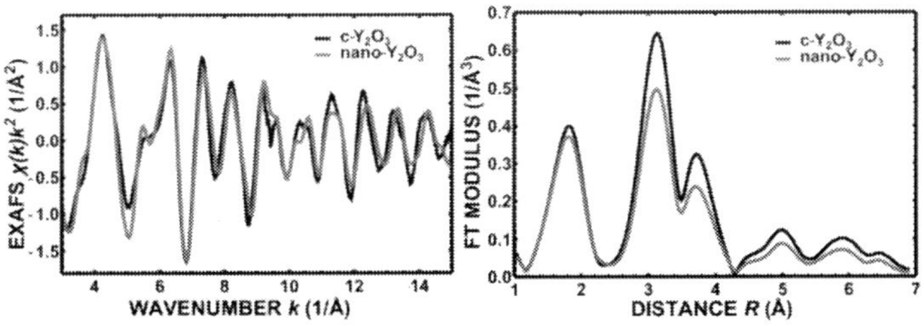

Figure 2. Room temperature Y K-edge EXAFS spectra $\chi(k)k^2$ and their Fourier transforms for crystalline Y_2O_3 (c-Y_2O_3) and nanocrystalline Y_2O_3 (nano-Y_2O_3).

2. Experimental and data analysis

X-ray absorption spectra of polycrystalline cubic Y_2O_3 (99.99%, Sigma-Aldrich) and nanocrystalline yttria (99.9%, size <50 nm (TEM), Sigma-Aldrich) were acquired at the Y K-edge (17038 eV) in transmission mode at the ELETTRA (Trieste, Italy) XAFS bending-magnet beamline [9]. The sample temperature was controlled in the range from 20 to 1000°C using the l'Aquila-Camerino vacuum glass furnace [9].

The experimental Y K-edge EXAFS spectra were extracted using the conventional procedure [10] and are shown together with their Fourier transforms (FT's) in Fig. 1. The first peak in FT at ~1.7 Å is due to the contribution of the first coordination shell (six closest oxygen atoms) around the absorbing yttrium atoms, whereas the second main peak at ~3.2 Å corresponds mainly to the outer coordination shells, formed by yttrium atoms.

Experimental Y K-edge EXAFS spectra and their Fourier transforms for crystalline Y_2O_3 and nanocrystalline Y_2O_3 are compared in Fig. 2. As one can see, the amplitude of EXAFS spectrum and the height of all peaks in FT for nanocrystalline yttria is slightly reduced due to the additional static disorder.

3. RMC/EA simulations

Reverse Monte Carlo (RMC) method [11, 12] allows one to analyze EXAFS data taking into account the influence of multiple-scattering effects and contributions of outer coordination shells, as well as structural and thermal disorder. In this approach a 3D structure model is constructed and atomic coordinates are changed in an iterative random process, aimed to minimize the difference between experimental and configuration-averaged EXAFS spectra. The theoretical EXAFS spectrum is obtained by ab initio multiple-scattering calculation using the FEFF8 code [13] for the given structure model. Recently we have demonstrated that the implementation of the so-called evolutionary algorithm (EA) in the conventional RMC scheme significantly improves the convergence properties of the simulations and makes feasible the analysis of EXAFS spectra for complex materials [8, 14].

The RMC/EA-EXAFS analysis of crystalline and nanocrystalline yttria was performed at the Y K-edge. In both cases the material was modelled as an infinite crystal employing periodic boundary conditions for the supercell composed of 2×2×2 unit cells of cubic Y_2O_3 (Fig. 3). The value of the lattice parameter $a_0 = 10.607$ Å was fixed at the experimental value for crystalline

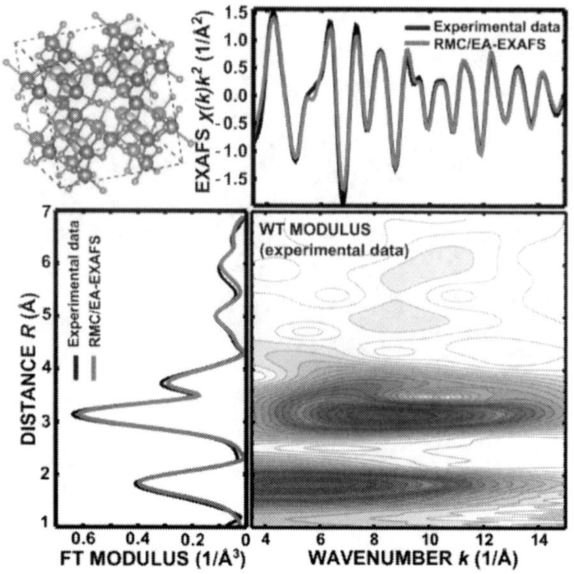

Figure 3. Unit cell of crystalline Y_2O_3 (upper left panel) and the results of RMC/EA–EXAFS calculations for crystalline Y_2O_3 at room temperature: the Y K-edge EXAFS spectra (upper right panel), their Fourier transforms (bottom left panel) and the wavelet transform (bottom right panel).

yttria obtained from neutron powder diffraction experiment [15].

In ab initio calculations of EXAFS spectra by the FEFF8 code [13] the multiple-scattering contributions were accounted up to the 4th order. Since yttrium ions occupy in Y_2O_3 two non-equivalent positions, the Y K-edge EXAFS spectra were calculated at each temperature for Y1 and Y2 sites separately, and the averaged EXAFS spectrum was calculated with the weights of 0.25 and 0.75 that correspond to the ratio of Y1 and Y2 atoms in Y_2O_3 structure. The averaged EXAFS spectrum then was used for the comparison with the experimental data.

The comparison of experimental and theoretical Y K-edge EXAFS spectra at each iteration was performed using Morlet wavelet transform (WT) [16, 17, 18] in the k-space range from 3.5 to 14.5 Å^{-1} and in the R-space range from 1.0 to 6.9 Å. Thus we were able to account for the features of EXAFS spectra in k and R spaces simultaneously.

Good agreement between theoretical and experimental EXAFS spectra was obtained after 1500 iterations (Fig. 3). The final structural model was used to estimate the values of required structure parameters.

4. Local structure of crystalline Y_2O_3

Two non-equivalent Y sites in cubic yttria are schematically shown in the inset in Fig. 4. Each of Y ions is surrounded by six oxygen atoms, located at the vertices of slightly distorted cubes. For Y1 ions, located at centrosymmetric site, two unoccupied cube vertices correspond to the ends of body diagonal and at equilibrium all six Y1–O distances are equal. For Y2 ions, located at noncentrosymmetric site, two unoccupied cube vertices correspond to the ends of face diagonal, and three groups of Y2–O distances are slightly different [5].

The atomic radial distribution functions (RDFs) for Y–O and Y–Y pairs around both Y1 and Y2 can be calculated straightforwardly from the structure model, obtained in RMC/EA-EXAFS simulations (Fig. 4). Note that the nearest peaks of RDFs around Y1 and Y2 have close shapes and positions, whereas the distribution of atoms in outer coordination shells differs significantly.

The width of RDF peaks can be characterized using mean-square relative displacement (MSRD) factors σ^2. Temperature dependencies of the MSRD factors for yttrium–oxygen (Y–O) and yttrium–yttrium (Y1–Y2, Y2–Y2) bonds are compared in Fig. 5. To estimate the uncertainties of obtained results, RMC/EA-EXAFS calculations were repeated several times

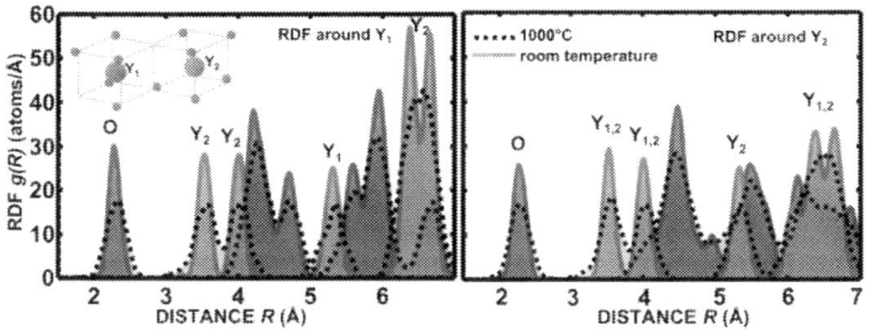

Figure 4. Radial distribution functions for Y–O and Y–Y pairs around Y1 and Y2 in crystalline Y_2O_3 at room temperature, reconstructed by RMC/EA method. RDFs, obtained for 1000°C data, are shown by dashed lines.

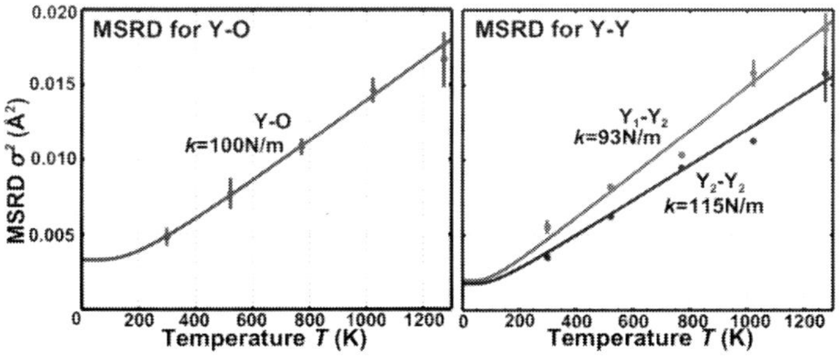

Figure 5. Temperature dependencies of the MSRD σ^2 factors for yttrium–oxygen (Y–O) and yttrium–yttrium (Y1–Y2, Y2–Y2) pairs in crystalline Y_2O_3. Best fitted correlated Einstein models are shown by solid lines.

with different sequences of pseudo-random numbers [8]. Within such uncertainties, the obtained MSRDs for all four non-equivalent Y–O bonds (one Y1–O and three Y2–O) have close values. Temperature-dependencies of MSRDs for Y1–Y2 and Y2–Y2 atom pairs are, in turn, quite different, although corresponding Y1–Y2 and Y2–Y2 distances are close: our results suggest that the MSRD for Y2–Y2 pair is smaller and increases slowly when the temperature rises, indicating that the atoms of Y2–Y2 pair interact strongly.

Temperature-dependencies of the MSRDs were best fitted using correlated Einstein model [19]. Using the values of Einstein frequency ω_E one can estimate effective bond-strength constants $\kappa = \omega_E^2\mu$, where μ is reduced mass of corresponding atomic pair. The obtained value for Y–O bond $\kappa = (100 \pm 4)$ N/m is in a good agreement with the value of bond-stretching constant for Y–O bond, determined by infra-red and Raman spectroscopies (103 N/m in [20] and 108 N/m in [21]).

The obtained effective bond-strength constants for Y–Y pairs are close to the value found for Y–O pair: $\kappa = (115 \pm 1)$ N/m for Y2–Y2 pair and is about 25% larger than the effective bond-strength constant $\kappa = (93 \pm 1)$ N/m for Y1–Y2 pair.

5. Local structure of nanocrystalline Y_2O_3

Reconstructed RDFs around Y atoms in nanocrystalline and crystalline Y_2O_3 are compared in Fig. 6. The peaks of RDFs for nanocrystalline sample are slightly broadened with respect to those for crystalline material. The MSRD values for Y–O, Y1–Y2 and Y2–Y2 pairs in nanocrystalline Y_2O_3 are equal to (0.0059 ± 0.0006), (0.0057 ± 0.0002), (0.0047 ± 0.0003) Å^2, respectively, and are close to corresponding values for crystalline sample at room temperature ((0.0049 ± 0.0005), (0.0056 ± 0.0005), (0.0036 ± 0.0004) Å^2, respectively). Note that the MSRD value for Y–O bond in nano-Y_2O_3 is close to that ((0.0059 ± 0.0016) Å^2), obtained in EXAFS studies of 40 nm Y_2O_3 particles doped with europium ions in [22]. One can conclude therefore that the size-related effects are not pronounced for our nanocrystalline yttria sample.

6. Conclusions

The analysis of temperature-dependent Y K-edge EXAFS spectra of crystalline and nanocrystalline Y_2O_3 has been performed using RMC/EA-EXAFS approach. This method

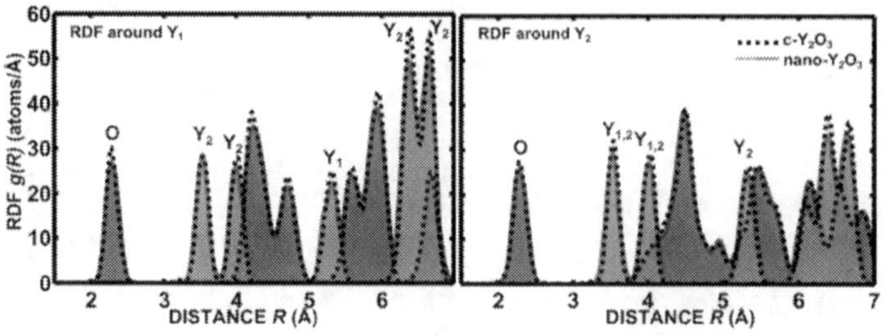

Figure 6. Radial distribution functions around Y1 and Y2 in nanocrystalline Y_2O_3 at room temperature, reconstructed by RMC/EA method. RDFs for crystalline Y_2O_3 at room temperature are shown by dashed lines.

allowed us to obtain and investigate in details a 3D structure model of Y_2O_3 and to follow its evolution upon temperature increase. It was shown that there are no distinguishable differences in dynamics of Y–O bonds for both centrosymmetric Y1 and noncentrosymmetric Y2 sites. Nevertheless, such differences were observed for Y–Y atom pairs: it was shown that the interaction between neighboring Y2 atoms is stiffer than the interaction between neighboring Y2 and Y1 atoms. The results, obtained for nanocrystalline yttria with particle size below 50 nm, showed only minor increase in structural disorder with respect to microcrystalline material, and no significant size-related effects were observed.

Acknowledgments

The work was supported by EURATOM project No. WP13-MAT-IREMEV-05-01 and European Social Fund project No. 2013/0015/1DP/1.1.1.2.0/13/APIA/VIAA/010. The experiment at ELETTRA synchrotron facility has received funding from the European Community's Seventh Framework Programme (FP7/2007-2013) CALIPSO under grant agreement No. 312284.

References

[1] Xu Y N, Gu Z Q and Ching W Y 1997 *Phys. Rev. B* **56** 14993–15000
[2] Lindau R, Möslang A, Rieth M, Klimiankou M, Materna-Morris E, Alamo A, Tavassoli A A F, Cayron C, Lancha A M, Fernandez P, Baluc N, Schäublin R, Diegele E, Filacchioni G, Rensman J, Schaaf B, Lucon E and Dietz W 2005 *Fusion Eng. Des.* **75** 989–996
[3] Klimiankou M, Lindau R and Möslang A 2007 *J. Nucl. Mater.* **367** 173–178
[4] Lindau R, Möslang A, Schirra M, Schlossmacher P and Klimenkov M 2002 *J. Nucl. Mater.* **307** 769–772
[5] Hanic F, Hartmanova M, Knab G G, Urusovskaya A A and Bagdasarov K S 1984 *Acta Crystallogr. B* **40** 76–82
[6] Gajović A, Tomašić N, Djerdj I, Su D and Furić K 2008 *J. Alloy Compd.* **456** 313–319
[7] Rehr J J and Albers R C 2000 *Rev. Mod. Phys.* **72** 621–654
[8] Timoshenko J, Kuzmin A and Purans J 2014 *J. Phys.: Condens. Matter* **26** 055401
[9] Di Cicco A, Aquilanti G, Minicucci M, Principi E, Novello N, Cognigni A and Olivi L 2009 *J. Phys.: Conf. Ser.* **190** 012043
[10] Aksenov V, Kovalchuk M, Kuzmin A, Purans Y and Tyutyunnikov S 2006 *Crystallogr. Rep.* **51** 908–935
[11] McGreevy R L and Pusztai L 1988 *Mol. Simul.* **1** 359–367
[12] Timoshenko J, Kuzmin A and Purans J 2012 *Comput. Phys. Commun.* **183** 1237–1245
[13] Ankudinov A L, Ravel B, Rehr J J and Conradson S D 1998 *Phys. Rev. B* **58** 7565–7576
[14] Timoshenko J, Anspoks A, Kalinko A and Kuzmin A 2014 *Phys. Scripta* **89** 044006
[15] Faucher M and Pannetier J 1980 *Acta Crystallogr. B* **36** 3209–3211
[16] Muñoz M, Argoul P and Farges F 2003 *Am. Mineral.* **88** 694–700
[17] Funke H, Scheinost A C and Chukalina M 2005 *Phys. Rev. B* **71** 094110
[18] Timoshenko J and Kuzmin A 2009 *Comp. Phys. Commun.* **180** 920–925
[19] Sevillano E, Meuth H and Rehr J J 1979 *Phys. Rev. B* **20** 4908–4911
[20] Repelin Y, Proust C, Husson E and Beny J M 1995 *J. Sol. State Chem.* **118** 163–169
[21] Ubaldini A and Carnasciali M M 2008 *J. Alloy Compd.* **454** 374–378
[22] Qi Z, Shi C, Zhang W, Zhang W and Hu T 2002 *Appl. Phys. Lett.* **81** 2857–2859

IOP Conf. Series: Materials Science and Engineering **77** (2015) 012031 doi:10.1088/1757-899X/77/1/012031

Local structure and lattice dynamics of cubic Y_2 $_3$: an x-ray absorption spectroscopy study

Karlis Lazdins and Alexei Kuzmin

Institute of Solid State Physics, University of Latvia, Kengaraga street 8, LV-1063 Riga, Latvia

E-mail: klazdins@inbox.lv

Abstract. Classical molecular dynamics was used to interpret the temperature dependent 300-1273 K Y K-edge x-ray absorption spectra of cubic c-Y_2O_3 and to validate two force-field models. It was shown that both models reproduce well thermal disorder effect, however are less accurate in description of static octahedra distortion around yttrium atoms. The temperature dependencies of mean-square displacements and mean-square relative displacements were obtained from molecular dynamics simulations for the first two shells of yttrium. The importance of anharmonic effects was demonstrated by comparison of molecular dynamics and lattice dynamics results.

1. Introduction

Cubic yttrium oxide, c-Y_2O_3 (space group $Ia\bar{3}$ (No. 206), $Z = 16$), is technologically important material with complex crystallographic structure [1]. In particular, an addition of nanosized yttria to steel matrix allows one to produce oxide dispersion strengthened (ODS) steels with improved mechanical properties and radiation resistance [2, 3, 4]. The optimization of ODS steel properties requires development of reliable theoretical models [5, 6]. Among different theoretical tools, molecular dynamics (MD) [6, 7] and Monte Carlo [8, 9] simulations open an access to in-depth understanding of materials properties at the atomic scale. However, their reliable use is challenging and demands the precise knowledge of interatomic potentials. Low concentration and nanosize of oxide particles in ODS steels makes x-ray absorption spectroscopy (XAS) a natural and highly e cient tool to probe their local structure [10, 11, 12, 13]. Besides, it was shown recently that XAS can be e ciently used as a validation tool for force-fields employed in the MD simulations [14, 15, 16].

In this study we have used the Y K-edge XAS to probe the local structure and lattice dynamics of yttria in a wide range of temperatures from 300 to 1273 K. The experimental results are analysed using classical molecular dynamics [17] and lattice dynamics [18] simulations.

2. Experimental

Temperature-dependent x-ray absorption spectroscopy of cubic c-Y_2O_3 (99.99 , Aldrich) was performed at the Y K-edge in transmission mode at the ELETTRA storage ring (Trieste, Italy) XAFS bending-magnet beamline [19]. The synchrotron radiation was monochromatized using a Si(111) double-crystal monochromator, and its intensity before and after the sample was measured by two ionization chambers. Polycrystalline c-Y_2O_3 powder was mixed with boron nitride and pressed into pellets. The sample thickness was optimized to obtain the absorption

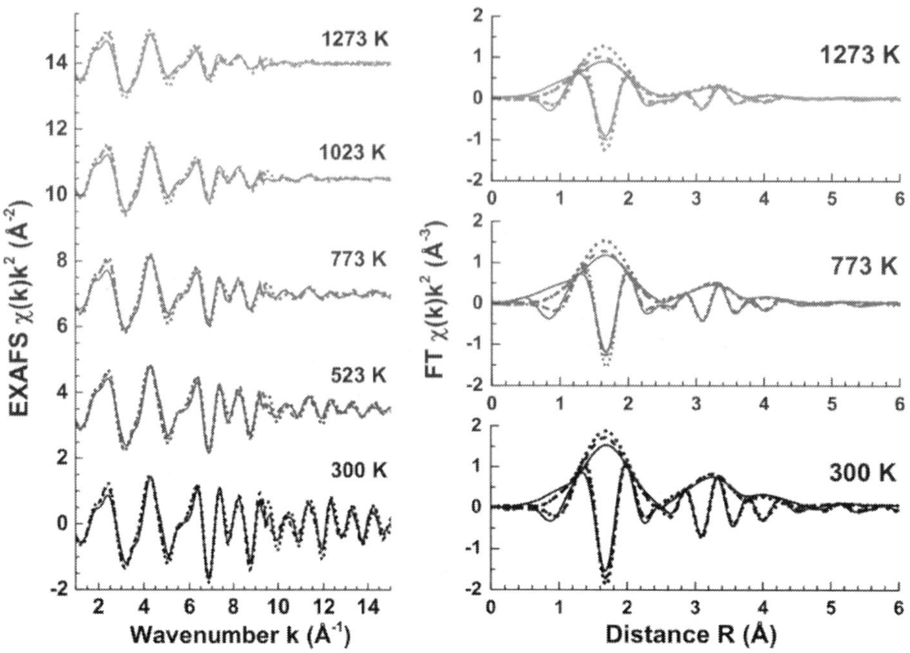

Figure 1. Comparison of the experimental (solid line) and configuration-averaged (dotted line - Model 1, dashed line - Model 2) Y K-edge EXAFS spectra $\chi(k)k^2$ and their Fourier transforms (modulus and imaginary parts) for c-Y_2O_3 at several selected temperatures.

Y K-edge jump value $\Delta\mu \approx 1$. The sample temperature was controlled using the vacuum glass furnace [19] in the range from 300 to 1273 K. The experimental Y K-edge extended x-ray absorption fine structure (EXAFS) (figure 1) was extracted using the conventional procedure [20].

3. Molecular and lattice dynamics simulations

Classical molecular dynamics (MD) and lattice dynamics (LD) simulations were performed using the GULP code [21, 22]. Note that classical MD does not include quantum effects, therefore it can be used only at relatively high temperatures. On the contrary, the LD simulations are limited by harmonic approximation and thus will fail upon increasing temperature. Therefore, the two methods complement each other in a description of material vibrational properties.

The MD simulation box was built up of eight unit cells with periodic boundary conditions ($2a_0 \times 2a_0 \times 2a_0$ supercell where a_0 is the lattice parameter). The crystallographic structure of c-Y_2O_3 [1] was used as starting configuration (figure 2). The unit cell contains two non-equivalent yttrium atoms, Y1 and Y2, which occupy 8(b) (1/4, 1/4, 1/4) and 24(d) (u, 0, 1/4) Wyckoff positions, respectively, and are octahedrally coordinated by oxygens. The octahedra around Y1 atoms are regular, thus all Y1–O distances are equal to 2.29 Å, whereas the octahedra around Y2 atoms are distorted with Y2–O bond lengths being in the range from 2.24 Å to 2.33 Å.

We have used the canonical (NVT) ensemble, so that the average temperature was kept constant during MD simulation run using Nosé-Hoover thermostat [23]. The Newton's equations of motion were integrated using the Verlet leapfrog algorithm [24] with a time step of 0.5 fs. The equilibration and production times were 20 ps each. A set of 4000 atomic configurations were accumulated during each production run, and the configuration-averaged Y K-edge EXAFS spectra were calculated [17], taking into account the presence of two non-equivalent yttrium

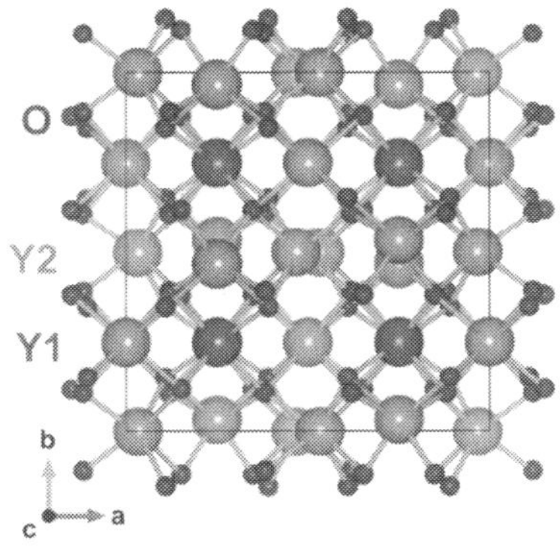

Figure 2. Crystal structure of cubic c-Y_2O_3 [1]. The unit cell is indicated. Two non-equivalent yttrium atoms (Y1 and Y2) are marked.

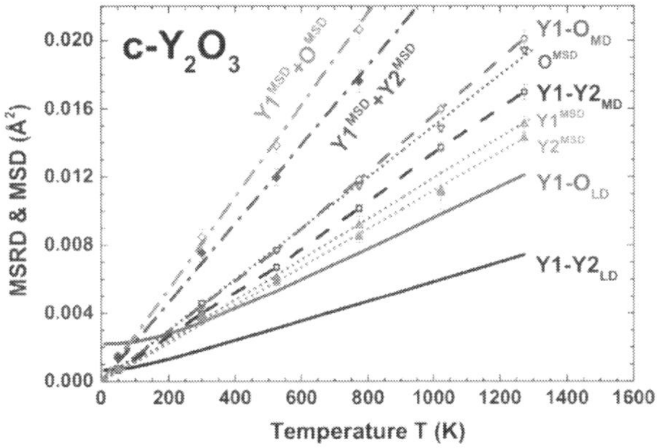

Figure 3. MSRDs for the Y1–O and Y1–Y2 atom pairs and MSDs for Y1, Y2 and O atoms. Dashed, dotted and dash-dotted lines - molecular dynamics (MD), solid lines - lattice dynamics (LD).

atoms (Y1 and Y2) [1]. This approach will be referred to as the MD-EXAFS method.

EXAFS calculations were done using ab initio real-space multiple-scattering (MS) FEFF8 code [25], including the MS contributions up to the 7th order. The self-consistent mu n-tin (MT)-type cluster potential was evaluated only once for the average crystallographic structure of c-Y_2O_3 [1], thus neglecting its small variation due to thermal disorder. The complex exchange-correlation Hedin–Lundqvist potential was used to describe the inelastic losses, and the MT radii ($R_{\rm mt}$(Y)=1.57 Å and $R_{\rm mt}$(O)=1.06 Å), evaluated by the FEFF8 code [25], were employed.

Two force-field models (Model 1 and Model 2) based on the rigid-ion Buckingham potential were used in both MD and LD simulations to describe interactions in c-Y_2O_3 (table 1)

$$U_{ij}(r_{ij}) = A_{ij}\exp(-r_{ij}/\rho_{ij}) - C_{ij}/r_{ij}^6 + q_iq_je^2/r_{ij} \qquad (1)$$

where e is the elementary charge and r_{ij} is the interatomic distance for the i–j atom pair. Model 1 is the re-optimized in the present work version of the model, originally developed in [26]: it includes three short-range Y–O, Y–Y and O–O interactions and long-range Coulomb interactions, defined by the effective ion charges q(Y)=+1.8 and q(O)=−1.2. In Model 2, taken from [27] excluding the shell, only two short-range interactions Y–O and O–O of the Buckingham-type were considered, and the effective ion charges were q(Y)=+2.4 and q(O)=−1.6.

Lattice dynamics simulations were performed using the second force-field model (Model 2). In the harmonic approximation, solving the eigenvalue problem of the dynamical matrix allows one to obtain its eigenvalues and eigenvectors, which in turn can be used to evaluate the mean-square displacements (MSDs) of Y and O atoms and the mean-square relative displacements (MSRDs) σ_{ij}^2 of the i–j atom pairs [28, 18]. Thus obtained MSRD values for Y1–O and Y1–Y2 bonds are compared with that determined from the MD simulations in figure 3.

4. Results and discussion

The experimental and calculated Y K-edge EXAFS spectra $\chi(k)k^2$ and their Fourier transforms at selected temperatures are compared in figure 1. The two force-field models (Model 1 and Model 2) give close EXAFS spectra in k-space, which are in good agreement with the experimental ones. However, the difference between theory and experiment is observed in R-space in the region of the first peak at ∼1.6 Å, which is due to the first coordination shell of

Table 1. Force-field parameters used in the MD and LD simulations.

Pair of atoms	A eV	ρ Å	C eVÅ^6	A eV	ρ Å	C eVÅ^6
		Model 1			Model 2	
Y–O	162401.94	0.172628	0.0	1822	0.309042	0.0
O–O	351366.27	0.187988	0.0	1822	0.305726	100.0
Y–Y	112925.98	0.232299	0.0			

yttrium atoms surrounded by six oxygens. While the observed difference is rather small, the second force-field model (Model 2), which neglects the Y–Y interactions, gives slightly better account for the thermal disorder upon increasing temperature from 300 to 1273 K. Note that both models reproduce well the atomic structure and disorder effects in the outer coordination shells above 2.5 Å.

Thermal vibrations of atoms play an important role in thermal and transport properties of materials, phase transitions, thermal expansion and interaction of radiation with matter, which are of significant relevance for ODS steels. Therefore, the knowledge of vibration amplitudes and correlation effects represent significant interest. This information can be experimentally obtained by direct structural tools as diffraction and EXAFS. While diffraction technique allows one to determine the mean-square displacements (MSD) of atoms relative to their mean positions [29], EXAFS contains, in principle, information on both MSRD and MSD. However, conventional analysis of EXAFS for the nearest coordination shell(s) around an absorbing atom allows one to determine only MSRD values [30]. Thus, an extraction of the MSD values from EXAFS is challenging and requires to perform the analysis of distant coordination shells [31, 32]. At the same time, both MSD and MSRD values can be directly obtained from the MD simulations, validated in advance using EXAFS data.

Classical molecular dynamics cannot be used at low temperatures because it ignores quantum effects. On the contrary, lattice dynamics simulations give accurate picture of thermal vibrations at zero temperature, but fail at high temperatures due to neglect of anharmonic effects. Therefore, these two approaches complement each other and should be utilized together. In figure 3 the MSRD values for Y1–O and Y1–Y2 atom pairs and the MSD values for Y1, Y2 and O atoms are shown: they were calculated by MD and LD using the same force-field model (Model 2). We have selected only regular Y1 site to exclude the need to account for additional static distortions of yttrium–oxygen octahedra around Y2 site. The LD results deviate strongly from MD already above ~200 K for yttrium–oxygen pairs and above ~50 K for yttrium–yttrium pairs, indicating the importance of anharmonic effects. Temperature dependence of MSDs at high temperatures, i.e. in the classical limit, is linear, and heavier yttrium atoms vibrate with smaller amplitudes than light oxygen atoms, as expected. MSD and MSRD for the i–j atom pair are connected by the equation $\mathrm{MSRD}_{ij} = \mathrm{MSD}_i + \mathrm{MSD}_j - 2\sqrt{\mathrm{MSD}_i}\sqrt{\mathrm{MSD}_j}\phi$, where ϕ is a dimensionless correlation parameter [33], which is equal to $+1$ for perfectly in-phase atom motion, to zero for completely independent motion and to -1 for perfectly opposite-phase motion. From data reported in figure 3 one can find that $\phi = 0.45$ at 50 K and increases slightly to $\phi = 0.60$ at 1273 K for both Y1–O and Y1–Y2 pairs, indicating prevalent in-phase atom motion in c-Y_2O_3.

5. Conclusions

We have shown that experimental Y K-edge EXAFS spectra of c-Y_2O_3 can be used to validate the force-field models (table 1) within the MD-EXAFS approach. The agreement between theory and experiment is restricted by the simplicity of the models, which are not able to describe accurately the distortion of yttrium–oxygen octahedra. However, the second model (Model 2) gives slightly better agreement for the first shell than the first one. Note that both models reproduce well thermal disorder in c-Y_2O_3 in a wide range of temperatures from 300 to 1273 K. A comparison between molecular dynamics and lattice dynamics simulations (figure 3) allows one to conclude on the importance of anharmonic effects and on the temperature range, where quantum effects can be neglected.

Acknowledgments
This work was supported by EURATOM WP13-MAT-IREMEV-05-01 pro ect and ESF Pro ect no. 2013/0015/1DP/1.1.1.2.0/13/APIA/VIAA/010. The experiment at ELETTRA synchrotron facility has received funding from the EC 7th Framework Programme FP7/2007-2013 CALIPSO under Grant Agreement no. 312284.

References
[1] Hanic F, Hartmanová M, Knab G G, Urusovskaya A A and Bagdasarov K S 1984 *Acta Crystallogr. B* **40** 76–82
[2] Ukai S and Fu iwara M 2002 *J. Nucl. Mater.* **307-311** 749–757
[3] Klueh R, Maziasz P, Kim I, Heatherly L, Hoelzer D, Hashimoto N, Kenik E and Miyahara K 2002 *J. Nucl. Mater.* **307-311** 773–777
[4] Schneibel J, Liu C, Hoelzer D, Mills M, Sarosi P, Hayashi T, Wendt U and Heyse H 2007 *Scripta Mater.* **57** 1040–1043
[5] Gope enko A, Zhukovskii Y F, Vladimirov P V, Kotomin E A and Möslang A 2010 *J. Nucl. Mater.* **406** 345–350
[6] Hirata A, Fu ita T, Wen Y R, Schneibel J H, Liu C T and Chen M W 2011 *Nat. Mater.* **10** 922–926
[7] Yashiro K, Yamaguchi A, Tanaka M, Okuda T, Koga K and Segi T 2012 *Mater. Trans.* **53** 401–406
[8] Alinger M, Wirth B, Lee H J and Odette G 2007 *J. Nucl. Mater.* **367-370** 153–159
[9] Hin C, Wirth B D and Neaton J B 2009 *Phys. Rev. B* **80** 134118
[10] Degueldre C, Conradson S and Hoffelner W 2005 *Comp. Mater. Sci.* **33** 3–12
[11] Pouchon M, Kropf A, Froideval A, Degueldre C and Hoffelner W 2007 *J. Nucl. Mater.* **362** 253–258
[12] Béchade J L, Menut D, Lescoat M L, Sitaud B, Schlutig S, Solari P, Llorens I, Hermange H, de Carlan Y, Ribis J and Toualbi L 2012 *J. Nucl. Mater.* **428** 183–191
[13] He P, Liu T, Möslang A, Lindau R, Ziegler R, Hoffmann J, Kurinskiy P, Commin L, Vladimirov P, Nikitenko S and Silveir M 2012 *Mater. Chem. Phys.* **136** 990–998
[14] Kuzmin A and Evarestov R A 2009 *J. Phys.: Conf. Ser.* **190** 012024
[15] Anspoks A, Kuzmin A, Kalinko A and Timoshenko J 2010 *Solid State Commun.* **150** 2270–2274
[16] Timoshenko J, Anspoks A, Kalinko A and Kuzmin A 2014 *Acta Mater.* **79** 194–202
[17] Kuzmin A and Evarestov R A 2009 *J. Phys.: Condens. Matter* **21** 055401
[18] Cope E R and Dove M T 2007 *J. Appl. Cryst.* **40** 589–593
[19] Di Cicco A, Aquilanti G, Minicucci M, Principi E, Novello N, Cognigni A and Olivi L 2009 *J. Phys.: Conf. Ser.* **190** 012043
[20] Aksenov V L, Kuzmin A Y, Purans J and Tyutyunnikov S I 2001 *Phys. Part. Nuclei.* **32** 675–707
[21] Gale J D 1996 *Phil. Mag. B* **73** 3–19
[22] Gale J D and Rohl A L 2003 *Mol. Simul.* **29** 291–341
[23] Hoover W G 1985 *Phys. Rev. A* **31** 1695–1697
[24] Hockney R W 1970 *Methods Comput. Phys.* **9** 136–211
[25] Ankudinov A L, Ravel B, Rehr J J and Conradson S D 1998 *Phys. Rev. B* **58** 7565–7576
[26] Belonoshko A B, Gutierrez G, Ahu a R and Johansson B 2001 *Phys. Rev. B* **64** 184103
[27] Bose P P, Gupta M K, Mittal R, Rols S, Achary S N, Tyagi A K and Chaplot S L 2011 *Phys. Rev. B* **84** 094301
[28] Chung J S and Thorpe M F 1999 *Phys. Rev. B* **59** 4807–4812
[29] Trueblood K N, Bürgi H B, Burzlaff H, Dunitz J D, Gramaccioli C M, Schulz H H, Shmueli U and Abrahams S C 1996 *Acta Crystallogr. A* **52** 770–781
[30] Beni G and Platzman P M 1976 *Phys. Rev. B* **14** 1514–1518
[31] Sapelkin A V and Bayliss S C 2002 *Phys. Rev. B* **65** 172104
[32] Jeong I K, Heffner R H, Graf M J and Bilinge S J L 2003 *Phys. Rev. B* **67** 104301
[33] Booth C H, Bridges F, Bauer E D, Li G G, Boyce J B, Claeson T, Chu C W and Xiong Q 1995 *Phys. Rev. B* **52** R15745–R15748

Vibrational spectra of tungsten oxytetrachloride

J Gabrusenoks
Institute of solid State Physics, University of Latvia, Riga, Latvia

E-mail: gabrusen@latnet.lv

Abstract. Raman spectroscopy has been applied to study the vibrational modes of $WOCl_4$. The symmetry of the vibrations is reported. Ab initio calculations are employed to determine the equilibrium lattice parameters and to calculate phonon frequencies in several Brillouin zone points.

1. Introduction

The tungsten-oxygen compounds have crystal lattices with different topologies. The $WOCl_4$, WO_2Cl_2 and WO_3 form crystal lattices with one-, two- and three-dimensional networks of W-O bonds respectively. It determines vibrational behaviour of the crystal lattice.

$WOCl_4$ crystalizes in a tetragonal body centered structure with space group of C_4^5(Fig.1) [1]. The crystal lattice is made by WO_2Cl_4 octahedra. Two oxygen atoms in the octahedra are in a trans-position with each other. W - O bonds are asymmetric. The chlorine atoms are in the plain perpendicular with the W – O bond. Octahedra are linked together by corners, where an oxygen atom is shared by two neighboring octahedra. The crystal structure consists of linear chains. These chains are parallel in the crystal lattice. In the chain the long bonds interchange with the short ones. Van der Waals forces attract chains together. The vibrational properties of $WOCl_4$ have not been widely studied. The Raman spectra were reported earlier in [2,3].

$WOCl_4$ is model crystal for materials with more complicated structure. $Tl_6[WOCl_4]_2$ [4], $Tl[NbOCl_4]$ and $Tl[NbOBr_4]$ [5] have linear metal-oxygen chains.

This work presents experimental and theoretical investigations of vibrational properties of $WOCl_4$.

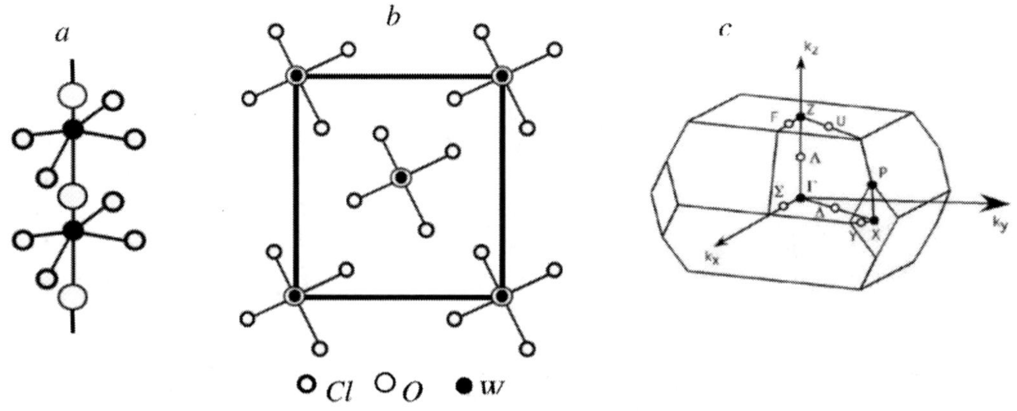

Figure 1. Crystal structure of $WOCl_4$. a) chain parallel to c-axe; b) projection on the plain (001); c) first Brillouin zone.

2. Experimental and computational details

Good optical quality $WOCl_4$ single crystals were grown by the reaction of WO_3 and $SOCl_2$ at 200°C in closed ampule. The dimensions of the crystals are nearly $0.5x0.5x10\ mm^3$.

IOP Conf. Series: Materials Science and Engineering **77** (2015) 012032 doi:10.1088/1757-899X/77/1/012032

The Raman spectra were measured with 1 cm^{-1} resolution using a 1403 double grating Spex Ramalog spectrophotometer with a conventional photon counting system. The spectra were excited with the 633 nm line of a He-Ne ion laser. The power of the excitation was 10mW.

Calculations were performed by using hybrid exchange density functional theory to determine equilibrium geometries and phonon frequencies. The Grimme dispersion correction for energy and gradient was used in combination with the B3LYP functional. The CRYSTAL09 computer code was used [6].

The group theoretical analysis have been applied to determine symmetries of vibrational modes.

3. Results and analysis

Relationship between symmetry of vibrations of different $WOCl_4$ lattices are given in table 1. The highest possible hypothetical symmetry of $WOCl_4$ chain is D_{4h}, as W – O bond lengths are equal. Actually existing $WOCl_4$ structure has asymmetric bonds, this leads to lowering of symmetry to C_{4v}. Symmetry lowers to C_4, if chains join in a crystal lattice.

The chain with symmetrical W – O bonds in WO_2Cl_4 octahedra has four Raman active optical vibrations A_{1g}+B_{1g}+B_{2g}+E_g. To the chain with asymmetrical W – O bonds the number of Raman active vibrations rise to 10: $3A_1$+$2B_1$+B_2+4E. Only one symmetry A_2 vibration stays inactive in the Raman scattering. Finally, if the chains joining in crystal lattice and having symmetry C_4 all optical vibrations 4A+3B+4E are Raman active. If the hypothetical $WOCl_4$ crystal is made by symmetrical W – O bonds, five optical vibrations $2A_g$+$2B_g$+E_g are Raman active.

Table 1. Relationship between symmetry of vibrations in different structures of $WOCl_4$. R and IR in the brackets show Raman and infrared activity respectively.

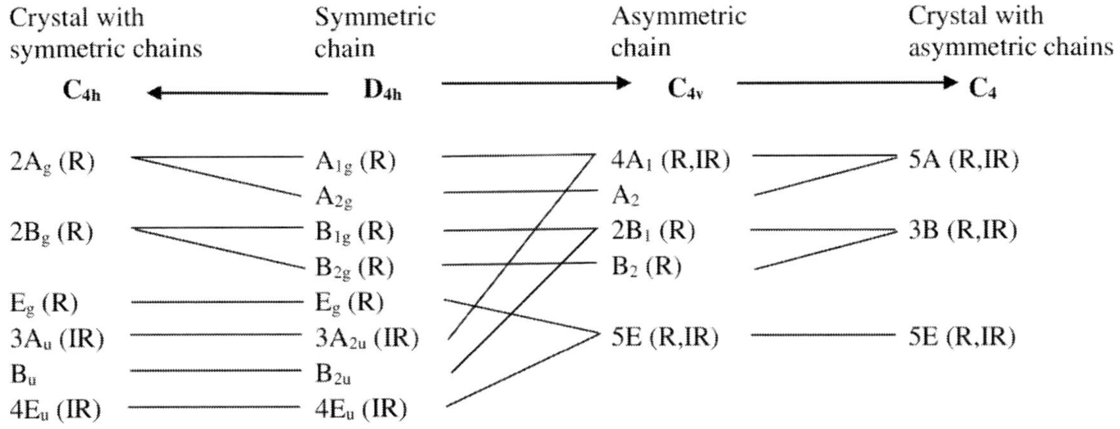

Raman spectrum of $WOCl_4$ crystal is shown in Fig.2. Thirteen bands are well observed in the Raman spectrum (Fig.2). Symmetry calculations of lattice vibrations provide for 11 Raman active vibrations (Tab.2).

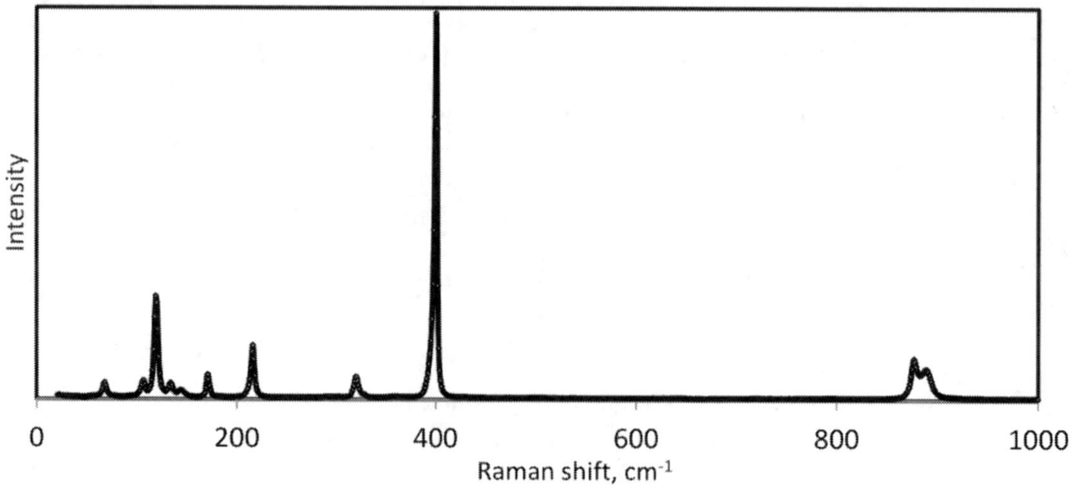

Figure 2. Raman spectrum of $WOCl_4$ at room temperature.

For further Raman spectra interpretation ab initio calculations of lattice vibrations have been performed. These calculations were performed for optimized crystal lattice. Optimized lattice parameters are shown in Table 3, where they are compared to the experimental parameters [1] of crystal lattice. The obtained sizes of a lattice cell do not differ from the experimental data of more than 1%.

Table 2. Experimental and calculated phonon frequencies of $WOCl_4$. Calculations were performed in the Brillouin zone points Γ, X and Z.

Γ				X	Z
Calculation		Experiment		Calculation	Calculation
Frequency,cm^{-1}	Symmetry	Frequency, cm^{-1}	Symmetry	Frequency,cm^{-1}	Frequency, cm^{-1}
				45; 56	63; 77
85	A	70	A	69	80
119	B	109	B	109	114
136	E	122	E	135	133
		136	A		
153	E	147	E	162	169
175	A	173	A	180	205
231	B	218	B	220	227
329	B	322	B	348	330
350	E	330	E	367	358
378	E	390	E	390	361
409	A	400	A	406	403
845	A	877	A	845	1013
		889	A		

Table 3. Ab initio optimized and experimental [1] lattice parameters of $WOCl_4$

	Calculation	Experiment [1]
a, Å	8,53	8,48
c, Å	3,95	3,99
$R_{W\text{-}O}$, Å	2.21	2,18
$R_{W=O}$, Å	1,74	1,81
$R_{W\text{-}Cl}$, Å	2,30	2,28

The frequencies of lattice vibrations were calculated in Brillouin zone center Γ and in two points X and Z on the Brillouin zone boundary. The calculations provide for one Raman active W-O stretching high frequency vibration in the Brillouin zone center Γ. Experimental Raman spectrum show doublet at 877-889 cm^{-1}. In [2] it was suggested that this is due to a multiphonon process in resonance with the fundamental W-O stretching vibration. This does not compare with computed and experimental vibrational frequencies. The highest possible two phonon frequency is equal to 800-820 cm^{-1}. It is too low for the experimentally observed 889 cm^{-1}.

If the sequence of experimentally obtained and theoretically calculated vibrational symmetry is compared, according to theoretical calculations 136 cm^{-1} Raman band should not be in Raman spectrum

4. Conclusions

Theoretical calculations describe lattice parameters and Raman spectra of $WOCl_4$. It is possible by using the Grimme dispersion correction for energy and gradient in computations. However, calculations do not explain existence of Raman bands 136 and 889 cm^{-1}. The issues of further investigations is to clarify the nature of additional Raman bands.

5. Acknowledgments

This work was supported by Latvian Science Council Grant No 374/2012

References

[1] H.Hess, H.Hartung, Z. Anorg. Allg. Chem. **344**, 157 (1966)
[2] I.R.Beattie, K.M.Livingston, D.J.Reynolds, G.A.Ozin, J. Chem.Soc. **A**, 1210 (1970)
[3] W.Bues, W.Brockner, F.Demiray, Spectrochim. Acta. **30A**, 579 (1974)
[4] J.Beck, Chem. Ber. **128**, 23 (1994)
[5] J.Beck, J.Bordinhao, Z. Anorg. Allg. Chem. **631**, 1261 (2005)
[6] Dovesi R, Saunders V R, Roetti C, et al, 2009 CRYSTAL09 User's Manual, University of Torino, Italy. URL: http://www.crystal.unito.it

IOP Conf. Series: Materials Science and Engineering **77** (2015) 012033 doi:10.1088/1757-899X/77/1/012033

Thermal deoxygenation of graphite oxide at low temperature

V Kampars, M Legzdina
Institute of Applied Chemistry of Riga Technical University, Paula Valdena 3, Riga, LV- 1048, Latvia

Abstract

. Synthesis of graphene via the deoxygenation of the graphite oxide (GO) is a method for the large-scale production of this nanomaterial possessing exceptional mechanical, electrical and translucent properties. Graphite oxide sheet contains at least four different oxygen atoms connected to the Csp^3 and Csp^2 atoms of the sheet in the form of hydroxyl, epoxy, carboxyl or carbonyl groups. Some of these functional groups are located at the surface but others situated at the edges of the platelets. To obtain the graphene nanoplatelets or the few-layer graphene the oxygen functionalities must be removed. Exfoliation and deoxygenation can be accomplished by the use of chemical reductants or heat. Thermal deoxygenation as greener and simpler approach is more preferable over chemical reduction approach. Usually a considerable mass loss of GO observed upon heating at temperatures starting at 200 °C and is attributed to the deoxygenation process. In order to avoid the defects of the obtained graphene sheets it is very important to find the methods for lowering the deoxygenation temperature of GO. Herein, we have investigated the way treatment of the Hummer's synthesis product with acetone and methyl *tert*-butyl ether under ultrasonication in order to lower the thermal stability of the graphite oxide and its deoxygenation temperature. The obtained results indicate that treatment of the graphite oxide with solvents mentioned above substantially reduces the reduction and exfoliation temperature (130 °C) under ambient atmosphere. The investigation of the composition of evolved gases by hyphenated Pyr/GC/MS method at different experimental conditions under helium atmosphere shows that without the expected H_2O, CO and CO_2 also sulphur dioxide and acetone has been released.

1.Introduction

Graphene is defined as sheets of sp^2-hybridized carbon, where the number of stacked sheets is lower than 10 [1]. The properties of graphene materials depend on the number of graphene layers in stack. Exceptional mechanical, electrical and translucent properties of the single and few layer graphene cause a great interest to this nanomaterial [2, 3] and make it attractive for a wide range of potential applications [4-8]. For example, the optical transmittance, high electrical conductivity and good stability allow to replaces the indium tin oxide by graphene for production of transparent electrodes of different devices [9]. Many fields of the potential application request to develop a market for this nanomaterial, therefore the elaboration of new methods for the commercial mass production of graphene is very important. Currently, there are few fundamentally different methods used for the graphene production: micromechanical cleavage of highly oriented graphite (top-down method), the growth the graphene sheets on different substrates (bottom-up methods), and thermolysis of structurally defined precursors and chemical exfoliation of graphite. Described methods have different variations, but remain suitable only for the fundamental investigation. High quality graphene films were grown on nickel and copper substrates as catalysts by chemical vapour deposition (CVD) at the temperature around 1000 °C by use of the reduced pressure and methane or other volatile carbon precursor [10]. Unfortunately, the CVD process is time and resource consuming and thus can't be commercially viable for the mass production [11]. Similar assessment can be done for the Arc-Discharge method [12]. Another process (chemical solid deposition, CSD) for the fabrication of high quality graphene sheets on SiC wafer can be done with the use of the epitaxial growth and annealation of SiC at temperatures around 2000 °C [13], but it is also considered as not commercially feasible. Different bottom-up approach for the synthesis of graphene can be done by the thermolysis of structurally-defined precursors [14], but the synthesis of the precursors is complex and time and resources consuming. Solvothermal-assisted exfoliation of the edge-selectively functionalized graphite (EFG) also has various problems for the practical use [15, 16]. Synthesis of graphene from graphite oxide (GO) is widely investigated and practically used top-down method [17], considered as one of the possible for the large-scale synthesis of graphene. The oxidation of graphite breaks up the sp^2-hybridized carbon sheets into nanoscale graphitic sp^2 domains surrounded by oxidised sp^3 domains as well as defects the carbon vacancies [18]. The oxidation product of the graphite has a layered morphology with -OH (phenols and alcohols) and 〉O (epoxide) functionalities on the hexagonal basal planes and carboxyl and carbonyl groups on the periphery. Although the oxygen functionalities wearing the attractive forces between layers, the graphite oxide retains stacked similar to the graphite but with much wider spacing, enabling facile exfoliation. The sonication in polar solvents is often used for the exfoliation of GO. After exfoliation or during the exfoliation the obtained GO shall be reduced in order to restore the sp^2-hybridized carbon sheets of graphene. It is known that the oxidation process of graphite followed by restoration of the sp^2-hybridized carbon sheets, proceeds with the introduction of some defects into the graphene structure [18, 19], therefore the obtained product usually is termed as reduced graphite oxide (rGO). Elaboration of the new methods of high quality rGO production produces a great interest [18]. Thermal reduction and exfoliation of GO is considered as most promising strategy for the employment in mass production of rGO materials and has potentially lower costs and higher quality material. During the thermal reduction of GO some of the partially oxidised sp^3 carbon atoms become fully oxidised to carbonaceous gases such as CO_2 and the rest is reduced to the sp^2 graphene product [18] The temperature for successful exfoliation is of great importance. Thermal analyses show that most of the functional groups can be removed at a temperature lower than 300 °C [20]. If the temperature is low, the thermal deoxygenation of GO is less complicated and more controlled. The low exfoliation temperature (as low as 200 °C) is a breakthrough for the mass production of rGO, which might result in low cost and sustainable energy consumption [20]. In order to lower the deoxygenation temperature we performed the additional treatment of GO prior the use, which later is synthesized accordingly to the modified Hummer's method [17], with MTBE and acetone by ultrasonification.

2.Experimental

The graphite flakes and other reagents were purchased from "Sigma-Aldrich" and used without further purification. IR-spectra were recorded on "Perkin-Elmer Spectrum 100" spectrometer by using "Universal attenuated total reflectance" accessory with 3 bounces. The X-ray powder diffraction (XRD) patterns were recorded on a „Bruker D-8 advanced" diffractometer with the "LynxEye" detector using the monochromatic CuKα radiation. The experimental data were collected over the 2θ range of 5 – 60 °. The thermal properties were characterised by thermogravimetric analyser "STA 6000" with sample mass 30 mg in a pure nitrogen flow of 20 ml/min and hyphenated Pyr-GC-MS system. In Pyr-GC-MS system the samples of graphite oxide were pyrolyzed using "CDS Analytical Inc. Pyroprobe 5200" pyrolysis rig at different experimental conditions. Gas chromatograph GC - 2010 Plus "Shimadzu" with FID detector was connected with mass spectrometer "GCMS-QP2010 Ultra, Shimadzu" to obtain the analysis results of the gases evolved. The chromatographic separation was carried out with the use of "Restek" capillary column – "Rtx-XLB" (60 m, 0.25 mm ID, film thickness 0.25 μm) with helium as a carrier gas. The ion source temperature was set at 220 °C, the injection temperature was 300 °C, the oven temperature program was 10 min at 30 °C, 40 °C/min to 270 °C, and 5 min at 270 °C.

The graphite oxide has been prepared according to the modified Hummer's method [21] by the use of 10 g of graphite flakes. Molten graphite flakes were mixed with conc. H_2SO_4, then $KMnO_4$ and $NaNO_3$ were gradually added with continuously stirring and cooling and the oxidation process at temperature 30 – 35 °C during 2h was performed. After them the 25% H_2O_2 solution was slowly added. The mixture was filtrated and then using centrifugation (9000 rpm) washed several times with 5 % HCl solution. The obtained by centrifugation tar-like graphite oxide was washed with de-ionized water repeatedly until the filtrate became neutral and don't contain the chloride ions. The obtained tar-like sediment after centrifugation was twice suspended in acetone and twice in MTBE by the use of the laboratory ultrasonic cleaner over 5 minutes. After centrifugation and drying at room temperature for 24 h following 4 h in vacuum at 50 °C, 22.8 g of a black coal-like product was obtained. CHNS microanalysis was performed using "Euro Vector EA 3000".

3.Results and discussion

FTIR spectrum of the synthesized GO significantly differs from the spectrum of the graphite and has absorption bands characteristic to oxygen functionalities (Fig.1).

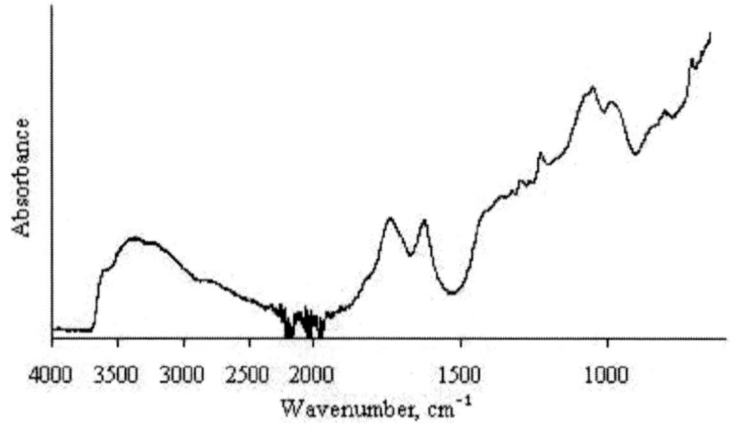

Figure 1. The FTIR spectrum of GO

IOP Conf. Series: Materials Science and Engineering **77** (2015) 012033 doi:10.1088/1757-899X/77/1/012033

There are absorption bands of stretching vibration of different OH and C-H groups in 3000 - 3400 and 2800 cm^{-1} region respectively. Stretching vibration of carboxyl groups and carbon-carbon double bonds at 1736-1710 and 1620 cm^{-1}, respectively. Also the characteristic C-O-C absorption and C-O bands at 1200 and 1045 cm^{-1}. The chemical oxidation of the graphite flakes causes the essential change of the XRD patterns (Fig. 2) of the graphite. The obtained sample has a new intensive peak at 2θ = 9.66 ° (d002) together with the weak peak at 20 °. The intensive peak at 2θ = 9.66 ° corresponds to the layered structure with an interlayer distance approximately at 0.915 nm and confirms the graphite-like stacked morphology of the GO. A bit larger basal spacing than usual [21, 22] by approximately 0.1 nm shall weaken the attractive forces between layers and enable the exfoliation.

It is known that the recovery of the graphene structure has been made by the deoxygenation of the graphite oxide after or simultaneously with the exfoliation. The removal of the oxygen functionalities from the graphite oxide and employment of the obtained colloidal suspension of the reduced graphene oxide is one of the most promising techniques to produce electrically conductive graphene-based platelets on a large scale [22]. The deoxygenation often has been performed utilizing chemical reductants [22, 23]. It is known that the reduced graphene oxide can also be produced from the GO by the thermal exfoliation and reduction [20]. As in the case of chemical reduction the properties of the product of thermal deoxygenation depends on the methods of preparation of the graphite oxide as well as of the realization of the thermal treatment, but the thermal deoxygenation is likely to be easier to implement and more sustainable process.

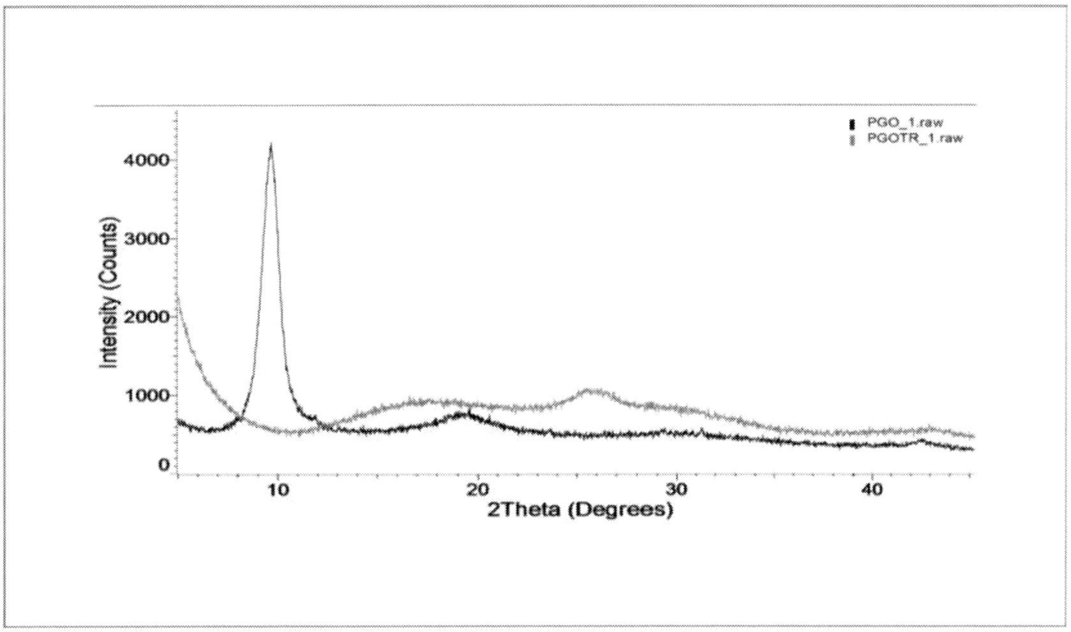

Figure 2. The XRD patterns of GO before and after the deoxygenation.

In order to investigate the thermal deoxygenation process we had begun with the thermogravimetric investigation of the synthesized GO. Thus, to the data reported [20, 24] considerable mass loss of the graphite oxide occurs at temperatures close to 200°C

and is accompanied by the release of CO_2, CO and H_2O. By employing solvents or dispersion in propylene carbonate the deoxygenation temperature has been decreased to 150 °C [25]. Our experiments showed that in all the cases when heating rate exceeds 1 °C/min an explosion like mass loss at 130 °C occurs indicating spontaneous exfoliation and deoxygenation of GO (Fig. 3). Analogous decomposition of GO obtained without the treating with acetone and MTBE proceed at 193 °C.
The data obtained from FTIR spectra of the products thrown out from the sample pan do not show the presence of oxygen containing functional group and the X-ray spectra show a weak and wide band in the region 22-26 °, while the peak at $2\theta = 10.5$ ° is absent (Fig. 2). The broad and weak peaks of reduced graphene oxide correspond to interlayer distances of 0.35 – 0.46 nm and can be related to the non-exfoliated part of reduced GO, obtained after the removing intercalated water molecules, carboxyl, hydroxyl and epoxide groups.

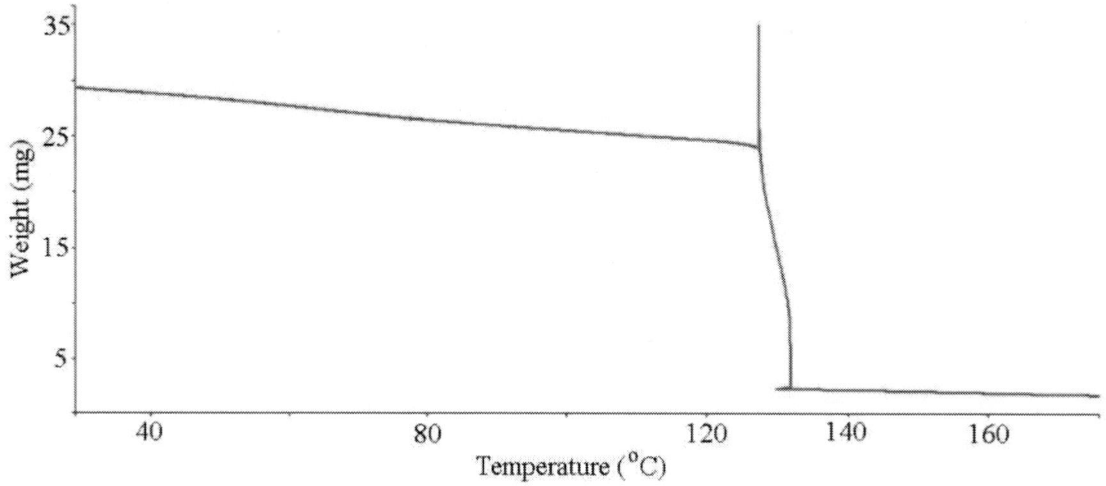

Figure 3. TG curve of mass loss of the synthesised GO at heating rate 5 °C/min.

Only when the heating rate is around 1 °C/min it was possible to register the mass loss curve (Table 1) what is with agreement with the reported earlier [25]. The mass loss curve consists of at least four stages of the anticipated H_2O, CO and CO_2 release, but the DTG curve allow established at least three steps of mass loss. The first step is through at 100 °C and can be attributed to the release of the absorbed water. Upon further heating over 100 °C the deoxygenation proceeds with the maximal rate at 158,4 °C.

Table 1. Mass loss of GO by heating rate 1 °C/min

No	Temperature interval, °C	Mass loss, %
1.	30-100	16.0
2.	100-185	27.4
3.	185-300	9.6
4.	30 - 750	60.6

Using the trapping option of "Pyroprobe 5200" it was possible to register the composition of the gas evolved upon heating at different temperatures. At the temperatures lower than 100 °C the evolved gas consisted mostly of water (78%), but also included considerable amount of carbon monoxide and carbon dioxide (21%), and acetone (0,7 %) what indicates that the decomposition of the labile oxygen containing functional group starts at unexpectedly low temperatures, however solvent isn't fully removed upon heating in vacuum at 50 °C. Fig. 4 shows that the evolution of carbon oxides and water proceeds until 800 °C and is part of the complex mechanism of deoxygenation process [23, 26]. The evolution of sulfur dioxide starts at approximately 200 °C reaches the highest rate at 280 °C and proceeds until reaches temperatures close to 800 °C.

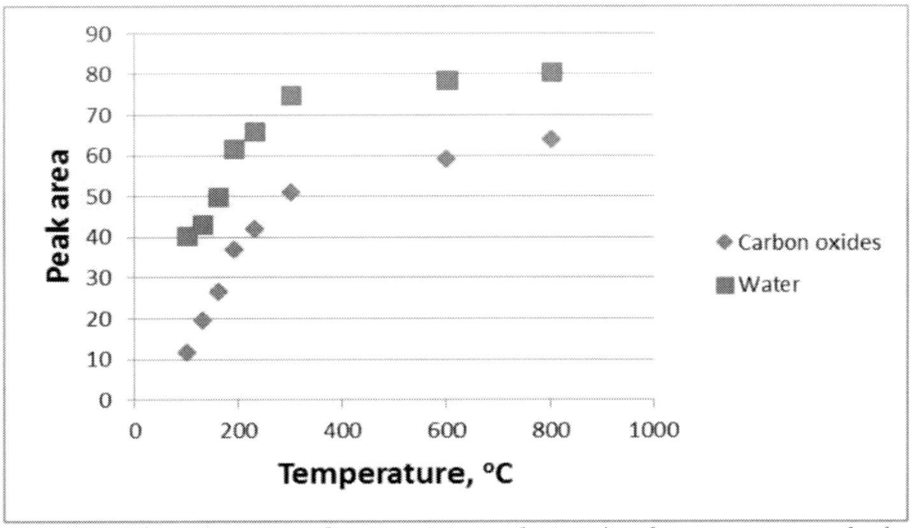

Figure 4. Integral peak areas of $CO + CO_2$ and H_2O in chromatograms obtained by step-by-step heating of GO in the temperature intervals 30-100, 100-130, 130-160, 160-190, 190-230, 230-300, 300-600 and 600-800 °C by heating rate 100 °C/min. Differential characteristics were obtained from thermal desorption and analysis of the gases evolved during each interval.

4.Conclusions

The obtained data indicates that the course of thermal deoxygenation is severely dependent on the pre-treatment of the graphite oxide with the volatile organic solvent. Such pre-treatment reduce the content of adsorbed water and decrease the temperature of thermal deoxygenation and exfoliation, what allows producing the reduced graphite oxide under milder conditions. The temperature of the deoxygenation protocol for the pre-treated graphite oxide is appreciably lower than the characteristic temperatures of deoxygenations of graphite oxide. During the thermal deoxygenation not only the H_2O, CO and CO_2, but also small amounts of sulphur dioxide and acetone has been released, what allow as consider that all parts of process (oxidation, treatment with organic solvents and thermal shock) influenced the chemical composition and structure of reduced graphene oxide.

5.Acknowledgements

The financial support from the program "IMIS" is gratefully acknowledged.

6.References

[1] A.K.Geim, K.S.Novoselov. The rise of graphene. Nat. Mater, 2007, 6, 183-191

[2] W.Choi, I.Lahiri, R.Seelaboyina, Y.S.Kang. Synthesis of graphene and its applications: a review. Critical Reviews in Solid State and Materials Sciences, 2010, 35, 52-71

[3] K.P.Loh, Q.Bao, P.K.Ang, J.Yang. The Chemistry of graphene. J.Mater.Chem., 2010, 20, 2277-2289

[4] Y.Zhu, S.Murali, W.Cai, X.Li, J.W.Suk, J.R. Potts, R.S. Ruoff. Graphene and Graphene Oxide: Synthesis, Properties, and Applications. Advanced Materials, 2010, 22 (35), 3906–3924

[5] M.D.Stoller, S.Park, Z.Yanwu, J.An, R.S.Ruoff. Graphene-Based Ultracapacitors. Nano Lett. 2008, 8, 3498–3502.

[6] N.Mohanty, V.Berry. Graphene-Based Single-Bacterium Resolution Biodevice and DNA Transistor: Interfacing Graphene Derivatives with Nanoscale and Microscale Biocomponents. Nano Lett. 2008, 8, 4469–4476.

[7] G.Jo, M.Choe, S.Lee, W.Park, Y.H. Kahng, T.Lee. The application of graphene as electrodes in electrical and optical devices. Nanotechnology, 2012, Volume 23 Number 11, doi:10.1088/0957-4484/23/11/112001

[8] Y.Sun, G.Shi. Graphene/polymer composites for energy applications. Journal of Polymer Science Part B: Polymer Physics, 2013, 51 (4), 231-253

[9] K.S.Kim, Y.Zhao, H.Jang, S.Y.Lee, J.M.Kim, J.H.Ahn, P.Kim, J.Y.Choi, B.H.Hong. Large-Scale Pattern Growth of Graphene Films for Stretchable Transparent Electrodes. Nature 2009, 457, 706–710.

[10] X.Li, W.Cai, J.An, S.Kim, J.Nah, D.Yang, R.Piner, A.Velamakanni, I.Jung, E.Tutuc, S.K.Banerjee, L.Colombo, R.S.Ruoff. Large-Area Synthesis of High-Quality and Uniform Graphene Films on Copper Foils, Science 2009, 324, 1312-1314

[11] S.De, J.N.Coleman. Are There Fundamental Limitations on the Sheet resistance and Transmittance of Thin Graphene Films? *ACS Nano*, 2010, 4 (5), 2713–2720.

[12] Y.Wu, B.Wang, Y.Ma, Y.Huang, N.Li, F.Zhang, Y.Chen. Efficient and Large-Scale Synthesis of Few-Layered Graphene Using an Arc-Discharge Method and Conductivity Studies of the Resulting Film. Nano Res., 2010, 3, 661-669

[13] J.A.Robinson, C.P.Puls, N.E.Staley, J.P.Stitt, M.A.Fanton, K.V.Emtsev, T.Seyller, Y.Liu. Raman Topography and Strain Uniformity of Large-Area Epitaxial Graphene, Nano Lett, 2009, 9, 964-968

[14] L.Zhi, K.Mullen. A bottom-up approach from molecular nanographenes to unconventional carbon materials. J. Mater. Chem., 2008, 18, 1472–1484

[15] X.Cui, C.Zhang, R.Hao, Y.Hou. Liquid-phase exfoliation, functionalization and applications of graphene. Nanoscale, **2011**, 3, **2118-2126**

[16] S.-Y.Bae, I.-Y.Jeon, J.Yang, N.Park, H.S.Shin, S.Park, R.S. Ruoff, L.Dai, J.-B.Baek. Large-Area Graphene Films by Simple Solution Casting of Edge-Selectively Functionalized Graphite. ACSNANO, 2011, 5 (6), 4974–4980

[16] X.Zhang , J.Xin, F. Ding. The edges of graphene. Nanoscale, 2013, 5, 2556-2569

[17] W.S.Hummers, R.E.Offeman. Preparation of Graphitic Oxide. J. Am. Chem. Soc. 1958, 80, 1339.

[18] D.Krishnan, F.Kim, J.Luo, R.Cruz-Silva,L.J. Cote, H.Dong Jang, J.Huang. Energetic graphene oxide: Challenges and opportunities. Nano Today, 2012, 7, 137-152

[19] D.W.Boukhvalov, M.I.Katsnelson. Modeling of Graphite Oxide. J. Am. Chem. Soc. 2008, 130, 10697–10701.

[20] H.Jeong, Y.P.Lee, M.H.Jin, E.S.Kim, J.J.Bae, Y.H.Lee. Thermal stability of graphene oxide, Chem.Phys.Lett., 2009, 470, 255-258

[21] K.Yin, H.Li, Y.Xia, H.Bi, J.Sun, Z.Liu, L.Sun. Thermodinamic and Kinetic Analysis of low-temperature thermal reduction of graphene oxide. Nano-Micro Lett., 2011, 3 (1), 51-55

[22] S.Park, J.An, J.R.Potts, A.Velamakanni, S.Murali, R.S.Ruoff. Hydrazine-reduction of graphite- and graphene oxide. Carbon, 2011, 49, 3019-3023

[23] S.Pei, H.Cheng. The reduction of graphene oxide. Carbon, 2012, 50 (9), 3210–3228

[24] H.L.Poh, F.Šaněk, A.Ambrosi, G.Zhao, Z.Sofer, M.Pumera. Graphenes prepared by Staudenmaier, Nanoscale, 2012, 4 (11), 3515-3522

[25] W.Chen, L.Yan. Preparation of graphene by a low-temperature thermal reduction at atmosphere pressure. Nanoscale, 2010, 2, 559–563

[26] M.Acik, Y.J. Chabal. A Review on Thermal Exfoliation of Graphene Oxide. Journal of Materials Science Research, 2013, 2 (1), 101-112

IOP Conf. Series: Materials Science and Engineering **77** (2015) 012034 doi:10.1088/1757-899X/77/1/012034

Optical Properties of Natural and Synthetic Beryl Crystals

V. Skvortsova[1], N. Mironova- Ulmane[1], L. Trinkler[1], V. Merkulov[2]

[1]Institute of Solid State Physics, University of Latvia
[2]Institute of Solid State and Semiconductor Physics NASB, Minsk, Belarus

e-mail: vera@cfi.lu.lv

Abstract. The results of investigation of photoluminescence and UV-Visible absorption spectra of natural beryl crystals from Ural Mountains before and after fast neutron irradiation and synthetic crystal grown in Belarus and Russia are presented. Photoluminescence (PL) spectra of synthetic beryl crystals contain a broad band with maxima 740 nm excited both by UV light (λ_{ex} = 260 nm, 271 nm) and laser excitation (λ_{ex} =263 nm). This band is connected with Fe^{2+} ions. The temperature lowering down to 8 K leads to appearance of narrow lines in the 680 - 720 nm regions. Emission lines observed in the luminescence spectra are connected with electron transition $^2E_g \rightarrow {}^4A_{2g}$ of the Cr^{3+} ions: R-lines (682.5 nm) arise from isolated Cr^{3+} ions occupying Al^{3+} sites; N-lines (691, 698, 703, 706 and 711 nm) arise from several types of exchange-coupled pairs of Cr^{3+} ions occupying first, second and third nearest and related neighbour Al^{3+} sites. It is shown that the absorption bands in the 690-580 nm region of natural pale blue beryl crystals caused by neutron irradiation belong to a complex center, which consists of Cr^{3+} ions and radiation defect - F or F^+- center. Presence of Fe^{2+} ions contributes to the persistence of the complex defect.

1. Introduction

Beryl ($Be_3Al_2Si_6O_{18}$) is a ring silicate with a hexagonal structure, space group P6/mcc ≡ D^2_{6h} [1, 2], in which Al^{3+} ions are located in octahedral coordination, and Si^{4+} and Be^{2+} ions - in tetrahedral coordination. The point symmetry of Si, Be and Al sites is C_s, D_2 and D_3, respectively. The dominant structural units are rings of Si_6O_{18} composition. The Si_6O_{18} rings lay one above the other along the c-axis forming intercommunicating channels of quite large dimensions. The colour of beryl $Be_3Al_2Si_6O_{18}$ is usually determined by its Fe^{2+} and Fe^{3+} content. The golden yellow colour is attributed to Fe^{3+} ions [3]. Presence of Fe^{2+} produces pale blue color. If Fe^{2+} and $Fe^{3+}$are present simultaneously, the color of crystal is dark blue. Green color in iron-containing beryl usually results from a mixture of blue and yellow components. Green color can also come from Cr^{3+} as it is found in various emeralds [4]. Beryl is naturally transparent; however inclusions and impurities may make it opaque.

Natural and synthetic minerals of beryl are used not only as gemstone but have also practical industrial application. Beryl crystals find application as laser and dosimetry materials [5, 6]. Small amount of impurities and structural defects have a significant effect on the physical properties of crystals, chemical and radiation resistance of products and devices manufactured on the basis of beryl crystals. Therefore it is important to estimate the present material defects and their effect on properties of material. Optical absorption (OA) spectra in the visible range were subject of several authors' reports on beryl [7-12].

The paper presents study of optical properties of natural crystals before and after fast neutron irradiation and synthetic beryl crystals that may provide a way to address the physical processes responsible for the creation of the colour centres.

2. Materials and Methods

Natural and synthetic beryl $Be_3Al_2Si_6O_{18}$ samples were obtained from different sources: natural crystals from Ural Mountains (Beryl Natural green-1, 2 and 3 and Beryl Pale blue) and synthetic crystals grown in Belarus (Beryl M) and Russia (Beryl Flux and Beryl Cr 0.1%). The samples thickness was in the range of 0.5 –2.15 mm for natural crystals and in the range of 0.475 to 5.75 mm for synthetic crystals.

The neutron irradiation was carried out in the Latvian 5 MW water-water research reactor. The fluence of fast neutrons with energy > 0.1 MeV was in the range 10^{14} – 10^{16} cm^{-2}. A cadmium filter was used for thermal neutrons absorption. During the neutron irradiation the samples temperature did not exceed 313 K.

A conventional experimental setup was used for measurements of photoluminescence (PL). Two types of UV light sources were used in the excitation channel of the setup: 1) a solid state laser DTL-389QT (λ =263 nm), with nominal pulse energy 4.7 mkJ at 1kHz, and 2) a deuterium lamp LDD-400 with an attached grating monochromator MDR-2. The PL signal was analysed by Shamrock monochromator SR-303i-B and detected by a CCD camera DV420A-BU2.

Absorption spectra were measured either with a "Specord 210" (Analytikjena) double-beam spectrophotometer (190-1100 nm operation range) or with a "Specord M-40" two-beam spectrometer (200-900 nm operating range). Optical measurements before and after irradiation were carried out at room temperature (RT). For photoluminescence measurements at low temperatures a sample was put into refregerator "Janis" providing temperatures from 8 to 300 K.

3. Results and Discussion

Fig. 1 and 2 show the photoluminescence spectra for different beryl crystals at various types of excitation. Photoluminescence spectra of beryl crystals contain a broad band with maxima 740 nm at UV light (λ_{ex} = 260 nm, 271 nm) as well as at laser excitation (λ_{ex} = 263 nm). This wide band is connected with Fe^{2+} ions [13]. When beryl crystals are exposed to UV light from the lamp, samples show a weak luminescence (Fig. 1, curve 3, 4 and Fig.2, curve 1, 3), use of more intense laser excitation allows additional observation of narrow line at 685 nm. The photoluminescence spectra of "Beryl Flux" at λ_{ex} = 260 nm and different temperatures are given in Fig. 3. The temperature lowering down to 8 K leads to appearance of narrow lines in the region from 680 to 720 nm. Some zero-phonon lines were observed in the electron transition $^2E_g \rightarrow ^4A_{2g}$ region of the Cr^{3+} ions: R-lines (682.5 nm) arising from isolated Cr^{3+} ions occupying Al^{3+} sites; N-lines (691, 698, 703, 706 and 711 nm) arising from several types of exchange-coupled pairs of Cr^{3+} ions occupying first, second and third nearest and related neighbour Al^{3+} sites [14]. Moreover in the photoluminescence spectra at low temperature there appears a band with maxima at 525 nm, its intensity increases with temperature decrease, which possibly could be explained by charge exchange between impurity ions [15]. PL measurements demonstrate that all studied beryl samples contain emission bands due to the presence of iron and chromium ions.

Besides the iron and chromium induced bands in PL spectra there are also additional bands observed in beryl samples. Excitation with light 260 nm leads to the appearance of 580 nm band in crystal "Beryl M" (Fig.2, curve 1) and 355 and 560 nm bands in crystal "Beryl Flus" (Fig.2, curve 3). The irradiated by fast neutron $\Phi=10^{14}cm^{-2}$ "Beryl Pale blue" crystal excitated by the laser shows a wide band with maxima 480 nm (Fig. 1, curve 2). This band probably is connected with peroxy-radical similar to irradiated topaz crystals [16].

The absorption spectra of beryl single crystals are given in Fig.4. The chromium containing beryl crystals spectra is characterized by two broad bands near 430 and 600 nm as well as by sharp R-lines (Fig. 4, curve 1, 3, 4). The observed absorption bands are ascribed to the electronic transitions of chromium ion: 630 nm transition $^4A_{2g}(F) \rightarrow ^4T_{2g}(F)$ and 431 nm transition $^4A_{2g}(F) \rightarrow ^4T_{1g}(F)$. The optical absorption band at 315–326 nm is caused by charge transfer from O^{2-} to Fe^{3+} in the tetrahedral Be^{2+} site [17]. The wide absorption band with peak in the near-infrared range at 813 nm is observed in natural beryl spectra (Fig.4, curve 2). The band is generally ascribed to internal electron transition of

$^5T_2$ (^5D) → ^5E (^5D) of Fe^{2+} ions localized in octahedral aluminium sites of beryl [18-21]. The absorption spectra of fast neutron irradiated Natural green beryl-2 crystal are shown in Fig. 5.

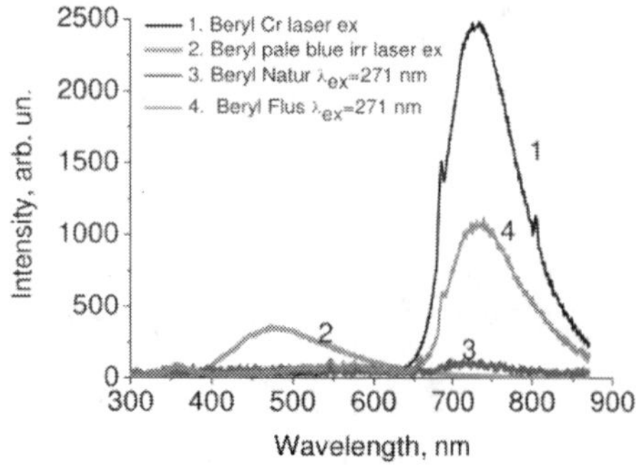

Figure 1. Photoluminescence spectra of: 1. Beryl Cr 0.1%, 2. Beryl pale blue after irradiation, 3. Beryl natural green-1, 4. Beryl flux. 1, 2 under laser excitation λ_{ex} =263 nm; 3, 4. under lamp excitation λ_{ex} =271 nm.

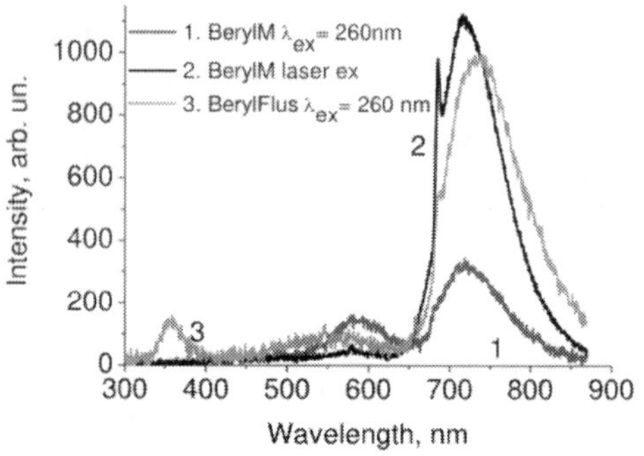

Figure 2. Photoluminescence spectra of: 1.Beryl M under lamp excitation λ_{ex} =260 nm, 2. Beryl M under laser excitation λ_{ex} =263 nm, 3. Beryl Flux under laser excitation λ_{ex} =263 nm.

Irradiation produces additional bands with maxima: 286, 370, 500 and 667 nm (Fig.5, curve 3). Most probably these bands are due to anion intrinsic defects (the F^+, F, F_2, F_2^+ and F_2^+ centers) similar to those observed in Al_2O_3 [22, 23]. The absorption spectra of natural pale blue beryl crystals before and after fast neutron irradiation $\Phi=10^{16}cm^{-2}$ are given in Fig. 6. Prior to irradiation there were no absorption bands associated with electronic transitions of chromium ions $^4A_{2g}(F) \rightarrow {}^4T_{2g}(F)$ (630 nm) and $^4A_{2g}(F) \rightarrow {}^4T_{1g}(F)$ (431 nm). Absorption spectra contain only the band 813 nm associated with Fe^{2+} ions. After the impact of fast neutrons the intensity of 813 nm band is reduced and the absorption edge is shifted due to appearance of an intense absorption band in the UV region. Besides, after irradiation there appear the additional bands with maxima 690, 645.9, 628.1, 606.2 and 588.7 nm (Fig. 6, curve 3). The crystal color is changed from pale blue to light green.

The so called "Maxixe" and "Maxixe-type" color centers observed at 600-700 nm were previously reported in paper [24-26]. Irradiation produces radicals NO_3^0 and CO_3^-, which seem to be related to blue colour in beryl, but crystals quickly lose colour after irradiation. The "nonfading" specimen examined in report [27] had not lost all of its colour over several years storage on a desk.

Green colour of beryl crystals caused by irradiation may be associated with changes in the valence state of iron and chromium ions. The Fe^{3+} impurity ion has the same valence as a substituted Al^{3+} ion. The Fe^{3+} impurity ions prefer to stay in the octahedral Al^{3+} sites in order to maintain the local charge neutrality of the beryl crystal. Consequently, the Fe^{2+} impurity ions are oxidized into the Fe^{3+} state. Possible reaction is: $Fe^{2+} \rightarrow Fe^{3+} + e^-$; $Fe^{2+} + V \rightarrow Fe^{3+} + X_s^-$, X_S^- being the constituent anions or radicals in regular lattice of beryl crystal.

Cr^{3+} impurity ions occupy only octahedral Al^{3+} sites in beryl crystal. Before the irradiation there is no absorption band corresponding to the Cr^{3+} impurity ions. Most likely, that the present chromium impurity ions are the chromium ions with the lower valence, namely bivalent Cr^{2+} impurity ions. Possible processes for valence change in chromium are the following: 1. $Cr^{2+} \rightarrow Cr^{3+} + e^-$; 2. $Cr^{2+} + V \rightarrow Cr^{3+} + X_s^-$ [26]. Besides, the addition reaction takes place: $Fe^{3+} + Cr^{2+} \rightarrow Fe^{2+} + Cr^{3+}$. This reaction is reversible. Position of the chromium bands observed in our irradiated Pale blue beryl crystals coincides with Cr^{3+} ions band position in emerald [1], only the bands for irradiated crystals are wider, than in emerald. Broadening of spectral bands is connected with decrease of Cr^{3+} local symmetry. Decrease of the local symmetry is associated with formation of a complex center. We suppose that the bands in region from 690 to 588 nm belong to a complex center, which consists of Cr^{3+} ion and a radiation defect (F or F^+- center – an anion vacancy, which captures one or two electrons). Concentration of F and F^+ centers increases with neutron fluence increasing. The number of perturbed chromium sites also increases and the band becomes more intense. Appearance of the additional bands due to exchange interaction between F or F^+-centers and transition metal impurity ions has been previously noted in MgF_2, $MgAl_2O_4$ and MgO [28-32]. In contrast to paper [24-26] our crystals retain the acquired color. We suppose that presence of Fe^{2+} ions in beryl crystals contributes to the color persistence.

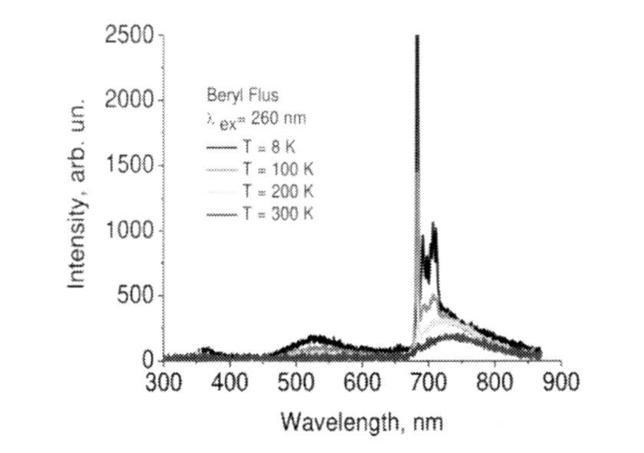

Figure 3. Photoluminescence spectra of Beryl Flux crystal at different temperatures, excitation wavelength 260 nm.

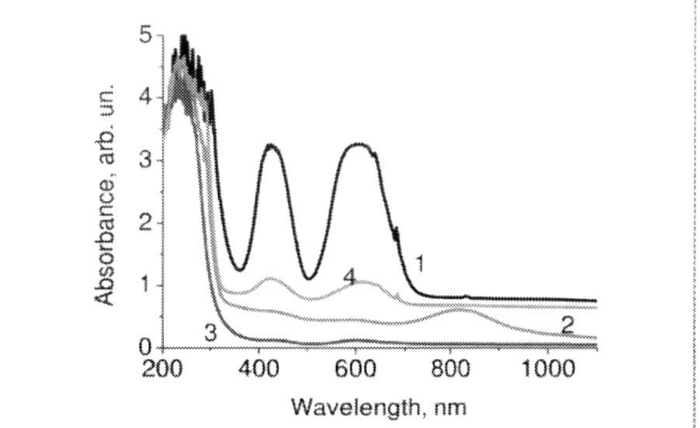

Figure 4. Beryl absorption spectra: 1. synthetic beryl Cr 0.1%, 2. Natural green beryl from Ural-1, 3. synthetic Beryl Flux, 4. synthetic Beryl M.

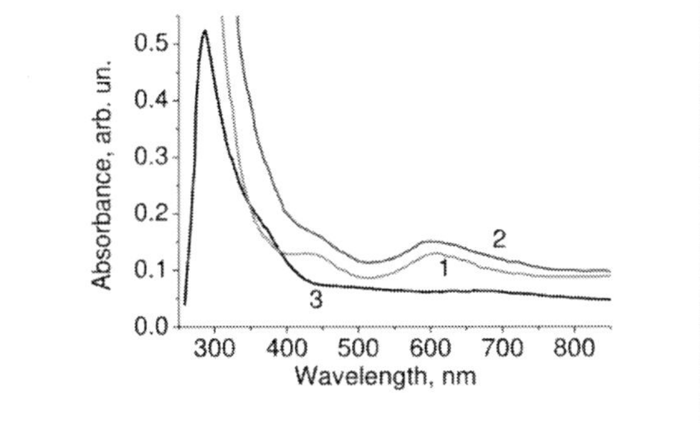

Figure 5. Absorption spectra of Natural green beryl-2: 1. before irradiation, 2. after fast neutron irradiation $\Phi=10^{16}cm^{-2}$, 3.additional absorption spectrum (2-1).

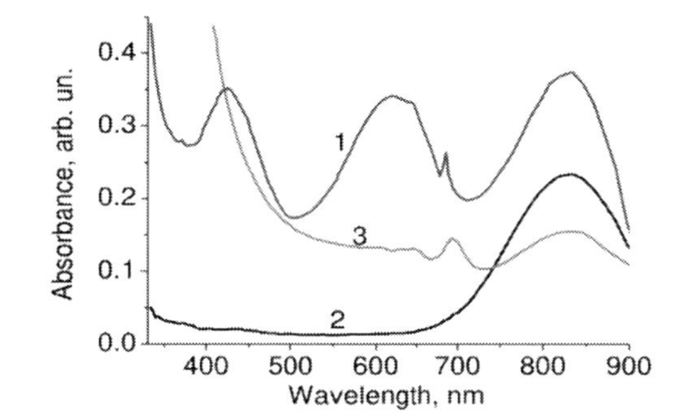

Figure 6. Absorption spectra of: 1. Natural green beryl-3, 2. Pale blue beryl before irradiation, 3. Pale blue beryl after fast neutron irradiation $\Phi=10^{16}cm^{-2}$.

4. Conclusions

The photoluminescence and optical absorption of natural and synthetic beryl $Be_3Al_2Si_6O_{18}$ crystals containing chromium and iron ions and defects produced by fast neutron irradiation fluence 10^{16} cm^{-2} (E> 0.1 MeV) has been investigated.

From the PL measurements it follows that all studied beryl samples contain emission bands due to the presence of iron and chromium ions. In the absorption spectra of natural beryl samples, which have no Fe^{2+} band, irradiation by fast neutrons causes appearance only of bands typical for radiation defects (F or F^+- centers), whereas in absorption spectrum of natural Pale blue beryl crystals, which have Fe^{2+} band, such irradiation leads to rise of the absorption bands in the region 690-580 nm, which belong to

IOP Conf. Series: Materials Science and Engineering **77** (2015) 012034 doi:10.1088/1757-899X/77/1/012034

a complex center consisting of Cr^{3+} ions and radiation defects. Presence of Fe^{2+} ions contributes to the persistence of the complex defect.

5. Acknowledgments
This work was supported by Latvian Science Council Grant No 402/2012

References
[1] Wood D L and Nassau K 1968 *Amer. Mineral.* **53** 777
[2] Morosin B 1972 *Acta Cryst* **B 28** 1899
[3] Ibragimova E M, Mukhamedshina N M, Islamov A Kh 2009 *Inorganic Materials* **45** 162
[4] Platonov A N, Taran M N and Balyatskii V S 1984 *Nature of color gems* (Moscow: Nedra) (in Russian).
[5] Fuxi Gan 1995 *Laser materials* (World Scientific)
[6] Moon P S 1974 *Nuclear Engineering and Technology* **6** 151
[7] Blak A R,.Isotani S, Watanabe S 1982 *Journal of Physical and Chemistry of Minerals* **8** 161
[8] Isotani S, Furtado W W, Antonini R, Dias O L 1989 *American Mineralogist* **74** 432
[9] Krambrock K, Pinheiro M V B, Guedes K J, Medeiros S M, Schweizer S, Castaneda C, BotelhoN F, Pedrosa-Soares A C 2002 *Nuclear Instruments and Methods* **B 191** 285
[10] Viana R R, Jordt-Evangelista H , Da Costa G M, Stern B 2002 *Physics and Chemistry of Minerals* **29** 668
[11] Mittani J C R, Watanabe S, Chubaci J F D, Matsuoka M, Baptista D L, Zawislak F C 2002 *Nuclear Instruments and Methods* **B 191** 281
[12] Skvortsova V, Mironova – Ulmane N, Trinkler L 2013 *Proceedings of the 9th International Scientific and Practical Conference Environment. Technology. Resources* **1** 195
[13] Gaft M, Reisfeld R, Panczer G 2005 *Modern Luminescence Spectroscopy of Minerals and Materials* (Springer-Verlag Berlin Heidelberg)
[14] Edgar A and Hutton D R 1978 *J. Phys. C: Solid State Phys*, **11** 5051
[15] Ivakin Yu D, Danchevskaya M N, Muravieva G P 2011 Vestnik of Moscow State University, ser Chemistry **52**. 18
[16] Priest V, Cowan D L, Yasar H and Ross F K 1991 *Phys. Rew.B* **44** 9877
[17] Edgar A and Vance E R 1977 *Phys. Chem. Mineral.***1** 165
[18] Schmetzer K, Berdesinski W, Bank H. 1974 *Zeitschrift der Deutschen Gemmologischen Gesellschaft* **23** 5
[19] Platonov A N, Taran M N, Minko O E, Polshyn E V (1978) *Physics and Chemistry of Minerals* **3** 87
[20] Bakhtin A I 1985 *Rock-Forming Silicates: Optical Spectra, Crystal Chemistry, Coloration Typomorphism, Kazan:* Kazan State University Press (in Russian)
[21] Marfunin A S 1979 *Spectroscopy, Luminescence and Radiation Centres in Minerals,* (Berlin: Springer)
[22] Atobe K, Nishimoto N, Nakagawa M.1985 *Phys. Stat. Sol. (a)* **89** 155
[23] Lee K H and Crawford J H Jr. 1977 *Phys. Rev.* **15** 4065
[24] Nassau K, Prescott B E, Wood D L 1976 *Am. Mineral* **61** 100
[25] Anderson L O 1979 *J. Gemmology* **16** 313
[26] Chen W, Gu H, Liu J, Wang F, Ma D, Zhu R 2010 *Physica B* **405** 331
[27] Nassau K and B. E. Prescott B E 1981 *Gems &* Gemology 21
[28] Kappers L A, Yun S I, Sibley W A 1972 *Phys. Rev. Lett*, **29** 943
[29] Shinn M D, Windscheif J C, Sardar D K, Sibley W A 1982 *Phys. Rev. B* **26** 2371
[30] Cain L S, Pogatshnik G J, Chen Y 1988 *Phys. Rev.*, **37** 2645
[31] Okada M, Kawakubo T, Seiyama T, Nakagawa M 1987 *Phys. Stat. sol.* (b) **144** 903
[32] Mironova N, Ulmanis U 1988 *Radiation defect and iron group metal ions in oxides* (Zinatne, Riga) (in Russian).

IOP Conf. Series: Materials Science and Engineering 77 (2015) 012035 doi:10.1088/1757-899X/77/1/012035

Raman, electron microscopy and electrical transport studies of x-ray amorphous Zn-Ir-O thin films deposited by reactive DC magnetron sputtering

M. Zubkins, R. Kalendarev, J. Gabrusenoks, K. Smits, K. Kundzins, K. Vilnis, A. Azens, J. Purans

Institute of Solid State Physics, University of Latvia, Riga, Latvia

E-mail: zubkins@cfi.lu.lv

Abstract. Zn-Ir-O thin films on glass and Ti substrates were deposited by reactive DC magnetron sputtering at room temperature. Structural and electrical properties were investigated as a function of iridium concentration in the films. Raman spectrum of Zn-Ir-O (61.5 at.% Ir) resembles the spectrum of rutile IrO_2, without any distinct features of wurtzite ZnO structure. SEM images indicated that morphology of the films surface improves with the iridium content. EDX spectroscopy and cross-section SEM images revealed that the films growing process is homogeneous. Crystallites with approximately 2 - 5 nm size were discovered in the TEM images. Thermally activated conductivity related to the variable range hopping changes to the non-thermally activated before iridium concentration reaches the 45 at.%.

1. Introduction

One of the obstacles to further developments of transparent electronics based on transparent conductive oxide (TCO) thin films is lack of p-type conductors [1]. Thin films with delafossite crystal structure (AMO_2) based on Cu are the most widely studied p-TCO material, but it suffers from low conductivity mainly due to low hole mobility [2]. In addition, elevated substrate temperature, which is required for thin film deposition [3], is not suitable for deposition on polymer substrates. From other p-TCO alternatives, such as binary oxides (ZnO:N, /As, /P, NiO [4]), mixed oxides (In_2O_3-Ag_2O [5]) or spinels ($ZnIr_2O_4$ [6], $NiCo_2O_4$ [7]), the amorphous thin films (Zn-Rh-O [8], Zn-Ir-O [9]) gained much interest due to low deposition temperature. No long range order is required for p-type conductivity in these materials and therefore the conductivity is independent of the film structure.

Earlier study [10] confirmed that the Zn-Ir-O films deposited by reactive DC magnetron sputtering do not show any observable XRD peak features, indicating that these films are x-ray amorphous. Electrical resistivity and transmittance of the films decreases considerably with the iridium content.

Magnetron sputtering is a widely used technique for thin film deposition, providing a high degree of composition control even for complex compound films. For binary metal oxides, either varied target composition or co-sputtering of two metal targets can be used to vary the metal-to-metal ratio in the films. The stoichiometry with respect to oxygen can be varied by varying the partial pressure of oxygen in the sputtering atmosphere.

This paper reports on properties of x-ray amorphous Zn-Ir-O thin films deposited by reactive DC magnetron sputtering. Structural and electrical properties were studied by Raman technique, SEM, TEM, EDX and DC conductivity measurements as a function of temperature.

2. Experimental details

Zn-Ir-O thin films were deposited on glass and Ti substrates, coated simultaneously by reactive DC magnetron sputtering from a metallic Zn (99.95 wt %) target with Ir (99.6 wt %) pieces on the target surface in an Ar+O_2 atmosphere. The substrates were washed in acetone and distilled water before deposition. Vacuum chamber was pumped to base pressure below 1×10^{-5} Torr by a turbo-molecular pump. During the deposition the substrate was kept at room temperature, the sputtering was conducted at 10 mTorr working pressure and 100 W sputtering power. The target to substrate distance was 5 cm. A set of samples was deposited at different oxygen to argon gas ratios (1/4, 1/2/, 1/1) and at different fractions of iridium on the zinc target erosion zone (1, 3, 5, 7, 10, 13, 15 %) to vary the atomic composition in the films. Pure c-ZnO_x and a-IrO_{2-x} thin films were deposited as a reference samples.

Elemental analysis of Zn-Ir-O films was done by X-ray fluorescence spectrometer Eagle III. The Raman shift from pulsed YAG laser (532 nm) was measured to further identify the structure of Zn-Ir-O thin films. Backscattering geometry was used to collect Raman spectra. Microstructural information on selected samples was obtained by high resolution scanning and transmission electron microscopy (SEM:Tescan Lyra and TEM: Tecnai GF20). The powder-like TEM samples were prepared by mechanical scraping from glass substrates. The electrical resistivity as a function of temperature was determined by Hall effect measurement system HMS 5000.

3. Results and discussion

3.1. Zn-Ir-O structural analysis

Normalized Raman spectra of Zn-Ir-O (6.7, 16.0, 24.1, 61.5 at.% Ir), x-ray amorphous iridium oxide (a-IrO_{2-x}) and crystalline zinc oxide (c-ZnO_x) films are shown in Fig. 1. Dashed vertical lines in Fig. 1 show the positions of the characteristic peaks of crystalline IrO_2 (rutile) and ZnO

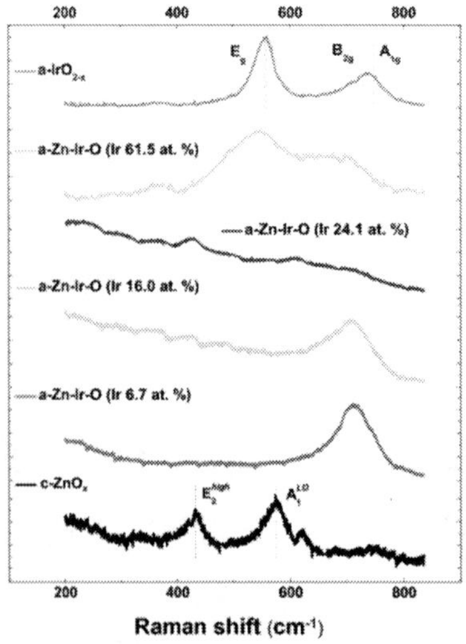

Figure 1. Raman spectra of Zn-Ir-O (6.7, 16.0, 24.1, 61.5 at.% Ir), a-IrO_{2-x} and c-ZnO_x.

(wurzite) [11,12]. Experimental configuration allows to detect only E_2^{high} and A_1^{LO} in ZnO case. In the Raman spectrum of Zn-Ir-O (6.7 at.% Ir) a band at 712 cm^{-1} is well observed. Intensity of this band decreases with the iridium concentration and disappears before concentration reaches 24.1 at.%. It is possible that this band corresponds to any solid solution of Zn and Ir oxides (for example Zn_2IrO_4 or Ir_2ZnO_4). We are not aware of any Raman spectra for these compounds available in the literature. Though Raman spectrum of similar compound Zn_2SnO_4 have strong A_g symmetry band at 668 cm^{-1}, with other Raman active bands being of low intensity [13]. Our ab initio calculation of lattice vibrations for crystalline Zn_2IrO_4 gives high frequency A_g vibration at 601 cm^{-1}. Raman spectrum of Zn-Ir-O (61.5 at.% Ir) mainly consists of IrO_2 phase and resembles the spectrum of a-IrO_{2-x}, without any distinct features of wurtzite ZnO. Due to amorphous Zn-Ir-O structure Raman peaks are extremely wide. The absolute intensity of c-ZnO_x spectrum is significantly lower compared to a-IrO_{2-x} spectrum, therefore, Raman data does not allow to draw definite conclusions about the presence or absence of ZnO phase in the films. Again, to our knowledge, there is no Raman literature data of amorphous ZnO thin films to compare with.

Zn-Ir-O thin films with low iridium concentration (6.7 at.%) have approximately 100 - 200 nm large particles on the surface (Fig. 2(a)). The cross-section SEM images of this film revealed that the bottom surface contains nano-sized holes of approximately 150 nm diameter (Fig. 2(d)).

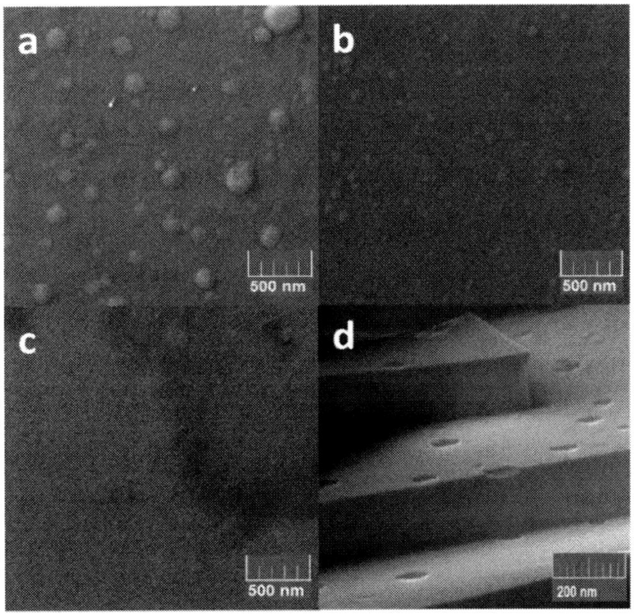

Figure 2. SEM images of Zn-Ir-O (6.7(a), 24.1(b), 40.6(c) at.% Ir and cross-section of 6.7(d) at.% Ir.

It could be speculated that the holes may be related to the way the islands grow and coalesce on the substrate in the beginning of the deposition. Size and concentration of the particles decreases with the amount of iridium (Fig 2(b)) and disappears before the iridium concentration reaches 40 at.% (Fig. 2(c)).

Fig. 3(a,b) shows the TEM images of the Zn-Ir-O (6.7, 67.4 at.% Ir) thin films. Atomic rows observed in the TEM images indicate that all the Zn-Ir-O thin films contain randomly oriented crystallites with approximately 2 - 5 nm size. These nucleation centers do not grow large enough to fill the whole volume of the film, amorphous atomic network is still observed around the nano-crystallites.

200 nm distance of the samples was scanned by energy dispersive x-ray spectroscopy (EDX). EDX results indicate that all the elements (Zn, Ir, O) are distributed uniformly across the film.

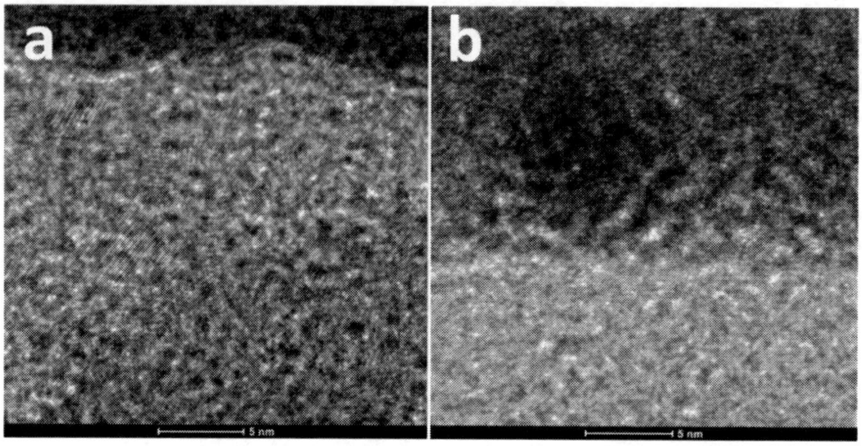

Figure 3. TEM images of Zn-Ir-O (6.7(a), 67.4(b) at.% Ir) thin films.

3.2. Zn-Ir-O electrical properties

The electrical transport properties of the samples were investigated by measuring the DC conductivity as a function of temperature between 90 K and 330 K. Fig 4(a) shows the lnρ as a function of 1000/T (K^{-1}) and T (K). Since the experimental data do not have linear behaviour against 1000/T axis, the experimental data cannot be fitted by the general conduction activation mechanism $\rho(T) = \rho_0 exp\left(-\frac{E_a}{k_B T}\right)$, which is commonly used to characterize the band conduction for semiconductors. Hopping conduction is one of the major features exhibited by

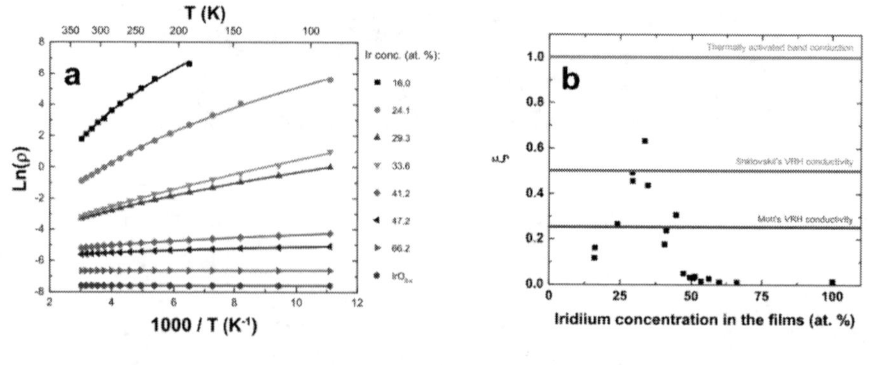

Figure 4. lnρ as a function of 1000/T (K^{-1}) and T (K) (a), and ξ as a function of iridium concentration in the films (b).

semiconductors with disorder structure [14] and this mechanism of conductivity is widely used for the interpretation of the experimental data. Generally, temperature dependence of the resistivity is given by $\rho(T) = \rho_0 exp\left(-\frac{T_0}{T}\right)^{\xi}$. Value of ξ depends on the model of conductivity (ξ=1/4 Motts variable range hopping (VRH) without Coulomb interaction [14], ξ=1/2 Shklovskiis VRH with long range electron-electron Coulomb interaction [15] and ξ=1 thermally activated band conduction). The movement of the charge carriers in electric field in the hopping conductivity is realized as tunnelling transitions with absorption or emission of phonons within a narrow strip of the localized states in the vicinity of the Fermi level. The solid lines in Fig. 4(a) are the theoretical fitting to the experimental data according to $\ln\rho = a + bT^{\xi}$, where a, b and ξ are fitting parameters. Exponent ξ from the fits to the experimental data of all samples as a function of iridium concentration is shown in Fig. 4(b). Zn-Ir-O thin films with the iridium concentration in the range from 0 to 16 at.% are not conductive. A thermal activated transport is found in the 16 – 45 at.% iridium concentration range indicating semiconductor behaviour. Values of ξ indicate that this thermal activated transport is related to the VRH. The almost constant resistivity i.e. not thermally activated conductivity is observed in the 45 – 100 at.% iridium concentration range.

4. Conclusions

Based on SEM and EDX studies, x-ray amorphous Zn-Ir-O thin films deposited by reactive DC magnetron sputtering are homogeneous in composition. However, high resolution TEM images revealed that structure contains randomly oriented nano-crystallites with approximately 2 - 5 nm size. Raman studies indicate that nano-crystallites are related to the IrO_2 phase at high iridium concentration (61.5 at.%). Phase of nano-crystallites at low iridium concentration has not been established.

Thermally activated variable range hopping is the dominated electrical mechanism in the Zn-Ir-O thin films with the iridium concentration from 16 to 45 at.%. Further increase of iridium content changes the electrical mechanism to the non-thermally activated conductivity.

Acknowledgments

This work has been supported by VBBKC L-KC-11-0005 project Nr.KC/2.1.2.1.1/10/01/006,5.3

References

[1] Szyszka B, Dewald W, Gurram S K, Pflug A, Schulz C, Siemers M, Sittinger V and Ulrich S 2012 Current Applied Physics 12 S2-S11
[2] Tate J, Jayaraj M K, Draeseke A D, Ulbrich T, Sleight A W, Vanaja K A, Nagarajan R, Wager J F and Hoffman R L 2002 Thin Solid Films 411 119-124
[3] Banerjee A N and Chattopadhyay K K 2005 Progress in Crystal Growth and Characterization of Materials 50 52-105
[4] Sato H, Minami T, Takata S and Yamada T 1993 Thin Solid Films 236 27
[5] Minami T, Shimokawa K and Miyata T 1998 J. Vac. Sci. Technol. A 16 1218
[6] Dekkers M, Rijnders G and Blank D G A 2007 Appl. Phys. Lett. 90 021903
[7] Windisch Jr. C F, Ferris K M and Exarhos G J 2001 J. Vac. Sci. Technol. A 19 1647
[8] Narushima S, Mizoguchi H, Shimizu K I, Ueda K, Ohta H, Hirano M, Kamiya T and Hosono H 2003 Advanced Materials 15 (17) 1409-1413
[9] Dekkers M 2007 Ph. D. Thesis, Universisty of Twente, Enschede p. 113
[10] Zubkins M, Kalendarev R, Gabrusenoks J, Vilnis K, Azens A and Purans J 2014 Phys. Status Solidi C 11 (9) 1493-1496
[11] Korotcov A V, Huang Y S, Tiong K K and Tsai D S 2007 J. Raman Spectrosc. 38 737
[12] Khan A 2010 J. Pak. Mater. Soc. 4(1) 5
[13] Shen X, Shen J. and You S. J. 2009 J. Appl. Phys. 106, 113523
[14] Mott N F 1968 J. Non-Crys. Solids 1 1
[15] Efros A L and Shklovskii B I 1975 J. Phys. C. 8 L49

IOP Conf. Series: Materials Science and Engineering **77** (2015) 012036 doi:10.1088/1757-899X/77/1/012036

Luminescence of Er/Yb and Tm/Yb doped FAp nanoparticles and ceramics

L. Grigorjeva[1], K. Smits[1], D.Millers[1], Dz.Jankoviča[2]

[1]Institute of Solid State Physics, University of Latvia,
Riga, 8 Kengaraga sr. LV-1063, Latvia
[2]RTU Institute of Inorganic Chemistry, Salaspils, 35 Miera str, LV-2169,
Latvia
L.Grigorjeva: lgrig@latnet.lv

Abstract. The nanoparticles of hydroxiapatite and fluorapatite doped with Er/Yb and Tm/Yb were synthesized and characterized by FTIR, XRD, SEM and TEM methods. The results of up-conversion luminescence studies were presented for the samples as prepared, annealed at 500°C and at 900-1000 °C. At annealing above 800°C the ceramic state was formed. It is shown that fluorapatite host is more appropriate than hydroxiapatite host for rare ions luminescence and up-conversion processes. The post preparing annealing of nanarticles significantly enhanced the luminescence intensity. The Tm/Yb doped fluorapatite shows intense up-conversion luminescence in 790-800 nm spectral region and is potentially useful for biomedical applications.

1. Introduction

Hydroxiapatite $Ca_{10}(PO_4)_6(OH)_2$ (HAp) and fluorapatite $Ca_{10}(PO_4)_6F_2$ (FAp) are well known biomaterials being the main components of bones and teeth. The synthetic HAp and FAp are bio-compatible materials and are studied mainly due to their biomedical applications, for example as coating material for metallic implants, dental implantology, drug delivery and deep tissue bio-imaging [1-3]. The energy levels and up-conversion luminescence mechanisms of Yb/Er and Yb/Tm ions in a number of hosts were described in details in the review article [3]. HAp and FAp are not mentioned among the typical host materials for up-converion luminescence the. The most studied up-conversion luminescence material is NaYF4:Yb:Er. It is shown in [4] that up-conversion luminescence of NaYF4:Yb:Er is applicable for cancer therapy and imaging.

The rare-earth (RE) doped HAp and FAp are prospective luminescent biomarkers, especially if the high efficiency of up-conversion luminescence in nanoparticles (NP) is be achieved.

However, up to date only few papers are devoted to up-conversion luminescence studies in FAp. In [5] the 543 nm and 654 nm luminescence bands were detected under 980 nm excitation due to up-conversion process of FAp:Yb/Ho. According to our knowledge the data of FAp: Yb/Er and FAp:Yb/Tm up-conversion luminescence are not available.

The FAp:Yb/Tm has a great potential for tissue bio-imaging since the Tm ion has one luminescence band at ~800 nm and it fits well in the window of optical transparency of biological tissues. This indicates the importance of present studies.

The FAp has similar to HAp hexagonal structure with space group *P63/m* [6] with only difference - the fluoride ions substitute OH groups in HAp lattice. The structure has two non equivalent Ca^{2+} positions: Ca(I) and Ca(II). The analysis of luminescence spectra and luminescence decay kinetics of Eu^{3+} doped HAp described in [7] showed that Eu^{3+} can substitute for Ca^{2+} in both non-equivalent sites in HAp.
The presence of residual molecular impurities can distort the symmetry of luminescence center, thus modifying the configuration coordinate and leading to luminescence quenching. Therefore in the present investigation the Fourier Transform Infrared absorption (FTIR) study was caried out for detection of some molecular impurities of as prepared samples and those after thermal treatment. Another possible channel for efficient quenching of luminescence is due to excitations trapping at surface states, followed by non radiative recombination. This channel effciency is NP size dependent.
In the present work the FAp codoped with Er/Yb and Tm/Yb NP were synthesized and characterized. The work is devoted to studies of luminescence processes under x-ray excitation and 980 nm laser line excitation.

2. Experimental

2.1. Sample synthesis and annealing.

The FAp synthesis procedure was performed similar to HAp synthesis described in our previous paper [8]. The starting materials were $(NH_4)_2HPO_4$, $Ca(NO_3)_2{\cdot}4H_2O$, Er_2O_3, Tm_2O_3, Yb_2O_3 and NH_4F. The RE oxides were dissolved in HNO_3 (1:3) and the obtained solutions of nitrate salts were added to $Ca(NO_3)_2{\cdot}4H_2O$ and stirred for 30 min at 80°C. The $(NH_4)_2HPO_4$ and NH_4F solutions were kept at constant pH=10 by adding NH_4OH. The obtained solution was drop-wise added to the solution of nitrate salts during 1h at 80°C. The prepared white suspension was stirred for 3h at 80°C and then cooled to room temperature and aged for 4h. The sediment was filtered through 3 μm membranes filter and washed in deionized water. The FAp powders were dried at 110°C 24h. The activator concentrations were calculated in at% units. For samples doped with Er/Yb and Tm/Yb the concentration of activators was 2 at% of Yb and 1 at% of Er or Tm.
The study was carried out on NP as prepared and after annealing at 500°C or 900°C in air. Above ~800°C the ceramic samples were obtained.

2.2. Experimental equipments.

The FTIR absorption spectra were registered with KBr pellet method using spectrometer Bruker EQUINOX55. The X-ray diffractometer (Bruker AXS D8 Advance) was used for structure identification; grain size was calculated from FWHM according to Scherrer's formula.
The Lyra (Tescan) SEM and Tecnai G20 (FEI) TEM equipments were used for samples morphology testing.
The luminescence was excited with X-rays (30 kV, 10 mA, W-target) through a Be window or with 980 nm laser diode irradiation (optical power 0.78 W). For the luminescence spectra registration the monochromator SHAMROC303 coupled with ANDOR iDUS DU401A-BV CCD camera was used. All experiments were carried out at 300K.

3. Results and discussion

3.1. TEM and SEM

The as prepared HAp and FAp NP consisted of cigar-shaped nanorods with diameters in the range of 30-50 nm and ~200 nm lengths (Fig.1a). After annealing the grains enlarged (Fig.1b), achieving diameters 40 – 60 nm and agglomerated; after sintering at high temperature the porous ceramic was obtained (Fig.1c).

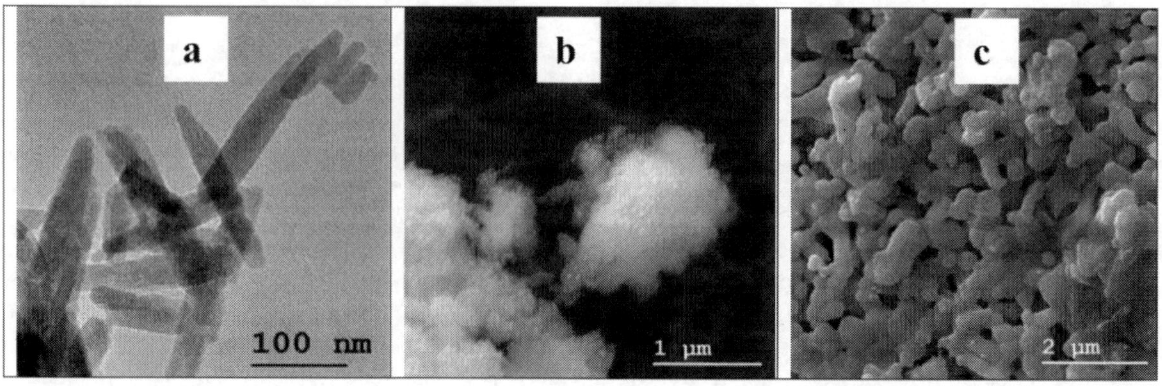

Fig.1. TEM micrograph of the undoped (as prepared) HAp (a), SEM images of FAp:Er/Yb sample annealed at 500 °C 2h (b) and FAp:Er/Yb ceramic sintered at 1000 °C 15 h (c).

3.2. XRD.

The HAp and FAP have hexagonal structure (P63/m). The XRD results for as prepared NP were compared with JCPDS cards number 00-009-0432 (HAp) and 00-15-0876 (FAp). Additional phases in RE doped NP were not detected. The grain sizes calculated according to Scherrer's formula are 20-30 nm for as prepared HAp and FAP and ~40-50 nm after 2h calcinations at 500 °C. These sizes are close to those obtained from TEM and SEM images for the corresponding samples.

3.3. FTIR

The FTIR absorption spectra of the prepared HAp and FAp NP show the absorption peaks associated with stretching and bending modes of the hydroxyl, PO_4^{3-} and carbonate groups [8,9]. The main aim of this study is to analyse the annealing effect on hydroxyl and carbonate group modes in FAp. For the comparison the HAp spectra are shown. The traces of carbonate group bands in as prepared NP were identified according to the [8,10,11]. Fig. 2a and b shows the FTIR spectra for as prepared and annealed HAp and FAp NPs.

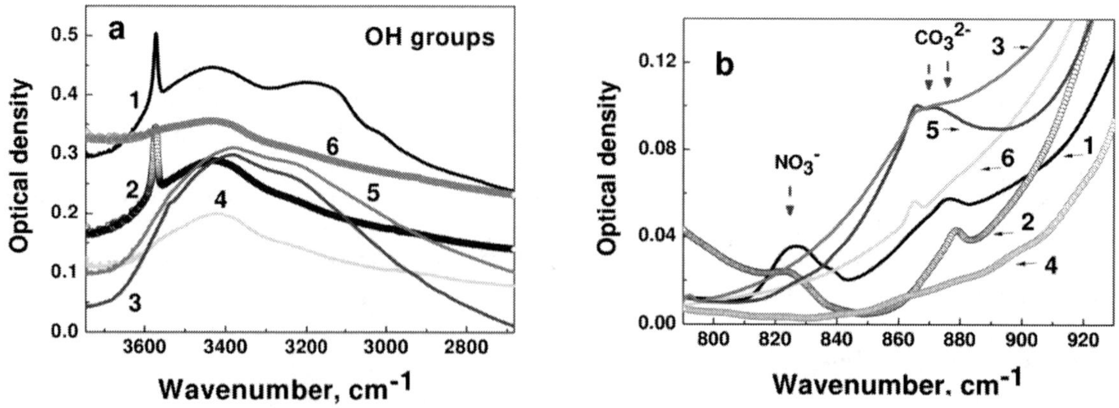

Fig.2. FTIR spectra of HAp and FAp NP in OH stretching (a) and CO_3^{2-}, NO_3^- (b) bending (ν_2) vibration modes region.

(1-HAp as prepared; 2 – HAp annealed at 800 °C 3h; 3 – FAp as prepared; 4 – FAp annealed at 500 °C 2h; 5 - FAp:Er/Yb as prepared; 6 - FAp:Er/Yb annealed at 500 °C 2h).

OH groups. In HAp the well resolved narrow OH band at 3572.8 cm^{-1} is not changing during annealing. The OH groups responsible for this narrow band are the components of HAp chemical composition. The wide bands at 3436.4 and 3170 cm^{-1} are partly reduced during annealing and the OH groups responsible for these bands are located on surface states.

In FAp only the wide OH bands were detected and these OH bands were annealed successfully. It is very important since the OH groups in FAp are the impurity ions and can affect the processes of charge and energy transfer. In FAp samples doped with RE ions the same wide OH bands are detected and annealing processes were observed similarly to undoped FAp.

CO_3^{2-} groups. In HAp NP the band at 878 cm^{-1} due to CO_3^{2-} group is well expressed and the annealing of this band is not efficient up to annealing temperature 800 °C. The CO_3^{2-} group band in FAp (867 cm^{-1}) is slightly shifted relatively to the peak position in HAp and is efficiently annealed even at 500 °C (Fig. 2 b).

NO_3^- groups. In HAp, the band at 825 cm^{-1} is due to nitrate ions [12] and this band is noticeably reduced during annealing. In FAp nanopowders this band was not detected even for the as prepared powders.

The molecular ions were incorporated during synthesis as the fragments from reagents or from the surrounding atmosphere. The presence of such ions can affect luminescence characteristics because these molecular ions can both have their own specific luminescence bands and interact with luminescence centers changing the luminescence yield.

3.4. Luminescence

The RE doped HAp and FAp NP are prospective materials as luminescent biomarkers, especially if the up-conversion luminescence is used. The radioluminescence and up-conversion luminescence under excitation in near infra-red spectral region were studied in the as prepared and annealed in air FAp:Er/Yb and FAp:Tm/Yb samples.

Our study of HAp:Er/Yb NP luminescence has revealed that the intensity of luminescence is low even for NP annealed at 900 °C in air. The reason of luminescence low intensity in RE doped HAp is possibly due to presence of OH^-, NO_3^- and CO_3^- groups with high thermal stability. Thus our attention was drawn to FAp samples since in FAp molecular impurities were annealed effectively (Fig. 2a and b) and FAp has a low phonon energy [10-11] – an important characteristic for up-conversion luminescence.

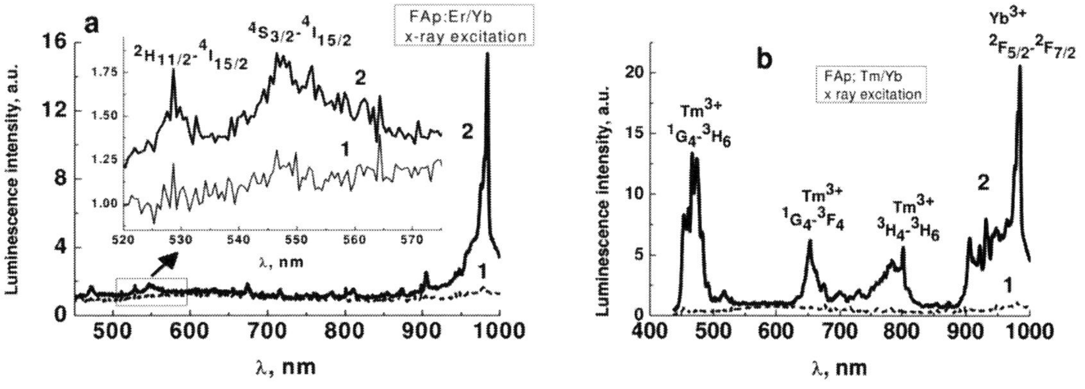

Fig.3. Luminescence spectra under x-ray excitation FAp:Er/Yb (a) and FAp:Tm/Yb (b) NP as prepared (1) and annealed 3h at 900 °C (2)

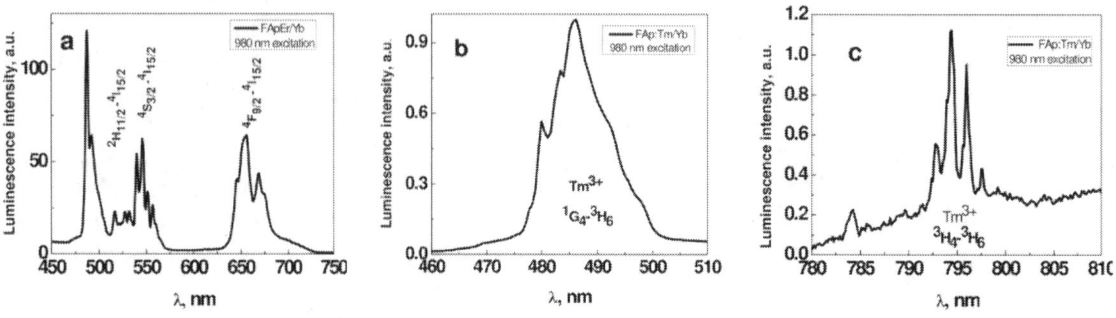

Fig.4. The up-conversion luminescence spectra of FAp:Er/Yb (a) and FAp:Tm/Yb (b,c) annealed 3h at 900°C.

Under X-ray excitation of annealed FAp NP the Yb^{3+} luminescence band (980 nm) is detected along with Er and Tm ion luminescence (Fig.3a and b). The luminescence intensity is negligible in the as prepared NP, but it increases considerably after annealing as well as after sintering (Fig.3a and b).
The up-conversion luminescence spectra of the annealed FAp:Er/Yb samples (Fig.4a) show the Er ion transitions at 525 nm ($^2H_{11/2}$-$^4I_{15/2}$), 545 nm ($^4S_{3/2}$- $^4I_{15/2}$) and 652 nm ($^4F_{9/2}$-$^4I_{15/2}$). The 490 nm band in the up-conversion spectra is the laser diode second harmonic line and is not detected under x-ray excitation. Attention should be drawn to the fact that spectra consist of red-green-blue lines, so white luminescence was obtained in the up-conversion process. This effect is important for new light sources and needs further research.
In FAp:Tm/Yb the ~479-495 nm, ~648 nm and ~795-800 nm luminescence bands are assigned to Tm^{3+} transitions under X-ray excitation. Note that the 648 nm luminescent transition (from $^1G_4$ to $^3F_4$ state) is detected under X-ray excitation and is not observed under 980 nm excitation. The $^3F_4$ state is one of excited states of Tm^{3+} ion involved in the up-conversion process. Therefore it is suggested that in the upconversion process the nonradiative transition from $^3H_5$ state to $^3F_4$ state provides the efficient population of $^3F_4$ state and thus the probability of radiative transition from $^1G_4$ to $^3F_4$ state becomes negligible. The radiative transition from $^3H_4$ state to the $^3H_6$ state is responsible for luminescence within 795 – 800 nm. The more detailed description of the upconversion processes can be found in [3].
The $^3H_4$-$^3H_6$ radiative transition (~800 nm) of Tm $^{3+}$ can be excited with 980 nm laser and is due to energy transfer from Yb^{3+} ion to Tm^{3+} ion in FAp:Tm/Yb. Since this luminescence falls in tissue transparency window, the further studies are important from biological point of view.

4. Conclusions

The single hexagonal crystal structure is confirmed in the as prepared HAp and FAp NP. RE ion doping does not create additional phases. Analysis of FTIR spectra shows that during the post preparing annealing at 500 °C the OH and CO_3^{2-} groups practically disappear in both undoped FAp and RE ions doped FAp.
The FAp NP is more suitable for the observation of RE luminescence and up-conversion processes than HAp NP. Our data of up-conversion luminescence reveal that the post preparing sintering at 900-1000°C of FAp NP significantly enhances the luminescence intensity. The FAP:Tm/Yb provides the intense up-conversion luminescence and potentially

can be used for biomedical applications, especially taking into account that the 800 nm luminescence band ($^3H_4$-$^3H_6$ transition of Tm^{3+} ion) is located close to optical transparency window of biological tissues.

Acknowledments. The work was supported by Latvian Material Science Program, LZP grant No. 302/2012.

References

1. V. J. Shirtliff V J and Hench L L 2003 *J.of Mater.Sci.* **38**, 4697
2. Chen F et al 2011 *Biomaterials* **32** (34) 9031
3. Xue C T 2013 *Laser Photonics Rev.* **7 (5)** 663
4. Wang C et al 2011 *Biomaterials* **32 (26)** 6145
5. Li X et al 2014 *Sci.Rep.* **4** 4446
6. Serret A, Cabanas M and Vallet-Regi M 2000 *Chem.Mater.* **12** 3836
7. Ternane R et al 2001 *Opt.Mater* **16** 291
8. Grigorjeva L et al 2013 *IOP Conf.Series: Mater.Sci and Engieneering* **47** 012005
9. Bakan F, Lacin O and Sarac H 2013 *Powder Technology* **233** 295
10. Lacout J L, Mikou M and Bonel G 1990 *J. of Solid State Chem.* **86** 323
11. Yi H et al 2014 *Phys. Chem.Minerals* **9** 0654
12. Miller F A and Wilkins C H 1952 *Analytical Chemistry* **24** 1253
13. Chen H et al 2006 *Crys.Grow.* **6** 1504
14. O'Donn M D et al. 2009 *J. of the Eur.Cer.Soc.* **29** 377

IOP Conf. Series: Materials Science and Engineering **77** (2015) 012037 doi:10.1088/1757-899X/77/1/012037

Novel method for feedstock production for high efficiency FZ c-Si PV

A Kravtsov[1]
Scientific Director,
SIA "KEPP EU", Riga LV-1034, Latvia

E-mail: doc@keppeu.lv

Abstract. The efficiency of solar cells manufactured from FZ wafers is significantly higher than the efficiency of cells made of CZ or multicrystalline wafers. The high cost of FZ silicon, however, limits the applicability of FZ wafers. The aim of the present work is to create a cost-effective method of FZ feedstock production based on pulling from skull with electron beam heating. This work covers the process of obtaining of raw silicon rods and FZ monocrystals from them.

1. Introduction

The efficiency of solar cells manufactured from float zone (FZ) wafers, thanks to their purity and structure perfection, is significantly higher than the efficiency of cells based on CZ or multicrystalline wafers. The applicability of such wafers is limited because of a) high deprecation costs, due to expensive FZ pullers, and b) the high price of polycrystalline silicon rods – FZ feedstock. The aim of the present work is to create a cost-effective method of FZ feedstock production. It requires crystal of cylindrical shape to be obtained from raw material applicable for PV applications, without further contamination during pulling.

2. Experiments

We based our work on an electron beam single crystals growing technique [1,2] and skull process[3]. Electron beam guns with cooled cathode were used. During the pulling process, two electron beam heaters were used. Each beam was moved over less than an 180 degrees arc with variable radius. The growth was conducted with the use of seed, which was pulled upwards while rotating. The crucible containing raw material stayed motionless. The appearance of the process is shown at figure 1.

[1] To whom any correspondence should be addressed.

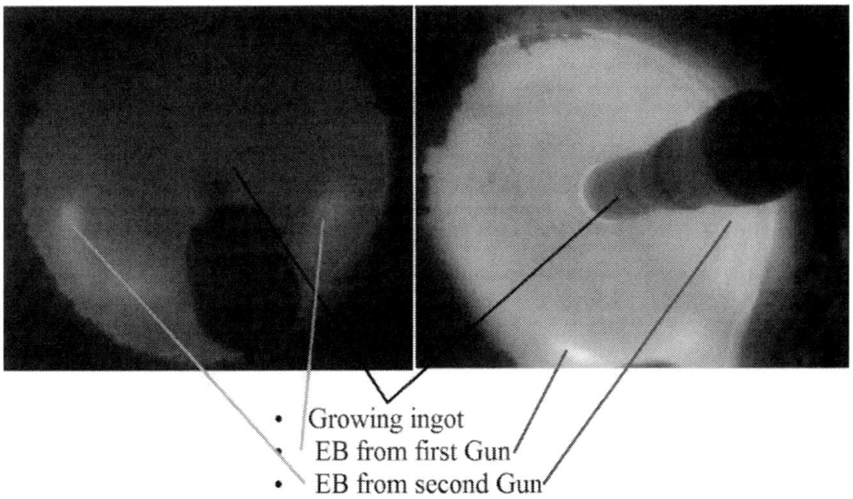

Figure 1. The process appearance.

We used leftover silicon doped by Sb and As to 100 ppma level as feedstock, then purified it prior to ingot pulling by technology described in [4]. After the purification process we pulled ingots as shown at figure 1, then we cut samples from the ends of ingots, and measured these samples by a glow-discharge mass-spectroscopy (GDMS) method. The results are shown in table 1.

Table 1. GDMS measurements of pulled rod in comparison with undoped Si.

Semiquantitative Concentrations (ppbw)	Undoped Si Chunk	Bottom Part of Pulled Rod
B	2.0	272.0
Al	0.7	1.8
Fe	0.4	0.7
Ni	5.2	0.7
Cu	24	10.0
As	1.0	1.8
Sb	0.5	1.4

According to GDMS results, the purity of pulled ingots is on the same level as in standard polysilicon, except for the concentration of boron, which can't be removed by the methods used.

A new generation of experimental furnace was created (shown at figure 2). It consists of melting and pulling (crystal) chambers, with a flap gate valve in between, a pulling mechanism and two electron beam guns, each equipped with its own power system. The process is conducted under residual pressure of not more than 0,05 mbar. The batch is placed in a copper water cooled mold, equipped with elements designed for thermal field management.

Figure 2. Appearance of experimental furnace.

Using this equipment, we have demonstrated the possibility of manually growing crystals of diameter up to 240 mm, and the purification of heavily doped silicon in this kind of process (figure 3). However, the ingots have a very volatile diameter, so for practical application grinding should be applied to ingot surface, thus a lot of weight is lost, driving the price of feedstock to an uncompetitive level. To address this issue, we've organized a new activity together with Elektronikas un datozinatnu instituts (EDI) under contract with SIA "LEO PETIJUMU CENTRS" N1.1.1.-5/-7-14. The first results of this activity (actually the second process from the beginning of cooperation) yielded a much better ingot surface, shown at figure 4. Enhancement was achieved due to the employment of PID-regulation of pulling rate, where the correction to rate value was calculated solely from the deviation between crystal-melt meniscus diameter and its preset value.

Figure 3. Rods pulled with electron beam heating with significant diameter fluctuations to +/- 10 mm.

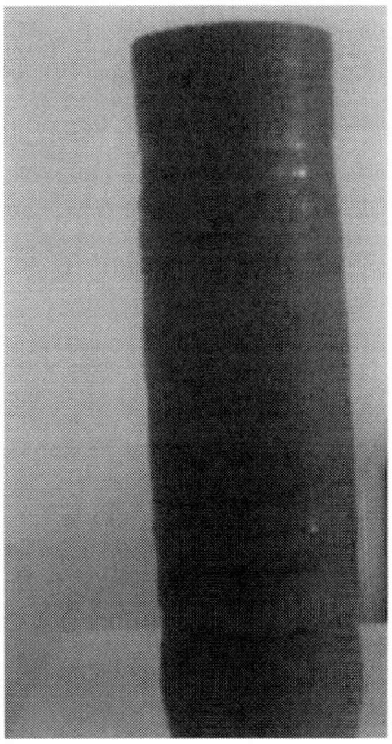

Figure 4. Silicon rod obtained with electron beam heating and enhanced diameter stabilization (with diameter fluctuation of less than +/-5 mm).

The rods obtained were tested for acceptability in FZ single crystal growth at the Leibniz Institute for Crystal Growth, Berlin, Germany (IKZ). Only one out of two rods tested yielded a small, defect-free single crystal with diameter under 80mm, shown at figure 5. In our opinion, the appearance of dislocations can be attributed to alien inclusions, described below.

Figure 5. Place of generation of dislocation at conical part of single crystal, grown from rod obtained on new generation of equipment.

Phase boundary measurements, using LPS of the rods obtained, were made in IKZ (figure 6). Measurements revealed that the crystallization front in rods pulled with the use of electron beam heating is curved towards melt (downwards) and has a substantial sag of ca 15mm for 100mm rod diameter, which can be attributed to the character of the surface heating.

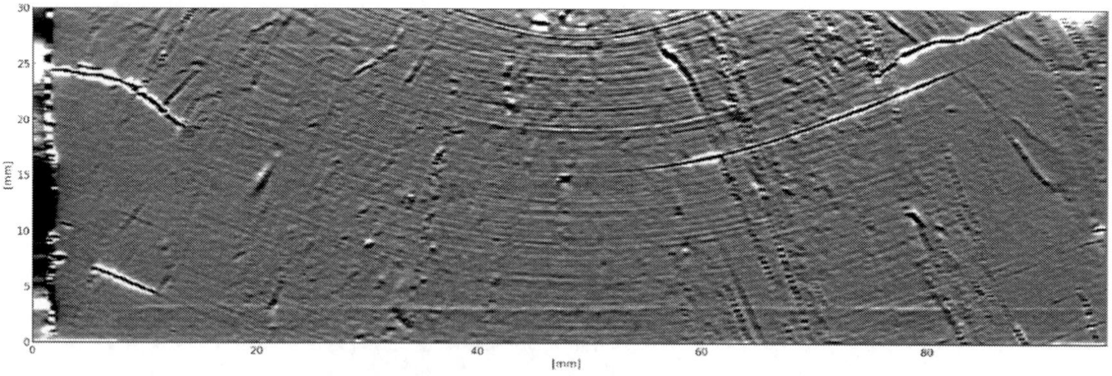

Figure 6. Crystallization front shape for rods grown with electron beam heating.

3. Results and discussions

Since trial growth did not result in obtaining defect-free monocrystals and the crystallization front study did not reveal the source of the problem, the initial research of pulled rods was made together with LU CFI, Riga, Latvia (CFI). Raman spectroscopy and SEM studies were performed at this stage.

Raman spectroscopy (figure 7, made by CFI) was conducted on a plate-shaped silicon sample mirror, polished on both sides with the use of two spectrometers in the 200..2000cm^{-1} range:

- Spex Ramalog (USA) (Resolution: 1 cm^{-1}, laser line: 532 cm^{-1});
- Advantage NIR (USA) (Res.: 3 cm^{-1}, laser line 785 cm^{-1}, objective: 90x).

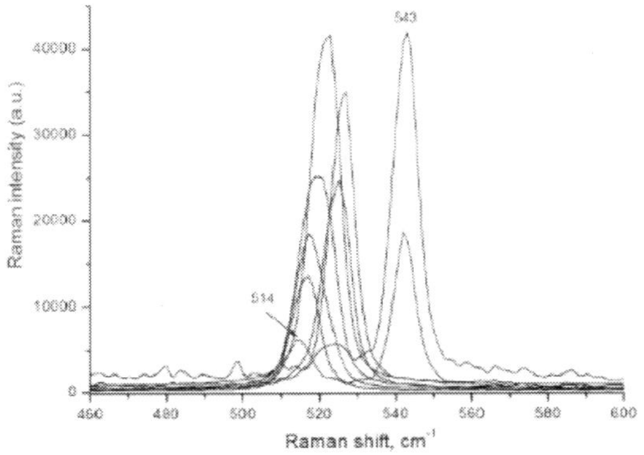

Figure 7. Raman-spectra for different points of silicon sample.

Surface contains monocrystalline (522 cm^{-1} shift) and polycristalline (517 cm^{-1} shift) parts. Also, as yet unidentified peaks with 543 cm^{-1} shift were observed.

Scanning electron microscopy (made by LU CFI, Riga, Latvia) revealed alien inclusions shown at figure 8. Possibly, this inclusion is silicon carbide, however, research is not finished yet and requires to be continued.

Previously, we have published [5] results where full-length, dislocation-free FZ single crystals were obtained from rods pulled in processes with a load weight less than 25kg. From our viewpoint, the difference between the results obtained for small and enlarged loads can be attributed for the most extent to the use of materials which were not previously used in a hot zone. In particular, while creating new hot zone carbon, containing materials were used, which if contacted by electron beam or molten silicon could produce SiC particles. To prevent an appearance of the alien particles described, we are making modifications in hot zone design with the use of carbon-free materials.

Figure 8. SEM analysis of silicon rods.

IOP Conf. Series: Materials Science and Engineering **77** (2015) 012037 doi:10.1088/1757-899X/77/1/012037

4. Conclusions

General results of this work are:

- An experimental furnace for pulling rods applicable in FZ process has been created.
- We successfully created the basis for technologies of:
  - production of competitive FZ feedstock from recycled materials in an arbitrary form, such as chunks, etc;
  - recycling leftovers of the FZ process.
- The possibility of manually growing crystals of diameter up to 240 mm has been revealed.
- The possibility of satisfactory diameter stabilization by means of PID-control of pulling rate, based on the measurement of the diameter of rod-melt meniscus, has been demonstrated.
- The possibility of purification of heavily doped silicon in this kind of process, has been demonstrated.
- The dislocation-free part of single crystal obtained proves the principal possibility of yielding dislocation-free single crystals from rods pulled from up to 100kg loads.
- It is necessary to proceed with working on hot zone design enhancement.

5. References

[1] Movchan B and Zlotin A 1963 Electron beam radial heater for furnaces for pulling monocrystals by Czochralsky method *Russian Patent No. 159293* (Moscow, USSR: Goscomizobreteniy)

[2] Cizek T F 1972 *J. Cryst. Growth* **12** 281-287

[3] Zumbrunnen A D 1979 High Frequency Resistance Melting Furnace *US Patent 4133969*, (Washington, D.C.: U.S. Patent and Trademark Office)

[4] Kravtsov An and Kravtsov Al 2011 Electron Beam Silicon Purification *26th EU PVSEC* (10.4229/26thEUPVSEC2011-2BV.4.7) pp 1814-1816

[5] Kravtsov A 2014 Ingots Pulled with Electron Beam Heating from Skull - a New Feedstock for FZ Crystals Applicable for Solar Cells *IEEE PVSC40* (Denver, CO, USA)

IOP Conf. Series: Materials Science and Engineering **77** (2015) 012038 doi:10.1088/1757-899X/77/1/012038

New Feedstock for c-Si Photovoltaics

Alexey Kravtsov[1], Alexander Shagun[2] and Anatoly Kravtsov[2]
[1]KEPP-service, Ltd, Moscow 117556, Russia
[2]SIA "KEPP EU", Riga LV-1034, Latvia

E-mail: alexeykravtsov@gmail.com

Abstract. Results from functional tests of highly doped silicon purified with electron beam melting, a new feedstock for photovoltaics are presented. Possibility of obtaining dislocation free single crystals from such feedstock in typical industrial processes (CZ and FZ) is shown, crystals' parameters are tested for coherence with requirements for PV silicon.

1. Introduction

Highly doped silicon feedstock, which can be purified by means of vacuum refining, exists on the market. Electron beam melting is an energy-efficient method of vacuum refinement and in [1] we have researched its applicability for refining such feedstock. A series of experiments on new materials for crystalline silicon photovoltaics contains data on the difficulties with using such new materials, unblended with virgin polysilicon, as crystals obtained from such feedstock possess unsatisfactory physical or structural properties. The purpose of the work was to demonstrate the possibility of obtaining dislocation-free single crystals from batches with 100 percent new feedstock in typical industrial Czochralski (CZ) and float zone (FZ) processes, and to test crystals' coherence with requirements for PV silicon.

2. Experiments and results

The furnace consists of a chamber with a cooled base plate in it. A graphite crucible with a quartz crucible inside are placed on the base plate. An electron beam gun is mounted on the top of the furnace to provide heating. The scheme is provided in figure 1[1]. The appearance of the process is shown in figure 2, and the basic principle of the technology was described previously in [2].

[1] To whom any correspondence should be addressed.

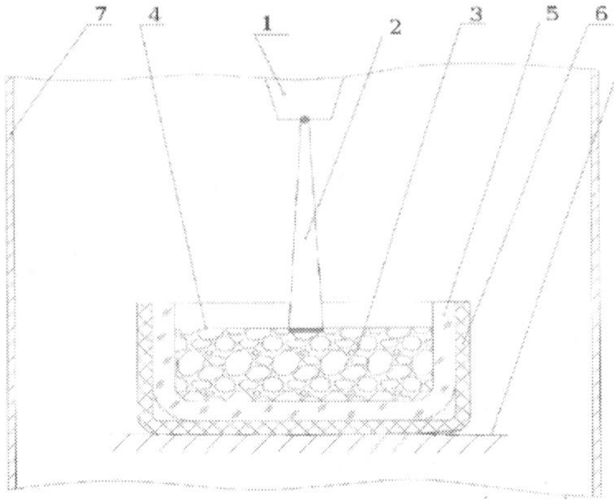

Figure 1. Furnace principal scheme: 1 — Electron beam emitter; 2 — electron beam; 3 — raw material; 4 — melt surface; 5 — quartz crucible; 6 — carbon crucible; 7 — chamber; 8 — cooled base plate.

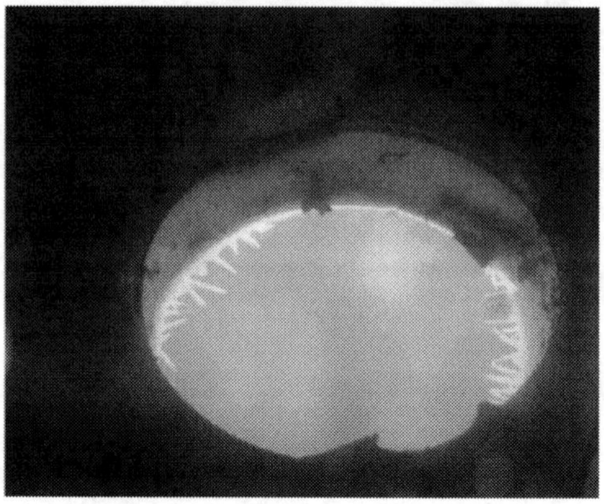

Figure 2. Photo of a process with 50 kgs load.

Three different companies tested rectified material and pulled CZ monocrystals from it. Each company used its standard process of obtaining CZ monocrystals for semiconductor or photovoltaic application. Each company applied chemical treatment before pulling, and all silicon chunks were etched in a mixture of HF and HNO_3 to clean potential impurities off the surface. No slug or similar contaminants were found on melt surface during the process. Seeds of <100> orientation were used for pulling. In all cases, excluding company A, after initial seeding an abrupt with the monocrystal occurred before reaching the target diameter, and the initial crystal was discarded. Then, at the second attempt, full length monocrystal including the end cone, was pulled.

2.1. Tests in company A

Company A used a 45kg load for CZ pulling. A 16" crucible was used and crystal was pulled at 0.8-1.2mm/min rate, with the resulting diameter in the 204-210 mm range. The resulting length was 520 mm (with both cones included), and weight 38.64kg. The ingot was cut into 2 parts, of 184 mm and 223 mm. A thick (3mm) wafer was cut from each part of the ingot, wafers were annealed, then resistivity measurements were made. The lifetime was measured at the bottom of the second part, only. Results are given in table 1.

Table 1. Resistivity, lifetime and shape of ingot parts grown at company A.

Ingot#	Length (mm)	Weight (kg)	Diameter (mm)	Conductivity	Resistivity (Ω*cm) top	Resistivity (Ω*cm) bottom	Lifetime (us)
7-16216-1	184	14.9	210.2	P	3.6-4.0	2.7-3.0	Not measured
7-16216	223	17.3	204-205.5	P	2.8-3.0	2.6-2.7	27-30

2.2. Tests in company B

Company B used a 60kg load for CZ pulling. A 16" crucible was used. After initial seeding an abrupt in the monocrystal occurred before reaching the target diameter, and the initial crystal was discarded. Then, at the second attempt, full-length monocrystal, including the end cone, was pulled. Crystal was pulled at a rate of 0.8-1.2mm/min, and the resulting diameter was in the 154-157mm range. The resulting length was 1777mm, and the weight 53.24kg. The ingot was cut into 3 parts: of 353mm, 356mm and 430mm. A thick (3mm) wafer was cut from each part of the ingot, and wafers were annealed, then resistivity measurements were made. The lifetime and concentrations of oxygen and carbon were also measured for the top and the bottom parts. The results are given in table 2.

Table 2. Resistivity, lifetime and concentration test results at company B.

Measurement point (mm)	Conductivity	Resistivity at centre (Ω*cm)	RRV (%) ASTM F81	Lifetime (us)	N_C (10^{17} at/ccm)	N_O (10^{17} at/ccm)
Top (0)	P	>190	-[a]	120	1.0	8.6
353	N	17.8	120	-[a]	-[a]	-[a]
709	N	5.5	35	-[a]	-[a]	-[a]
1139	N	1.14	11.6	12	9.8	6.3

[a] Not measured.

2.3. Tests in company C

Company C used a 40kg load for CZ pulling. A Saint Gobain Quartz S.A.S. 16x12,6" crucible was used. The material underwent chemical treatment before pulling, and all silicon chunks were etched. However, a piece of quartz was found on the melt surface during the process, thus the initial crystal (shoulder and 20mm of the body, with total weight of 1.55kg) was discarded. Then, at the second attempt, full-length crystal, including the end cone, was pulled. The resulting diameter was in the 155-158mm range. The resulting length was 630mm, and the weight 33kg (figure 3).

The path of growth ended at the 60mm level, from the tip of the end cone at diameter ca 90mm, dislocation withdrawal amounted to 150mm, thus 480mm of resulting crystal were free of dislocations. The ingot was cut into two main parts: A (diameter 154-157mm, length 300mm, and weight 12.08 kg) and B (diameter 157-157.5mm, length 267mm, and weight 13.38 kg), each of them preceded and followed by three thick wafers of 2.2, 3.0 and 3.0 thickness. The thick wafers were then annealed and used for resistance and concentrations of C and O FTIR measurements. The results of the measurements are provided in table 3.

Table 3. Resistivity, lifetime and concentration test results at company C.

Measurement point (mm)	Conductivity		Resistivity (Ω*cm)		RRV (%) ASTM F81	FTIR measurement point (mm)	N_C (10^{17} at/ccm)	N_O (10^{17} at/ccm)
	Before anneal	After anneal	Before anneal	After anneal				
Shoulder (0)	N	P	3.85	7.0-7.1	3.9	2.2	8.75	<0.2
316.4	P	P	8.83	7.3-7.4	-4.1	308.2	7.62	<0.2
613.6	P	P	10.9	9.5-9.6	-5.5	589.4	7.31	<0.2
682.6 (end cone)	P	-[a]	14.5	-[a]	-[a]	-[a]	-[a]	-[a]

[a] Not measured

Then glow-discharge mass spectroscopy (GDMS) measurements of CZ crystals top and end, crucible leftovers and a piece of raw material were carried out at two different laboratories. The results are shown in table 4.

Table 4. GDMS measurements of CZ crystals top and end, crucible leftovers (potscrap) and a piece of raw material by two laboratories (concentrations given in ppbw).

	Raw material		Crystal top		Crystal end		Potscrap	
Lab	1	2	1	2	1	2	1	2
[B]	10	2	20	12	40	35	40	30
[P]	80	6	12	12	40	28	90	75
[As]	320	90	200	60	190	170	450	110
[Sb]	<10	2	<10	<1	<10	4-6	20	12
[Ti]	5	<5	0	0	<2	10	5	15
[Fe]	<150	10	<150	0	<150	30-60	<150	250
[Ni]	100	<5	<30	0	<30	8	200	540
[Cu]	65	<5	18	0	15	11	<10	7

The tail part of the ingot was divided into four thick wafers, which were then tested for dislocations. During etching in a Sirtl bath, no dislocations were found. Only standard microdefects for CZ method (square forms as in figure 4) with density 800 per cm2 appeared. Etching in a Secco bath did not reveal dislocation or square forms at <100> surface. The top part of the ingot was also tested for dislocations, and none were found.

Figure 3. 6" crystal grown at Company C.

Figure 4. Square forms at Si after Sirtl bath, illuminated.

2.4. Tests of FZ rods

In FZ experimental processes, two silicon rods were pulled from the crucible during the rectification process. The rods had a volatile diameter in the 100-130mm range, and weight of ca 15kg. Mechanical and then chemical treatment was used to bring rods to a constant diameter for further use in the standard FZ process. Subsequently, two FZ monocrystals, of 65 and 100 mm diameter, were grown. The resistivity and lifetime were measured at seed and non-seed ends of the crystals, and the results are given in table 5.

Table 5. Resistivity and lifetime of FZ crystals.

Parameter	B-3-10		B-3-20	
	Seed end	Non-seed end	Seed end	Non-seed end
Diameter (mm)	64-66		100	
Conductivity	P	P	P	N
Resistivity (Ω*cm)	2.4-2.5	2.1-2.3	180	17
Lifetime (us)	1780	1254	2279	300

3. Discussions

Our research [1] for individual chunks of silicon demonstrated an acceptable dopant level for further usage of the material for crystal growth. However, we have not yet performed tests on a big batch of new feedstock and dislocation-free single crystal pulling.

All growth experiments shared feedstock - heavily doped electronic grade material rectified with electron beam. Despite the fact that the same technological procedure was employed, the resistivity of the feedstock obtained varied between processes. In particular, P-type feedstock with resistivity in the 2-10 Ω*cm range was used by companies A and C and for growth of B-3-10 FZ crystal. The single crystals obtained possess acceptable uniformity in parameter distribution. The lifetime could not be measured in the crystal obtained by company C, for reasons that remain unknown.

Meanwhile, when N-type material with resistivity above 1 Ω*cm was used by company B, and high resistivity N-type feedstock was used for producing B-3-20 crystal, crystals with highly uneven distribution of resistivity along crystal length were obtained. An increase in N-type dopants' concentrations towards the end of crystal led to a decrease in the lifetime from 120 to 12 us and 2300 and 300 us, respectively.

To increase yields, the N-type dopants' content is to be minimized and the boron level is to be stabilized. To minimize the former to <10 ppba level, the dependence between dopant concentration and refinement time for the purification process needs to be found to decrease the level of impurities to an acceptable level, while keeping the process energy-efficient. To obtain crystals with resistivity in the P-type 1.5-3 Ω*cm range, boron content may be increased to a level of 70 ppbw in the existing process, but to increase yields or use this feedstock for higher resistivity applications, inner sources of boron should be found.

It is unlikely that source of boron is highly doped EG-Si, since it is produced from "virgin" polysilicon processed in CZ furnaces used for "N-processes", which implies very low levels of boron content. Quartz crucibles used for the typical CZ process generally do not introduce impurities at levels of more than 10 ppbw. Also, in previous work it has been shown that the background impurities of the rectification furnace are at level of ~1 ppbw if boron is involved.

To solve both tasks - to assuredly decrease the level of N-type dopants below 10 ppba and to prevent unsystematic introduction of boron in amounts more than 1ppbw - additional studies are necessary.

4. Conclusions

General results of this work are:

- Electron beam rectification of heavily doped electronic grade material enables one to receive feedstock that can be used directly, without blending for successful production of solar grade silicon by well-known methods such as CZ and FZ and without any modifications. These

IOP Conf. Series: Materials Science and Engineering **77** (2015) 012038 doi:10.1088/1757-899X/77/1/012038

methods, applied to the given feedstock, yield reasonable resistivity and lifetime levels in dislocation-free crystals.

- Testing of purified silicon in various independent companies showed the same results – dislocation-free single crystals were obtained without difficulties.
- The electrical parameters of ingots showed the possibility to use them for solar application.
- However, one of the key steps in obtaining material of good quality is thorough control of the initial raw material before rectification.

5. References

[1] Kravtsov A 2010 Method of vacuum refinement of silicon *Russian Patent No. 2381990* (Moscow, Russia: Ruspatent)

[2] Kravtsov An and Kravtsov Al 2011 Electron Beam Silicon Purification *26th EU PVSEC* (10.4229/26thEUPVSEC2011-2BV.4.7) pp 1814-1816

Acknowledgments

A part of the work was funded by the Latvian Ministry of Environmental Protection and Regional Development and SIA "Vides investīciju fonds" during the KPFI-14/1 project. We thank Topsil Semiconductor Materials A/S (Denmark), PCMP (Russia) and Prolog-Semicor (Ukraine) for assistance in conducting experiments.

IOP Conf. Series: Materials Science and Engineering **77** (2015) 012039 doi:10.1088/1757-899X/77/1/012039

Structure and Photocatalytic Properties of TiO_2-WO_3 Composites Prepared by Electrophoretic Deposition

Ineta Liepina[1], Gunars Bajars[1], Marcis Rublans[2], Janis Kleperis[1], Andrejs Lusis[1], Evalds Pentjuss[1]

[1]Institute of Solid State Physics, University of Latvia, Latvia
[2]Faculty of Physics and Mathematics, University of Latvia, Latvia

liepina.ineta@gmail.com

Abstract. In this work TiO_2–WO_3 composite films containing different oxide concentrations were prepared by electrophoretic deposition on steel substrates. Composite coating structures were analyzed by X -ray diffraction, Raman spectra and scanning electron microscopy. The results showed an even distribution of WO_3 particles in the entire composite layer. Light absorption measurements were used for photocatalytic properties evaluation. It was found that the removal ratio of methylene blue depends on the (TiO_2):(WO_3) concentration ratio. The most effective photodegradation was determined for the sample that was electrophoretically deposited from the suspension with the molar content ratio $n(TiO_2)$:$n(WO_3)$ 2:1.

1. Introduction

Due to its photocatalytic capability, low cost and chemically inert properties, TiO_2 is a promising material for water and air contamination treatment technologies [1-3]. Most of investigations have been devoted to the pure TiO_2 films because of its reusability. However, practical application of such films is limited by recombination of photo-induced hole-electron pairs. Coupling TiO_2 with other inorganic oxides such as SiO_2, SnO_2, WO_3, In_2O_3, $(Sr,La)TiO_{3+\delta}$ and $ZnFe_2O_4$ can increase the energy range of photo-excitation and this will change the photocatalytic efficiency [4].

Electrophoretic deposition (EPD) method has been widely applied to obtain pure TiO_2 thin films as well as doped TiO_2 thin films with transition elements. Advantages of electrophoresis over other coating methods are the homogeneity of thin films on large substrates, as well as better control of particle size and shape [5, 6].

In this work we focus on the preparation of TiO_2-WO_3 composite film using EPD method. WO_3 is a suitable photo-electron storing material to couple with TiO_2 because WO_3 has a more narrow band gap (3.0 eV) than that of TiO_2 (3.25 eV for anatase) to allow the transfer of photogenerated electrons from TiO_2 to ensure effective charge separation. Photocatalytic properties of coatings were determined with measurements of methylene blue decomposition under UV irradiation.

2. Experimental

The first part of this work includes electrophoretic deposition that was carried out using suspension of pure TiO_2 (*Sigma*-Aldrich, 98%) or TiO_2:WO_3 mixtures (molar ratio in suspension 1:1, 2:1, 3:1 and 1:2) in isopropanol (Fluka Analytical, 97%). To improve conductivity of dispersion media, pH of the suspension was adjusted with 1 mL HCl (Fluka Analytical, 98%). Each suspension was sonificated for 30 min. For the electrophoretic growth two 3x3 cm^2 stainless steel (grade 316) electrodes were used. Distance between electrodes was 1 cm. A constant voltage of 100 V was applied between the electrodes by DC Power Supply (Agilent Technologies N5772A) and held for 10 min. The deposited coatings were gradually heated (5 °C/min) and annealed at 500 °C for 2 h.

Photocatalyitic measurements were carried out using methylene blue (MB) solution under UV irradiation. Each sample was placed in 80 mL of 0.43 mg/L MB solution in a quartz beaker and illuminated with Hg lamp from 30 cm distance. Each composite was used to obtain MB solution sample series after 0; 0.5; 1; 3 and 6 hours of photodegradation. Light absorption measurements (Jenway 6300, 1 cm quartz cell) were used to calculate MB concentration after photodegradation had occurred.

3. Results and Discussion

TiO_2–WO_3 composite films were obtained from suspension with the following n(TiO_2):n(WO_3) ratio: 1:1 (a), 2:1 (b), 3:1 (c), 1:2 (d) and pure TiO_2 (e). The X-ray diffraction (PANalytical X"Pert PRO Cu Kα, λ = 0.15418 nm) pattern for sample d is shown in Figure 1. Characteristic peaks at 2θ 25.35°, 48.05°, 54.05° and 55.10° of the composite coating indicate that after thermal treatment TiO_2 has crystallized in anatase form with average crystallite size 200 nm (Scherrer equation) and WO_3 has monocline structure (characteristic peaks at 2θ 23.72° and 33.65°) with crystallite size 80–90 nm.

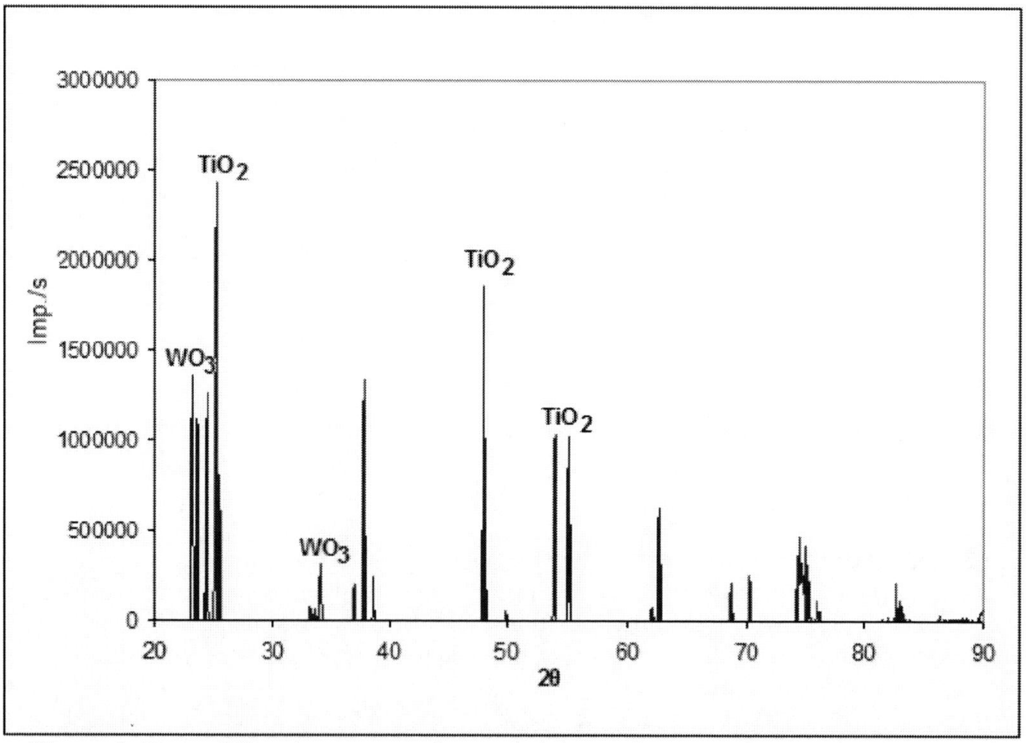

Figure 1. X-ray diffraction pattern of annealed d sample composite on a steel substrate

The concentration of WO_3 in the obtained composites was also determined from X-ray diffraction patterns - 18.2 wt.% (a), 17.8 wt.% (b), 16.6 wt.% (c) and 19.9 wt.% (d). As it is seen, the coating compositions don't have the same TiO_2:WO_3 ratio as their corresponding suspensions. This might be caused by the van der Waals force that affected coagulation of WO_3 particles near the electrode.

Figure 2 shows the Raman spectrum of annealed composite coating. The vibrational bands at 150 cm^{-1} (E_g), 400 cm^{-1} (B_{1g}), 515 cm^{-1} (A_{1g}) and 633 cm^{-1} (E_g) confirm phase transition of electrophoretically deposited TiO_2 to anatase phase [6]. The bands sited at 274, 327, 716 and 806 cm^{-1} are typical Raman peaks of crystalline WO_3 of monoclinic phase [7].

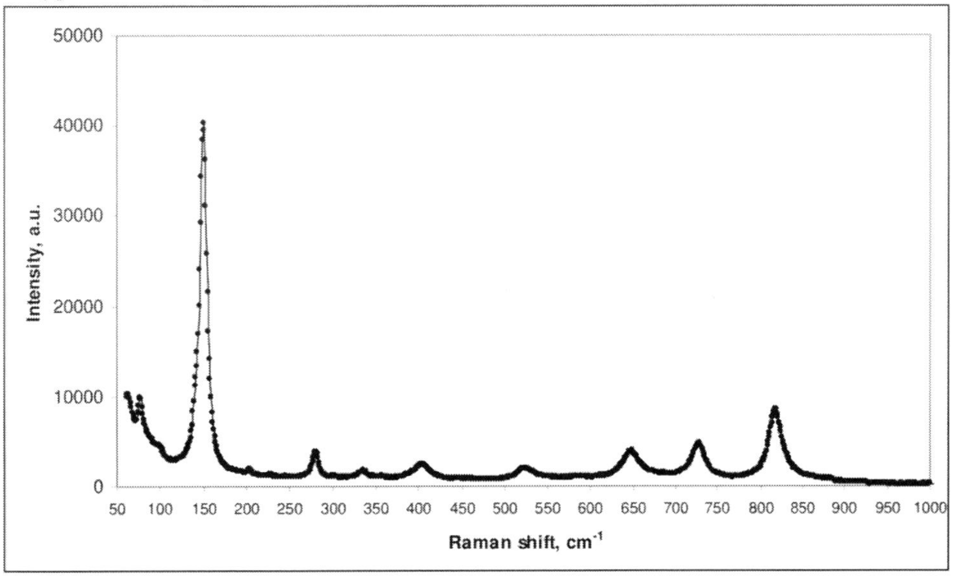

Figure 2. Raman spectrum of annealed composite coating of sample b

Scanning electron microscopy (SEM) images in Figure 3 show the cross-section and surface of c (top row) and d (bottom row) using both – scattered and back scattered detectors for imaging. As it is seen, c and d samples show an even distribution of WO_3 particles (seen as bright spots because of stronger backscattering) in the entire composite layer. This was observed as a common trend for all the composite samples.

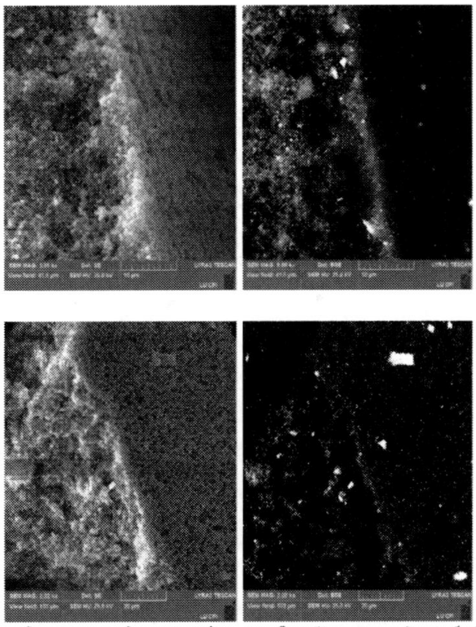

Figure 3. SEM images of composite coatings of c (top row) and d (bottom row) samples annealed at 500 °C

IOP Conf. Series: Materials Science and Engineering **77** (2015) 012039 doi:10.1088/1757-899X/77/1/012039

Figure 4 shows that the size of WO_3 crystallites can be estimated about 200–400 nm, while X-ray diffraction showed notably lower crystalline size.

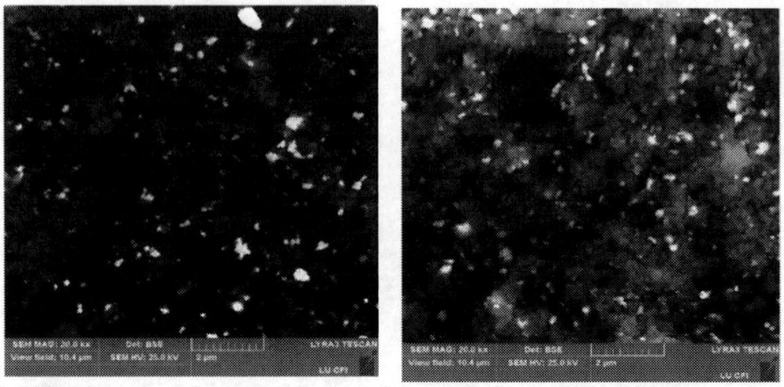

Figure 4. SEM images of composite coatings c (left) and d (right) annealed at 500 °C

Photocatalysis graph (Figure 5) shows that in the first hour photo degradation occurs at a slower rate. This can be explained with the absorption and dissociation of water molecules in contact with oxygen bridges in TiO_2 structures. Only after reduced oxygen was formed on composite surface the noticeable photodegradation of MB was observed. As expected a sample of clear stainless steal substrate (control) has no effect on MB decomposition. Most effective photodegedation occurred with the sample b ($n(TiO_2)$:$n(WO_3)$ ratio = 2:1) with the resulting content of WO_3 17.8 wt.% - after 6 hours of UV irradiation MB had decomposed to less than 10% of its initial concentration. Slightly slower photodegradation was observed for the samples a and c where the concentration of WO_3 in films were somewhat higher or lower than that for the sample b (18.2 wt.% and 16.6 wt.%) respectively.

The sample d ($n(TiO_2)$:$n(WO_3)$ ratio = 1:2) with the highest resulting content of WO_3 (19.9 wt.%) showed a considerably slower photodegradation rate of MB. Even a sample of pure TiO_2 (e) showed more effective photocatalytic properties than d sample. This can be explained with the instability of WO_3 – unlike the inert TiO_2, tungsten trioxide undergoes a transformation to WO_4^{2-} in a solution that has pH over 4 [8].

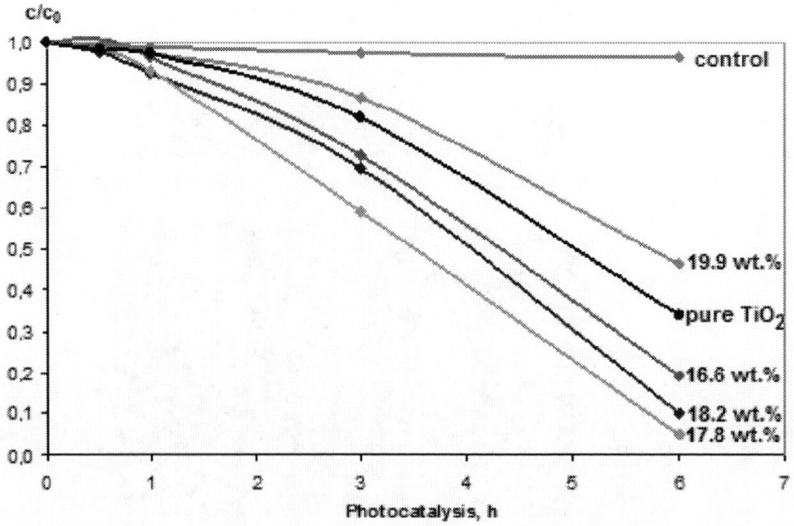

Figure 5. Photodegradation rate of MB using composite coatings under UV irradiation

Conclusions

Uniform TiO_2-WO_3 coatings have been successfully obtained by electrophoretic deposition method on stainless steel substrates. X-ray diffraction patterns confirmed the crystallization of anatase TiO_2 with average particle size 200 nm and monocline WO_3 with particle size 80-90 nm, while scanning electron micrographs showed WO_3 size distribution of 200–400 nm. The removal ratio of methylene blue depends on the (TiO_2):(WO_3) concentration ratio. The most effective photodegradation was determined for the sample that was electrophoretically deposited from the suspension with the molar content ratio $n(TiO_2)$:$n(WO_3)$ 2:1 with the resulting content of WO_3 17.8% - after 6 hours of UV irradiation MB had decomposed to less than 10% of its initial concentration.

Acknowledgments

This work was financially supported by Latvian project of scientific cooperation 666/2014. The authors thank K. Kundzins for SEM observations and J. Gabrusenoks for Raman spectroscopy measurements.

References

[1] Daude N, Gout C and Jouanin C 1977 *Phys. Rev. B* **15** 3229
[2] Wold A 1993 *Chem. Mater.* **5** 280
[3] Kim S B and Hong S C 2002 *Appl. Catal. B Environ.* **35** 305
[4] Georgieva J, Valova E, Armyanov S, Philippidis N, Poulios I and Sotiropoulos S 2012 *J. Hazardous Matter.* **211-212** 30
[5] Landmann M, Rauls E and Schmidt W G 2012 *J. Phys.: Condens. Matter* **24** 195
[6] Liepina I, Bajars G, Lusis A, Mezinskis G and Vanags M 2013 *IPO Conf. Series: Material Science and Engeneering* **49** 012060
[7] Diaz-Reyes J, Delgado-Macuil R J, Dorantes-Garcia V, Perez-Benitez A, Balderas-Lopez J A and Ariza-Ortega J A 2010 *Material Science and Engeneering B* **174** 182
[8] Wang G, Ling Y, Wang H, Yang X, Wang C, Zhang J Z and Li Y 2012 *Energy Environ. Sci.* **5** 6180

IOP Conf. Series: Materials Science and Engineering **77** (2015) 012040 doi:10.1088/1757-899X/77/1/012040

Microwave assisted synthesis of nanocomposites in Zn-Sn-O-Ag system and their photocatalytic activity

J Grabis, A Letlena, Dz Rasmane, A Krumina

Institute of Inorganic Chemistry RTU, Miera 34, Salaspils, LV-2169, Latvia

E-mail: grabis@nki.lv

Abstract. The crystalline Zn_2SnO_4, Zn_2SnO_4/Ag and $ZnO\text{-}Zn_2SnO_4$ nanoparticles with specific surface area in the range of 24.1–28.5 m^2/g were prepared by combining microwave assisted synthesis with additional calcination at 900 °C. The parameters of the powders were compared with those of ZnO and ZnO/Ag nanoparticles prepared by microwave synthesis at 150 °C. The crystallite size of oxide and silver nanoparticles was in the range of 30–52 nm and 18–60 nm, respectively, depending on composition. The prepared nanoparticles showed high photocatalytic activity in degradation of methylene blue under ultraviolet and sunlight irradiation.

Key words: microwave synthesis, zinc oxide, stannate, silver, photocatalytic activity.

1. Introduction

Zinc oxide and zinc stannate are promising candidates for application in solar cells, sensors, medicine, cosmetic, photocatalysis for degradation of organic pollutants due to their attractive electrical and optical parameters [1, 2]. The effective use of nanoparticles as photocatalysts depends on the particle size, morphology, crystallinity, and used dopants [3]. Besides this ZnO/Zn_2SnO_4 particulate composites show improved photocatalytic activity with respect to lone ZnO or Zn_2SnO_4 [4].

Due to the wast area of application of zinc oxide based nanoparticles, various their preparation methods, such as sol-gel, hydrothermal, combustion, precipitation, microwave synthesis have been elaborated [1, 2]. The hydrothermal synthesis is recognized well for preparation ZnO and Zn_2SnO_4 nanoparticles but particles size is relatively large [1, 5]. The promising fast and simple microwave synthesis ensures preparation of ZnO rod-like or quazi-spherical nanoparticles with size about 10–50 nm [6] or nanorods with length about 1.1 μm and diameter about 0.63 μm [7].

The aim of the present work was to extend microwave synthesis to preparation of nanoparticles in the Zn-Sn-O-Ag system with different content of the components and to compare their parameters and photocatalytic activity.

2. Experimental

Experiments were performed by using analytical grade chemicals. In the typical synthesis process separate solutions of 0.05 M of $Zn(CH_3COO)_2{\cdot}2H_2O$ in 200 mL deionized water, 0.025 M of $SnCl_2{\cdot}2H_2O$ in 100 mL mixture of ethyl alcohol and methyl alcohol, and $AgNO_3$ in water (100 mL) were prepared. The appropriate amounts of the solutions were mixed and stirred. Water solution of formaldehide for reducing silver was added. Afterwards NH_4OH or NH_4HCO_3 was added dropwise to reach pH 10. The mixture (500 mL) was stirred at room temperature and then transferred into a microwave reactor (Mastewave BTR (Anton Paar GmbH)) and treated at 120–160 °C and 9–11 bars for 5–30 min. The precipitates were separated by filtration and washed with deionized water. The obtained products were

IOP Conf. Series: Materials Science and Engineering **77** (2015) 012040 doi:10.1088/1757-899X/77/1/012040

dried at 80 °C for 6 h. Afterwards Zn_2SnO_4 containing powders were calcinated at 850–900 °C in air in a furnace.

Phase and chemical composition of nanopowders was determined by X-ray diffraction (XRD) analysis (AdvanceD8, Bruker AXS), X-ray spectroscopy (S4Pioneer, Bruker AXS). The crystallite size was calculated from broadening of diffraction maxima using software Topas3 (Bruker AXS). Specific surface area (SSA) of powders was determined by the argon absorption-desorption method. Morphology and size of particles were determined by SEM.

Photocatalytic activity of the samples was evaluated by the degradation of methylene blue (MB) under direct ultraviolet (UV) irradiation of Hg lamp (120 W) at distance 110 mm and sunlight irradiation (18200 Lux, Luxmeter Lx-101B) in summer between 1 and 3 p.m. The aqueous solution of MB ($7.6 \cdot 10^{-3}$ g/L) with catalyst (2g/L) was introduced in a quartz vessel and treated by ultrasound for 10 min in the dark to obtain homogeneous solution. During irradiation the solution was stirred mechanically and cooled by air flow.

The treatment of MB solution by UV and sunlight irradiation without catalyst showed that during 2 h the degradation degree MB reached 11 and 6% respectively. The samples of irradiated suspension were taken each 10 min for analysis. The degradation of MB was determined by measuring the light absorption by solution after removing solid particles at the wavelength of 622 nm using Jenway-6300 spectrometer.

3. Results and discussion

The XRD patterns of the as-prepared samples confirmed that microwave synthesis at 150 °C allows produce pure crystalline ZnO and ZnO/Ag but Zn_2SnO_4 powders contained admixture of ZnO and SnO_2. Pure crystalline Zn_2SnO_4 or Zn_2SnO_4/Ag nanoparticles were obtained after post-synthesis heat treatment at 850–900 °C for 2 h (Figure 1). After additional calcination up to 900 °C the crystallite size of Zn_2SnO_4 and Ag was the range of 32–52 nm and of 60 nm, respectively.

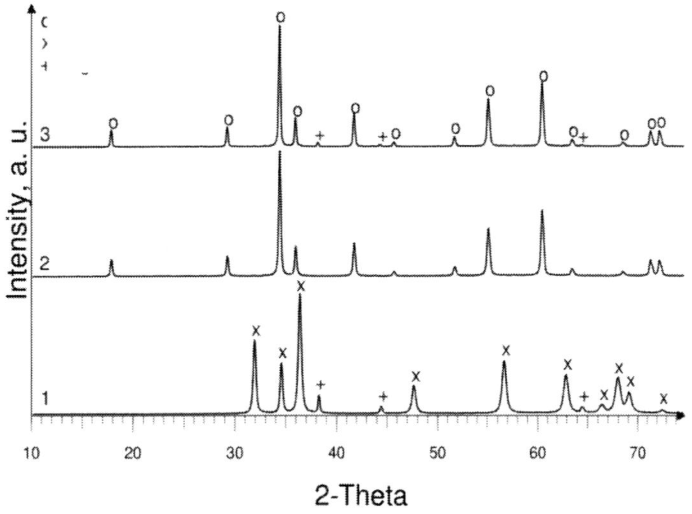

Figure 1. X-ray diffraction patterns of the prepared nanoparticles: 1 – ZnO/Ag; 2 – Zn_2SnO_4; 3 – Zn_2SnO_4/Ag 2; o – Zn_2SnO_4; x – ZnO; + – Ag.

The parameters of samples prepared by microwave synthesis at 150 °C for 20 min and with additional calcination Zn_2SnO_4 containing samples at 900 °C for 2 hours are shown in Table 1.

The SSA of the prepared powders was in the range of 21.3–28.5 m^2/g and it was a little higher for Zn_2SnO_4 containing samples despite to high calcination temperature (Table 1). The determined crystallite size of ZnO and Zn_2SnO_4 nanoparticles was in the range of 30–33 nm and 35–52 nm. The results provided evidence that presence of Ag had weak influence on crystallite size of ZnO.

Table 1. Parameters of the prepared nanoparticles

Sample	Ag content, mol%	SSA, m^2/g	Crystallite size, nm		
			ZnO	Zn_2SnO_4	Ag
ZnO	-	22.8	33	-	-
ZnO/Ag 1	1.0	22.9	32	-	18
ZnO/Ag 2	2.0	23.1	30	-	20
ZnO/Ag 3	3.0	21.3	31	-	50
Zn_2SnO_4	-	24.1	-	40	-
Zn_2SnO_4/Ag 2	2.0	25.9	-	35	60
Zn_2SnO_4-20ZnO*	-	28.5	-	52	-

* Zn_2SnO_4 sample contained 20 wt.% ZnO

The crystallite size of Ag nanoparticles was in the range of 18–60 nm depending on concentration of Ag. The determined typical particle size of ZnO in the range of 40–100 nm, especially particle size of Zn_2SnO_4 in the range of 30–80 nm, was close to the calculated crystallite size (Figure 2).

The morphology and particle size of the prepared ZnO powders differed strongly from nanotubes and nanosheets (0.5–2.0 µm) prepared by hydrothermal synthesis for 9–4 h at 90 °C [5] or from nanorods with a diameter 0.63 µm and lenght about 1.1 µm or flower-like ZnO particles prepared by microwave synthesis from zinc chloride solution [7]. The shape and size of the prepared stannate nanoparticles differed from cubic stannate nanoparticles with crystallite size about 500 nm [8] or from rod-like Zn_2SnO_4 nanoparticles with lenght about 300 nm and thickness of 50 nm prepared by hydrothermal synthesis [9]. Therefore applied synthesis ensures preparation of ZnO, ZnO/Ag and Zn_2SnO_4 as well as Zn_2SnO_4/Ag, Zn_2SnO_4-ZnO nanoparticles with relatively low crystallite and particle size.

Figure 2. SEM micrographs of ZnO (a); Zn_2SnO_4 (b); Zn_2SnO_4/Ag (c).

The crystallinity of prepared samples, their particle size, phase composition and presence of silver determined their photocatalytic activity under UV Hg lamp and sunlight. The photodegradation of MB under UV irradiation is shown in Figure 3.

All tested powders showed good photocatalytic activity – they degradated MB up to 90–98% during 60–100 min depending on their composition. ZnO/Ag nanoparticles showed the highest photocatalytic activity. The low influence of silver on improvement of the catalytic acitivity of Zn_2SnO_4 can be connected with a large crystallite size of Ag particles. The photocatalytic activity of the prepared nanoparticles under UV irradiation of Hg lamp decreased in the following order: ZnO/Ag2>ZnO>Zn_2SnO_4-20ZnO>Zn_2SnO_4/Ag>Zn_2SnO_4.

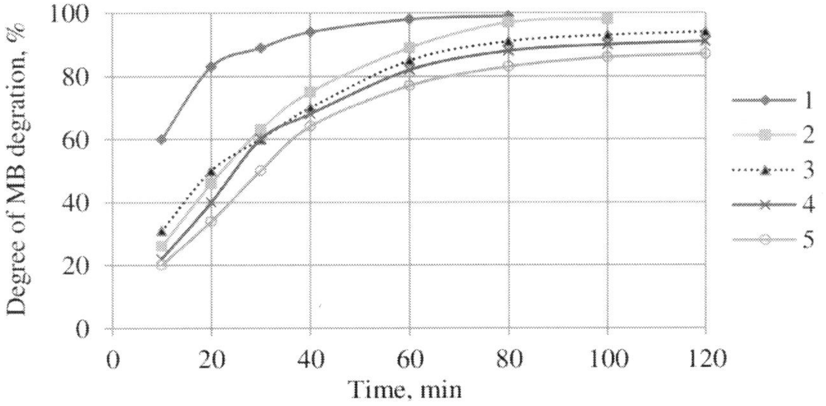

Figure 3. Photodegradation of MB under UV irradiation in dependence on used photocatalysts: ZnO/Ag 2 (1); ZnO (2); Zn_2SnO_4-20ZnO (3); Zn_2SnO_4/Ag 2 (4); Zn_2SnO_4 (5).

The important feature of the prepared nanopowders is their high photocatalytic activity under sunlight irradiation (Figure 4). The ZnO/Ag nanoparticles exibited the highest photocatalytic activity again depending on content of silver. According to our studies the silver content of about 2% was optimal. The higher Ag content led to increase of the particle size causing decrease of the photocatalytic activity. The photocatalytic activity of the prepared nanoparticles under the sunlight irradiation decreased in the following order: ZnO/Ag 2>ZnO/Ag 1>Zn_2SnO_4/Ag 2>Zn_2SnO_4.>ZnO.

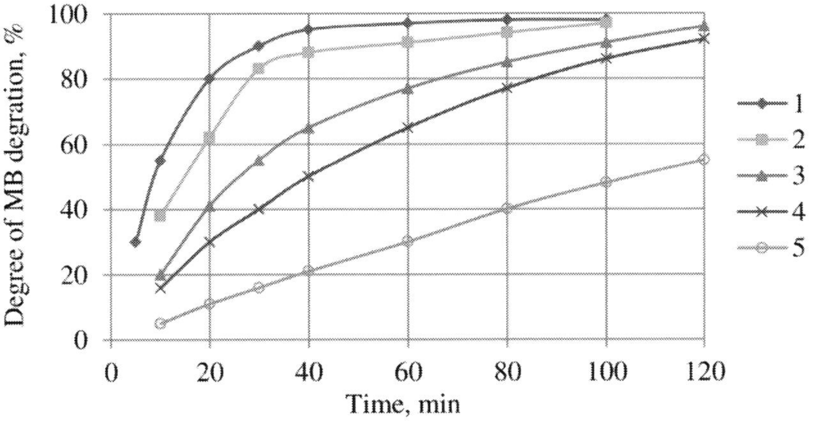

Figure 4. Photodegradation of MB under the sunlight irradiation in dependence on used photocatalysts: ZnO/Ag 2 (1); ZnO/Ag 1 (2); Zn_2SnO_4/Ag 2 (3); Zn_2SnO_4 (4); ZnO (5).

4. Conclusions

The combining microwave synthesis with additional calcination at 900 °C for 2 h ensures preparation of crystalline Zn_2SnO_4, ZnO-Zn_2SnO_4 and Zn_2SnO_4/Ag nanoparticles with specific surface area 24.1–28.5 m^2/g close to that of ZnO and ZnO/Ag (21.3–22.8 m^2/g) prepared by microwave synthesis at 150 °C for 20 min.

Crystalline size of ZnO and Zn_2SnO_4 nanoparticles was in the range of 30–33 nm and 35–52 nm, respectively, but crystalline size of Ag nanoparticles was in the range of 18–60 nm depending on the content of silver.

The prepared nanoparticles exhibited high photocatalytic activity in degradation of MB under UV and sunlight irradiation. ZnO and Zn_2SnO_4 doping with Ag enhanced their photocatalytic activity.

Acknowledgments

The research was supported by Latvian National Research Program IMIS^2.

5. References

[1] Baruah S and Dutta J 2011 *Sci. Technol. Adv. Mater.* **12** 1
[2] Vaseem M, Umar A, Hahn V B 2010 *Metal oxide nanostructures and their application* **5** 1
[3] Seki S, Sekizawa T, Haga K, Sato T, Takida M, Seku Y, Sawada Y,Yubula K, Siskido T 2010 *J. Vac. Sci. Technol.* **188** 3268133
[4] Khan M N, Al-Hinnai M, Al-Hinnai A, Dutta J 2014 *Ceram. Int.* **40** 8743
[5] Elamin N, Elsanousi A 2013 *J. Appl. Ind. Sci.* **1** 32
[6] Tran Thi Ha, Ta Dinh Canh, Nquyen Viet Tuyen. 2013 *ISRN Nanotechnology* article ID 497873 **92** 3683
[7] Zhu Jian-yu, Zhang Jing-xia, Zhou Hui-fen, Qin Wen-ging, Chai Li-yuan, Hu Yua-hua 2009 *Trans. Noferrous Met. Sci. China* **19** 1578
[8] Miyauchi M, Liu Z, Zhao Z G, Anandan S, Hara K 2010 *Chem. Commun.* **46** 1529
[9] Ji X, Huang X, Liu J, Jiang J, Li X, Ding R, Hu Y, Wu F, Li Q 2010 *J. Alloys Compd.* **503** L21

IOP Conf. Series: Materials Science and Engineering **77** (2015) 012041 doi:10.1088/1757-899X/77/1/012041

Synthesis and thermal decomposition of $Mg(BH_4)_2$–TMO ($TMO=TiO_2;ZrO_2;Nb_2O_5;MoO_3$) composites

I. Saldan[1,2], I. Llamas-Jansa[1], S. Hino[1], C. Frommen[1], B.C. Hauback[1]*
[1]Institute for Energy Technology, Physics Department, Box 40, NO-2027 Kjeller, Norway
[2]Ivan Franko National University of Lviv, Department of Physical and Colloid Chemistry, 6 Kyryla and Mefodiya Str., UA-79005, Lviv, Ukraine

E-mail: Bjorn.Hauback@ife.no

Abstract
Ball-milling has been applied to prepare composites between γ-$Mg(BH_4)_2$ and several transition metal oxides ($TMO=TiO_2;ZrO_2;Nb_2O_5;MoO_3$). No chemical interaction between $Mg(BH_4)_2$ and the TMO was observed during composite preparation and only a partial amophization of γ–$Mg(BH_4)_2$ was detected. Complete decomposition of $Mg(BH_4)_2$ into crystalline MgB_2 was found above 500 °C for all composites. The used TMOs with exception of ZrO_2 were chemically converted to oxides with lower oxidation states and the corresponding borides during the thermal decomposition of $Mg(BH_4)_2$.

1. Introduction
Magnesium borohydride is one of the most promising hydrogen storage materials though its kinetics and reversibility are still main challenges. The decomposition of $Mg(BH_4)_2$ occurs via several stages [1-4]. Different magnesium polyboranes appear depending on the process conditions. The most stable intermediate is magnesium dodecaborane, $MgB_{12}H_{12}$, which eventually transforms into MgB_2. The geometrical similarity of the boron icosahedral frame-work in $MgB_{12}H_{12}$ and that of bulk boron can be a reason for the kinetic stability of the dodecaborane [5] that by-turn may hamper rehydrogenation. Apparently, the synthesis of $Mg(BH_4)_2$ from lower polyboranes might be possible at reasonable hydrogen pressure and temperature [6]. A catalyst that prevents the dodecaborane formation may be one of the possibilities to improve the reversibility of $Mg(BH_4)_2$ dehydrogenation. Therefore development of effective additives for the improved decomposition/formation of $Mg(BH_4)_2$ is one of the most important problems to solve before its practical application. High valence transition metal compounds with carbon, nitrogen, oxygen or halogenide ligands have the ability to form bonds to hydrogen with varying stoichiometries. This encourages the fast dissociation into atomic hydrogen or its recombination to hydrogen molecules [7]. An optimum stability of the transition metal compounds can be found in some oxides which are less stable than corresponding carbides or nitrides but more stable than corresponding halogenides.

In the present paper transition metal oxides (TMO) are studied as a possible catalyst for the $Mg(BH_4)_2$ decomposition in the prepared $Mg(BH_4)_2$–TMO ($TMO=TiO_2;ZrO_2;Nb_2O_5;MoO_3$) composites.

2. Experimental part

Commercial γ-$Mg(BH_4)_2$ (≥95%, Sigma Aldrich) was used to prepare $Mg(BH_4)_2$–TMO (TMO=TiO_2;ZrO_2;Nb_2O_5;MoO_3) composites in molar ratios of 1:0.02. Rutile (TiO_2) (99.90%, Sigma Aldrich), ZrO_2 (97.7%, Merck), Nb_2O_5 (99.99%, Sigma Aldrich) and MoO_3 (99.5%, Sigma Aldrich) were chosen as additives. The as-received powders were milled for 1 h using a planetary micro mill (Fritsch pulverisette 7) in argon atmosphere (speed 300 rpm). New stainless steel balls 10 mm in size, with a 40:1 balls- to-powder ratio, were used. The balls were cleaned by standard procedure and dried by technical fan. All handling and preparation of the samples took place in a MBraun UniLab glove box with continuously purified argon atmosphere where oxygen and moisture values were less than 1 ppm.

Powder X-ray diffraction (PXD) was performed in a Bruker AXS D8 Advance instrument fitted with a Göbel mirror and a LynxEye™ 1D detector. The samples were encapsulated in 0.8 glass capillaries sealed with glue in the glovebox. Data acquisition was performed with a step size of 0.02° and 2s/step measurement time in range of 5-65° for 2Θ. The structure data base implemented in the Eva program was used for phase analysis.

Differential scanning calorimetry (DSC) and thermo-gravimetrical analysis (TGA) was performed simultaneously using a Netzsch STA 449 F3 Jupiter instrument. Samples were measured inside aluminum crucibles fitted with pierced lids. Argon with a flow rate of 50 ml/min was used as a purge gas. Experiments were performed between room temperature (RT) and 500 °C using a heating rate of 2 °C/min.

Temperature-programmed desorption (TPD) coupled with residual gas analysis (RGA) was used to estimate gas release. The samples were installed in a sealed copper sample holder. Experiments were carried out under vacuum (~10^{-4}mbar) with a heating rate of 2 °C/min from RT to 500 °C.

3. Results and discussions

The prepared $Mg(BH_4)_2$–TMO (TMO=TiO_2;ZrO_2;Nb_2O_5;MoO_3) composites were analyzed by PXD, DSC and TPD (Figs. 1-4). Based on PXD experiments the ball milled composites were similar to a physical mixture of $Mg(BH_4)_2$ and corresponding oxides powder (Fig.1). No indication for a chemical interaction between $Mg(BH_4)_2$ and the additives was found under the current milling conditions. Partial amophization of magnesium borohydride was observed after 1 h of ball-milling through an increase of the background below 25° in 2Θ (Fig. 1). Similar findings have been reported in [8] where it was concluded that $Mg(BH_4)_2$ decomposition is hardly affected by ball milling. In addition we observe that the level of amorphization depends on the nature of the additives. PXD analysis confirms that the smallest changes in the background is observed for $Mg(BH_4)_2$–ZrO_2 composite. This finding might be explained by the high hardness of ZrO_2 (8 in Mohs scale) and thermodynamic stability (ΔH^o_f = –1080 kJ/mol) [8]. Changes of diffraction peaks (lower intensity and peak broadening) might be the result of a reduced particle size in the prepared composites.

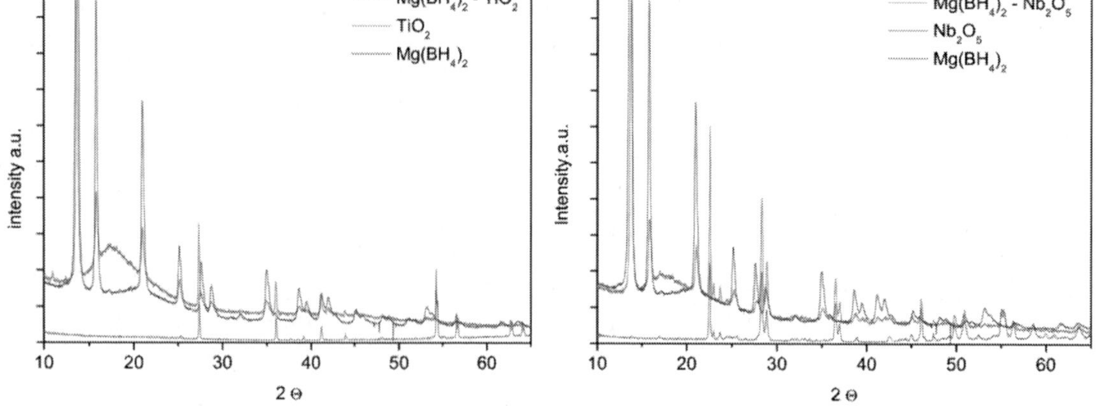

IOP Conf. Series: Materials Science and Engineering **77** (2015) 012041 doi:10.1088/1757-899X/77/1/012041

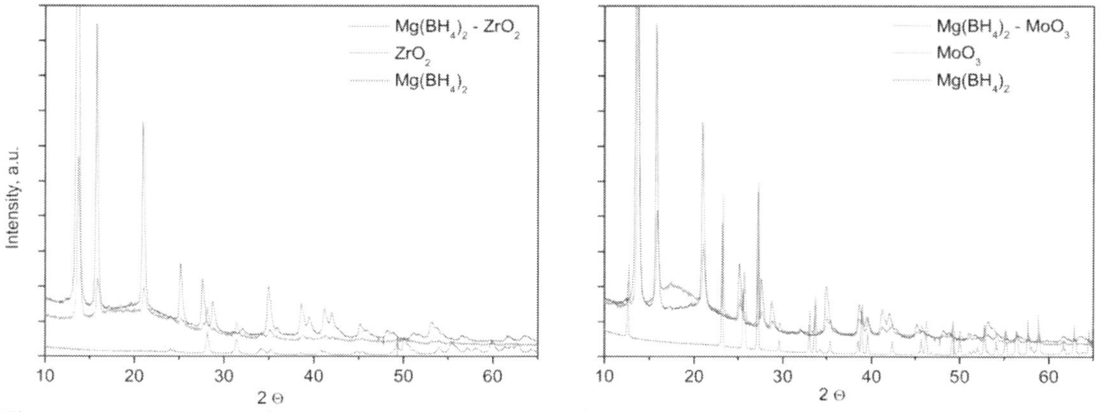

Fig. 1. PXD patterns of $Mg(BH_4)_2$–TMO (TMO=TiO_2;ZrO_2;Nb_2O_5;MoO_3) composites together with commercial $Mg(BH_4)_2$ and corresponding additive.

DSC experiments for the $Mg(BH_4)_2$–TMO (TMO=TiO_2;ZrO_2;Nb_2O_5;MoO_3) composites and pure $Mg(BH_4)_2$ are presented in Fig. 2. Because of the partial amorphization of $Mg(BH_4)_2$ in the composites their DSC curves show a small exothermic peak at above 100 °C due to recrystallization. For pure γ-$Mg(BH_4)_2$ two endothermic peaks are observed at ~165 °C and ~195 °C which can be attributed to the $\gamma \rightarrow \varepsilon$ and $\varepsilon \rightarrow \beta$ phase transformations [9]. The exothermic peak at ~320 °C corresponds to the crystallization of amorphous MgH_2 and the two endothermic signals at ~309 and ~333 °C can be assigned to the decomposition of $Mg(BH_4)_2$, and the decomposition of MgH_2 into magnesium metal and hydrogen gas, respectively [10]:

$$Mg(BH_4)_2 \rightarrow 3H_2 + 2B + MgH_{2\,(amorph)} \quad (309\,^{\circ}C) \qquad (1)$$
$$MgH_{2\,(amorph)} \rightarrow MgH_{2\,(cryst)} \quad (320\,^{\circ}C) \qquad (2)$$
$$MgH_{2\,(cryst)} \rightarrow Mg + H_2 \quad (333\,^{\circ}C) \qquad (3)$$

The $Mg(BH_4)_2$–ZrO_2 composite shows a thermal behavior similar to that of pure $Mg(BH_4)_2$. The other prepared composites show a small shift in the order of ~10-20 °C towards higher temperatures for decomposition of MgH_2 (Eq. 3).

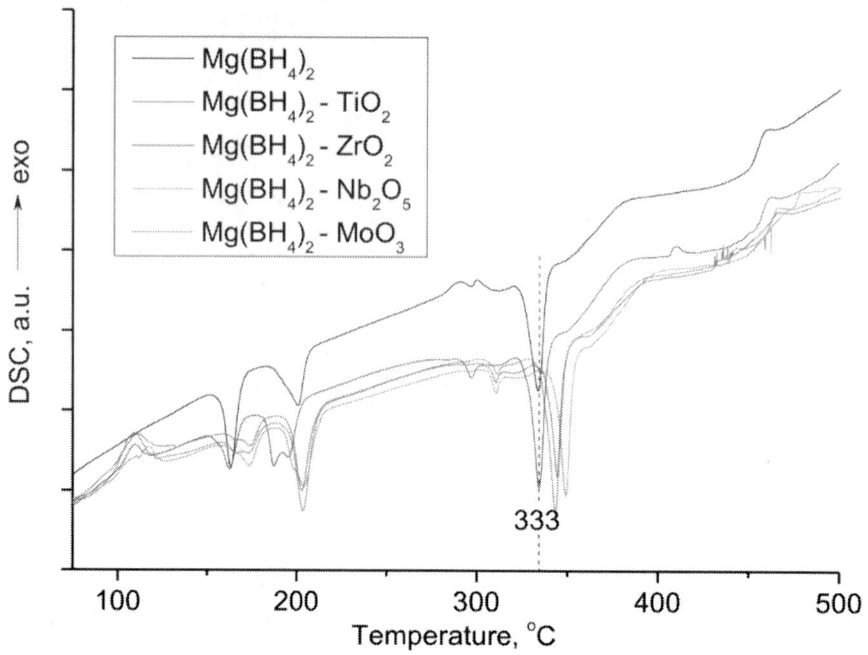

Fig. 2. DSC traces of $Mg(BH_4)_2$–TMO (TMO=TiO_2;ZrO_2;Nb_2O_5;MoO_3) composites and commercial $Mg(BH_4)_2$ as reference.

TPD analysis of $Mg(BH_4)_2$–TMO (TMO=TiO_2;ZrO_2;Nb_2O_5;MoO_3) composites showed gas release during temperature development (Fig. 3). There are three main events of hydrogen release for pure $Mg(BH_4)_2$. The most intense peak is observed at ~300 °C and other at ~333 and ~355 °C. Only for $Mg(BH_4)_2$–ZrO_2 composite these three events had similar intensity but were shifted to lower temperature by ~10 °C. Since the signal at ~165 °C is attributed to a small hydrogen release during the $\gamma \rightarrow \varepsilon$ phase transformation, it might be concluded that for $Mg(BH_4)_2$–ZrO_2 composite the $Mg(BH_4)_2$ phase undergoes a low amorphization. The TPD curves for $Mg(BH_4)_2$–TMO (TMO=TiO_2;MoO_3) showed a different intensity distribution as compared to pure $Mg(BH_4)_2$ and $Mg(BH_4)_2$–ZrO_2 although the temperatures were similar to those of pure $Mg(BH_4)_2$. Gas release of $Mg(BH_4)_2$–Nb_2O_5 composite was combined in one main signal but shifted to higher temperature.

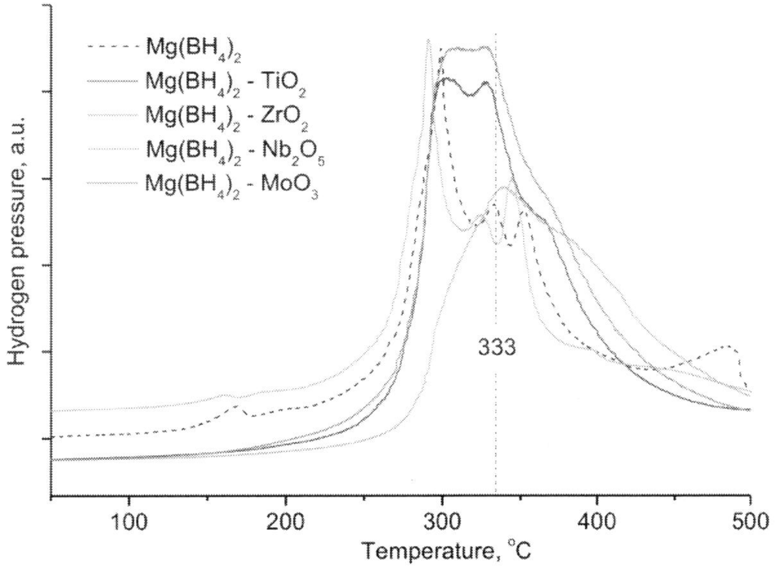

Fig. 3. TPD-RGA data for $Mg(BH_4)_2$–TMO (TMO=TiO_2;ZrO_2;Nb_2O_5;MoO_3) composites and commercial $Mg(BH_4)_2$ for comparison.

The $Mg(BH_4)_2$–TMO (TMO=TiO_2;ZrO_2;Nb_2O_5;MoO_3) composites and pure $Mg(BH_4)_2$ were studied again by PXD after thermal decomposition (Fig. 4). For all the samples PXD experiments confirm the complete decomposition of $Mg(BH_4)_2$ into crystalline MgB_2 after TPD carried out till ~500 °C. A small amount of CuO as an impurity phase (most probably from the copper sample holder) was detected, and its peak positions have been indicated by tickmarks in Fig. 4. In addition, all of the used TMOs with the exception of ZrO_2 were chemically converted into oxides with lower oxidation states and the corresponding borides. This implies that a chemical reaction between $Mg(BH_4)_2$ and the TMOs has occurred during heating and only ZrO_2 remains unaffected by this reaction.

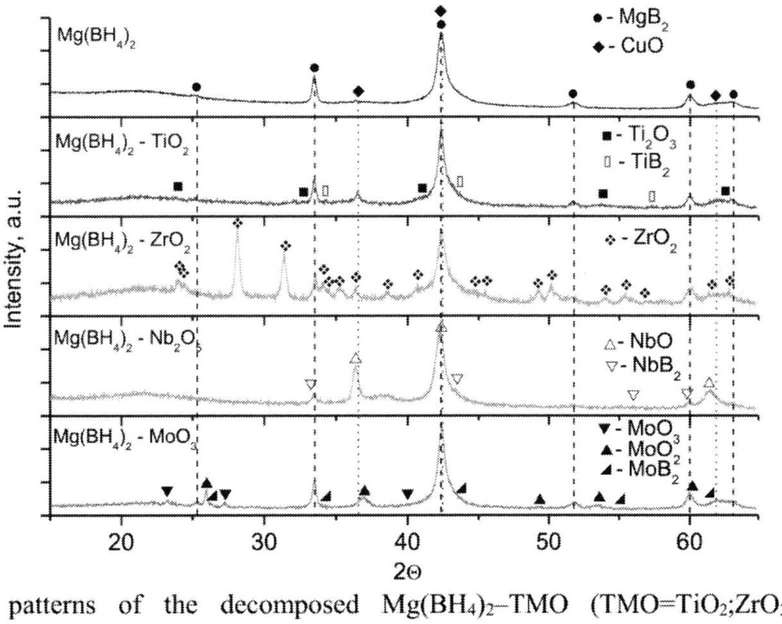

Fig. 4. PXD patterns of the decomposed $Mg(BH_4)_2$–TMO (TMO=TiO_2;ZrO_2;Nb_2O_5;MoO_3) composites and commercial $Mg(BH_4)_2$.

IOP Conf. Series: Materials Science and Engineering **77** (2015) 012041 doi:10.1088/1757-899X/77/1/012041

4. Conclusions

The current experimental work on the synthesis and thermal decomposition of $Mg(BH_4)_2$–TMO (TMO=TiO_2;ZrO_2;Nb_2O_5;MoO_3) composites leads to the following main conclusions:

1. There is no chemical interaction between γ-$Mg(BH_4)_2$ and the additives under the applied milling procedure while some amophization for γ-$Mg(BH_4)_2$ is observed by an increase of the background intensity at $2\Theta<25°$. The smallest change in the background is observed for $Mg(BH_4)_2$–ZrO_2 composite;
2. Positions of DSC peaks of $Mg(BH_4)_2$–ZrO_2 composite are very similar to that of pure γ-$Mg(BH_4)_2$. A small shift by ~10 °C to higher temperatures is observed for the DSC traces of the others composites;
3. There are three main gas release events (~300; 333 and 355 °C) for pure γ-$Mg(BH_4)_2$ that are shifted to lower temperatures by ~10 °C for the $Mg(BH_4)_2$–ZrO_2 composite. The composites containing TiO_2, Nb_2O_5 and MoO_3 show changed TPD curves that are furthermore shifted to higher temperatures;
4. All transition metal oxides with the exception of ZrO_2 react with γ-$Mg(BH_4)_2$ at high temperatures and are chemically converted into oxides with lower oxidation state and the corresponding transition metal borides.

The present results suggest that of all the TMOs tested only ZrO_2 might act as a catalyst during $Mg(BH_4)_2$ decomposition since it is not chemically altered.

Acknowledgement

The authors thank to The Research Council of Norway for financial support.

References:

[1] G. Soloveychik, Y. Gao, J. Rijssenbeek, M. Andrus, S. Kniajanski, R.C. Bowman, Jr, S.J. Hwang, Ji.Ch. Zhao. Magnesium borohydride as hydrogen storage material: properties and dehydrogenation pathway of unsolvated $Mg(BH_4)_2$. J. Hydrogen Energy 34 (2009) 916-928.

[2] H.W. Li, K. Kikuchi, Y. Nakamori, N. Ohba, K. Miwa, S. Towata, S.I. Orimo. Dehydriding and rehydriding process of well-crystallized $Mg(BH_4)_2$ accompanying with formation of intermediate compounds. Acta Materialia 56 (2008) 1342-1347.

[3] K. Chłopek, C. Frommen, A. Leon, O. Zabara, M. Fichtner. Synthesis and properties of magnesium tetrahydroborate, $Mg(BH_4)_2$. J. Mater. Chem . 17 (2007) 1396-3503.

[4] E. Rönnebro Development of group II borohydrides as hydrogen storage materials. Currrent Opinion in Solid State and Materials Science 15 (2011) 44-51.

[5] H.W. Li. K. Miwa, N. Ohba, T. Fujita, T. Sato, Y. Yan, S. Towata, M.W. Chen, S.I. Orimo. Formation of an intermediate compound with a $B_{12}H_{12}$ cluster: experimental and theoretical studies on magnesium borohydride $Mg(BH_4)_2$. Nanotechology 20 (2009) 204013 (7pp)

[6] M. Chong, A. Karkamkar, T. Autrey, S.I. Orimo, S. Jalisatgi, C.M. Jensen. Reversible dehydrogenation of magnesium borohydride to magnesium triborane in the solid state under moderate conditions. Chem. Comm. 47, (2011) 1330-1332.

[7] G. Barkhordarian, T. Klassen, R. Borman. Catalytic mechanism of transition-metal compounds on Mg hydrogen sorption reaction. J. Phys. Chem. C 110, (2006) 11020-11024.

[8] Handbook of Chemistry and Physics, 74th ed.; Editor, Lide D. R.; 1993–1994.

[9] S. Hino, J.E. Fonneløp, M. Corno, O. Zavorotynska, A. Damin, B. Richter, M. Baricco, T.R. Jensen, M.H. Sørby, B.C. Hauback. Halide Substitution in Magnesium Borohydride, J. Phys. Chem. C, 116 (2012) 12482-12488.

[10] J. Rijssenbeek, Y. Gao, G. Soloveichik, J.C. Hanson. In 2006 MRS Fall Meeting. Symposium Z: Hydrogen Storage Technologies: Boston, MA, 2006 p Z2.4

IOP Conf. Series: Materials Science and Engineering **77** (2015) 012042 doi:10.1088/1757-899X/77/1/012042

Electrophoretic Nanocrystalline Graphene Film Electrode for Lithium Ion Battery

Kaspars Kaprans, Gunars Bajars, Gints Kucinskis, Anna Dorondo, Janis Mateuss, Jevgenijs Gabrusenoks, Janis Kleperis, Andrejs Lusis

Institute of Solid State Physics, University of Latvia, 8 Kengaraga street, LV-1063, Riga, Latvia

E-mail: k.kaprans@gmail.com

Abstract. Graphene sheets were fabricated by electrophoretic deposition method from water suspension of graphene oxide followed by thermal reduction. The formation of nanocrystalline graphene sheets has been confirmed by scanning electron microscopy, X-ray diffraction and Raman spectroscopy. The electrochemical performance of graphene sheets as anode material for lithium ion batteries was evaluated by cycling voltammetry, galvanostatic charge-discharge cycling, and electrochemical impedance spectroscopy. Fabricated graphene sheets exhibited high discharge capacity of about 1120 $mAh \cdot g^{-1}$ and demonstrated good reversibility of lithium intercalation and deintercalation in graphene sheet film with capacity retention over 85 % after 50 cycles. Results show that nanocrystalline graphene sheets prepared by EPD demonstrated a high potential for application as anode material in lithium ion batteries.

1. Introduction

Rechargeable lithium ion batteries (LIBs) have attracted much attention as power sources for portable electronic devices, electric vehicles, and energy storage systems due to the high energy density, high operating voltage and low self-discharge rate. To meet the increasing demand for LIBs with high reversible capacity and energy density, as well as more durable cycling performance, an innovative anode material is required. Graphite is commonly used as an anode material for LIBs because of its high columbic efficiency, acceptable specific capacity and cyclic performance by forming intercalation compounds LiC_6 [1-5]. However theoretical specific capacity of graphite is 372 $mAh \cdot g^{-1}$ because of the limited Li-ion storage sites within the sp^2 carbon structure [6-8]. One alternative to graphite has recently been reported – graphene nanosheets (GNS) obtained by various fabrication routes [9-15]. Graphene, two-dimensional graphite, is a rapidly rising star in material science. It has extremely low thickness, high aspect ratio (the ratio of lateral size to thickness), excellent electrical conductivity and good mechanical properties, which qualify it a potentially promising anode material for lithium ion batteries [14-16]. To date, various graphene sheet (GS) fabrication methods such as chemical, liquid phase and anodic electrochemical exfoliation have been developed. Unfortunately these "top down" methods have drawbacks such as toxic chemical usage, the generation of graphene defects in sheets and they are also time consuming [17, 18]. While "bottom-up" methods for graphene sheet production, such as epitaxial growth and chemical vapour deposition, enable to produce high quality and large area graphene sheets [19, 20], these methods demand sophisticated equipment and it is complicated to control the fabrication process [18]. Electrophoretic deposition (EPD) is an economical and versatile processing technique that has been applied for deposition of coatings and films. It has many advantages in the preparation of thin films from suspensions, such as high deposition rate and throughput, good uniformity and controlled thickness of the obtained films, no need of binders, and simplicity of scaling up [21]. Fabrication of graphene and CNT/graphene composite nanosheets by electrophoretic deposition technique was

IOP Conf. Series: Materials Science and Engineering **77** (2015) 012042 doi:10.1088/1757-899X/77/1/012042

demonstrated by Seo, Lee et. al. [22, 23]. However the specific capacity values of obtained nanosheets by EPD from isopropanol have been dramatically reduced after the first discharge.

In this study direct EPD method is used to prepare free standing porous graphene sheets from water suspension of graphene oxide (GO) and subsequent thermal reduction of as-deposited multilayers. Structure and morphology of graphene sheets were examined by X-ray diffraction analysis (XRD), Raman spectroscopy and scanning electron microscopy (SEM). The electrochemical performance of graphene sheets as LIB anode materials were evaluated by cycling voltammetry (CV), galvanostatic charge-discharge cycling, and electrochemical impedance spectroscopy (EIS).

2. Experimental

2.1. Fabrication of graphene sheets

Water suspension of graphene oxide was used for electrophoretic deposition of films under potentiostatic mode. GO was dispersed in de-ionized water at the concentration 10 $mg{\cdot}mL^{-1}$ and diluted to the concentration 1 $mg{\cdot}mL^{-1}$ afterwards. Before the EPD process GO suspension was sonicated for 1 h in order to ensure homogenous dispersion. Electrodes used for EPD process were 1 mm thick 316 stainless steel substrates with a size of 1 x 4 cm. The distance between the two electrodes was 10 mm, and the applied electric field was 30 $V{\cdot}cm^{-1}$ (DC power supply Agilent Technologies N5772A). Under the applied voltage, the negatively charged graphene oxide particles migrated toward the positive electrode and were subsequently orderly deposited on it. Thermal reduction of as-deposited GO films was performed by heating at 700 ºC in argon/hydrogen flow (Ar:H_2 volume ratio – 95:5). The thickness of the graphene films was tuned ranging from several nanometers to a few micrometers by varying the deposition conditions, including the concentration of graphene oxide, the applied voltage, and the deposition time. The thickness of the films was evaluated using a Veeco Dektak 150 profilometer.

2.2. Characterization of graphene sheets

Obtained graphene layers were analyzed by X-ray diffraction (Philips X'Pert Pro MPD diffractometer, Cu K_α radiation) and Raman spectroscopy (Advantage-785 spectrometer). Morphologies of graphene nanosheets were characterized by scanning electron microscopy (Hitachi S4800). The electrochemical properties of these films as an electrode for lithium ion batteries were tested by various methods such a cyclic voltammetry, galvanostatic charge-discharge cycling and electrochemical impedance spectroscopy (VoltaLab 40). The measurements were performed in Swagelok type two electrode cell with metallic lithium as a reference and a counter electrode, and $LiPF_6$ in ethylene carbonate and dimethyl carbonate mixture (volume ratio 1:1) as an electrolyte. The cell was assembled in an argon filled glove box. The active area of the GS electrode is 0.8 cm^2, the thickness of graphene film ~ 1.5 μm.

3. Results and discussion

3.1. Structure and morphology of graphene sheets

The structural changes from GO to graphene during the thermal reduction of electrophoretically prepared films were investigated by XRD measurements. The XRD patterns of GO film, stainless steel substrate and graphene sheet obtained at electric field 30 $V{\cdot}cm^{-1}$ are shown in Figure 1. The XRD pattern of graphene exhibits a characteristic peak (002) of graphite at 26.5°, which was not observed for GO [5, 12]. The characteristic peak of GO at 11° was not observed for graphene sheets. It means that most of the GO was successfully converted to graphene by thermal reduction of the as-deposited GO films.

Raman spectroscopy was performed to indicate the structures of GO and graphene sheets by analysing carbon D and G bands. The D band at 1350 cm^{-1} is ascribed to defects and disordered carbon, while G band at 1585 cm^{-1} arises from the zone centre E_{2g} mode, corresponding to ordered sp^2 bonded carbon [5, 23]. As seen from the Figure 2 relative intensities of D and G bands are comparable which means that both GO and graphene sheets exhibit the presence of a significant number of defects and high disorder.

IOP Conf. Series: Materials Science and Engineering **77** (2015) 012042 doi:10.1088/1757-899X/77/1/012042

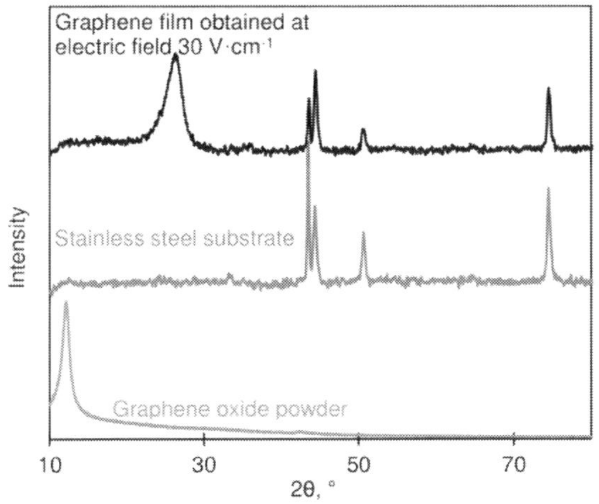

Figure 1. XRD pattern for graphene oxide, stainless steel substrate and graphene film obtained at electric field 30 V·cm^{-1}

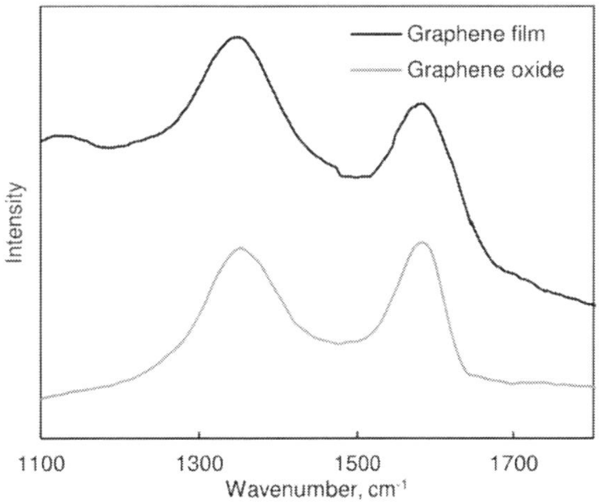

Figure 2. Raman spectra of graphene oxide (red) and graphene film (black)

The morphology of graphene sheets obtained by EPD with the electric field 30 V·cm^{-1} were characterized by SEM and their images are shown in Figure 3. SEM images of graphene sheets exhibit a nanocrystalline structure with the particle size in the range of 200 – 400 nm. Such nanocrystalline structure has been obtained due to the water decomposition during the EPD process at high voltage followed by the intensive evolution of gases on the electrodes. Therefore obtained graphene films possess not only layered but also nanocrystalline structure which is advantageous for enhanced Li-ion transport in films.

Fig. 3. SEM images of graphene films

3.2. Electrochemical performance of graphene sheets

Electrochemical properties of graphene sheets were first characterized by CV method. Typical cyclic voltammogram of GS electrode at 1 mV·s^{-1} from 0.01 to 4 V vs. Li/Li^{+} is shown in Figure 4. The CV profile is almost similar to the graphitic carbon with two cathodic and two anodic peaks [24, 25]. The wide reductive peak at around 1.2 V can be attributed to the formation of SEI film. The visible cathodic peak observed at around the potential 0.5 V and corresponding anodic peak at around 1.8 V complies to the Li-ion intercalation and deintercalation during the potential scan. This result is consistent with the charge-discharge voltage profiles (Figure 5).

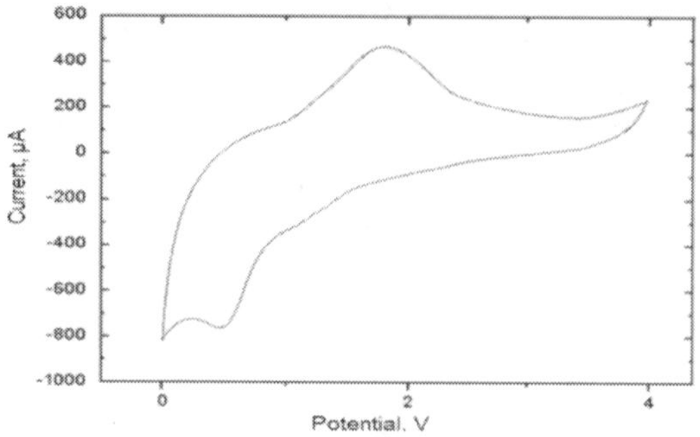

Figure 4. CV curve of graphene sheet electrode at the potential scan rate 1 mV·s^{-1}

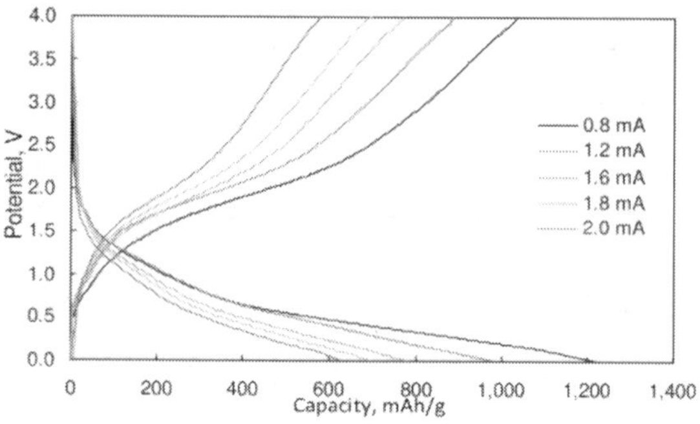

Figure 5. Galvanostatic charge-discharge profiles of graphene sheet electrode at the potential range 0.01 – 4 V

The gravimetric charge capacity dependence on discharge current is shown in Figure 6. At slow discharge rate (I = 0.8 mA) gravimetric charge capacity of obtained GS anode reaches high value ~ 1120 mAh·g^{-1} which is comparable or even higher than reported by other authors [5, 12, 15, 24-26]. The increasing of discharge rate causes the gradual decrease of capacity. At much higher discharge rate (I = 2 mA) capacity falls to ~ 620 mAh·g^{-1}. The cycling behaviour of obtained GS electrodes at I = 2 mA within the voltage range of 0.01 to 4 V is shown in Figure 7. It was found that after 50 cycles the reversible capacity of GS electrode still maintains ~ 530 mAh·g^{-1}. The Coulombic efficiency for first 50 cycles of GS electrode exceeded over 85 %. The further decrease of capacity during the cycling is negligible.

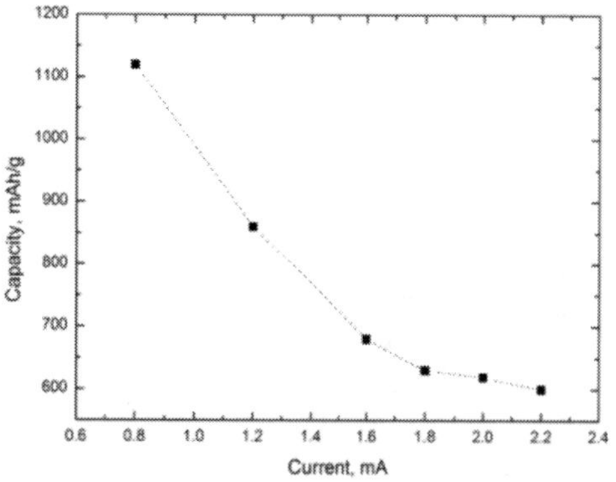

Figure 6. Discharge rate performance of graphene sheet electrode

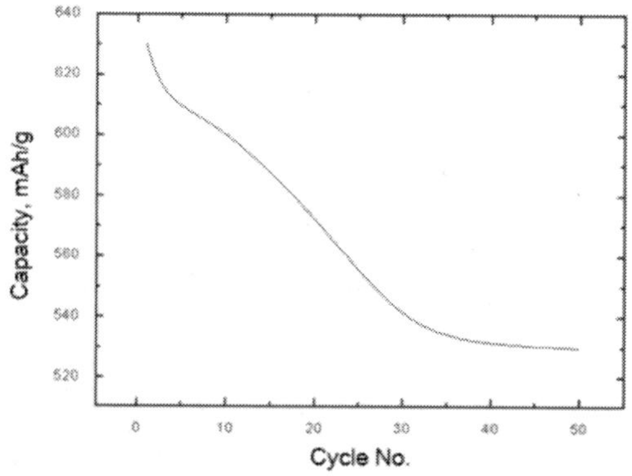

Figure 7. Cycling performance of graphene sheet electrode

4. Conclusions

Graphene sheets were fabricated by electrophoretic deposition method followed by thermal reduction. Results of scanning electron microscopy, X-ray diffraction and Raman spectroscopy confirm the formation of nanocrystalline graphene sheet films. The profile of voltammetric curves indicated smooth lithiation and delithiation processes of obtained graphene films. The initial discharge capacity was about 1120 mAh·g^{-1} at a constant current density of 1 mA·cm^{-2}. Extended cycling was performed that demonstrated good reversibility of lithium intercalation and deintercalation in graphene sheet film with capacity retention over 85 % after 50 cycles. Results show that nanocrystalline graphene sheets prepared by EPD exhibited a great potential for application as an anode material in lithium ion batteries because of the very nanocrystalline structure of GS, which promotes the Li-ion intercalation through the graphene layers. The use of light-weight graphene and lithium metal provides a high gravimetric capacity and energy density. Experimental results indicate that EPD is a highly powerful tool for the deposition of nanocrystalline graphene films from water suspension of graphene oxide.

Acknowledgements
The financial support of Latvian Council of Science Cooperation Project No. 666/2014 is greatly acknowledged.

References

[1] J-R Dahn, T Zheng, Y Liu, J S. Xue 1995 *Science* **270** 590.
[2] K Fukuda, K Kikuya, K Isono, M Yoshio 1997 *J. Power Sources* **69** 165.
[3] H Nozaki, K Nagaoka, K Hoshi, N Ohta, M Inagaki 2009 *J. Power Sources* **194** 486.
[4] F Ding, W Xu, D Choi, W Wang, X Li, M H Enqelhard, X Chen, Z Yang, J-G Zhang 2012 *J. Mater. Chem.* **22** 5.
[5] L-Z Bai, D-L Zhao, T-M Zhang, W-G Xie, J-M Zhang, Z-M Shen 2013 *Electrochim. Acta* **107** 555.
[6] J O Besenhard, J Yang, M Winter 1997 *J. Power Sources* **68** 87.
[7] J-M Tarascon, M Armand 2001 *Nature* **414** 359.
[8] K S Novoselov, A K Geim, S V Morozov, D Jiang, Y Zhang, S V Dubonos, I V Grigorieva, A A Firsov 2004 *Science* **306** 666.
[9] E Yoo, J Kim, E Hosono, H Zhou, T Kudo, I Honma 2008 *Nano Lett.* **8** 2277.
[10] M-H Liang, L J Zhi 2009 *J. Mater. Chem.* **19** 5871.
[11] D Pan, S Wang, B Zhao, M Wu, H Zhang, Y Wang, Z Jiao 2009 *Chem. Mater.* **21** 3136.
[12] P Guo, H Song, X Chen 2009 *Electrochem. Commun.* **11** 1320.
[13] T Li, L J Gao 2012 *J. Solid State Electrochem.* **16** 557.
[14] X Li, Y H Liu, A Lushington, R Li, X Sun 2013 *Nanoscale* **5**12607.
[15] T D Dao, J-E Hong, K S Ryu, H M Jeong 2014 *Chem. Engineer. J.* **250** 257.
[16] G Kucinskis, G Bajars, J Kleperis 2013 *J. Power Sources* **240** 66.
[17] X Huang, X Qi F Boey, H Zhang 2012 *Chem. Soc. Rev.* **41** 666.
[18] V Singh, D Joung, L Zhai, S Das, S I Khondaker, S Seal 2011 *Prog. Mater. Sci.* **56** 1178.
[19] C Berger, Z Song, X Li, X Wu, N Brown, C Naud, D Mayou, T Li, J Hass, A N Marchenkov, E H Conrad, P N First, W A de Heer 2006 *Science* **312** 1191.
[20] X Li, W Cai, J An, S Kim, J Nah, D Yang, R Piner, A Velamakanni, I Jung, E Tutuc, S K Banerjee, L Colombo, R S Ruoff 2009 *Science* **324** 1312.
[21] Z-S Wu, S Pei, W Ren, D Tang, L Gao, B Liu, F Li, C Liu, H-M Cheng 2009 *Adv. Mater.* **21** 1756.
[22] S-D Seo, I-S Hwang, S-H Lee, H-W Shim, D-W Kim 2012 *Ceram. Internat.* **38** 3017.
[23] S-H Lee, S-D Seo, K-S Park, H-W Shim, D-W Kim 2012 *Mater. Chem. Phys.* **135** 309.
[24] Y Yang, F Lu, Z Zhou, W Song, Q Chen, X Ji 2013 *Electrochim. Acta* **113** 9.
[25] J Zhang, H Cao, X Tang, W Fan, G Peng, M Qu 2013 *J. Power Sources* **241** 619.
[26] Y Hu, X Li, D Geng, M Cai, R Li, X Sun 2013 *Electrochim. Acta* **91** 227.

IOP Conf. Series: Materials Science and Engineering **77** (2015) 012043 doi:10.1088/1757-899X/77/1/012043

Mechanical Properties of Composite SPEEK Polymer Membranes Modified with Ionic Liquids

E Sprugis[1], **I Reinholds**[1], **G Vaivars**[1,2]

[1]Department of Chemistry, University of Latvia, Latvia

[2]Institute of Solid State Physics, University of Latvia, Latvia

e-mail: Guntars.Vaivars@cfi.lu.lv

Abstract. In this work, the mechanical properties of sulphonated polyetheretherketone (SPEEK) membranes impregnated with 3 different ionic liquids (1-butyl-2,3-dimethyl-imidazolium dimethylphosphate ([BMMIM][Me_2PO_4])), 1,2,3-trimethylimidazolium dimethylphosphate ([MMMIM][Me_2PO_4])), 1,3-dimethylimidazolium dimethylphosphate ([MMIM][Me_2PO_4])) have been investigated.
Prepared SPEEK/ionic liquid composite membranes are characterized by mechanical testing both in room and elevated temperatures. It was found that the stiffness and tensile strength of composites decreased by increasing the content of ionic liquid and the length of alkyl radical in ionic liquid as well as by increasing the temperature.

1. Introduction

Many of today's environmental problems are due to the world's dependence on fossil fuel. The fuel cell is one of the most promising alternative solution, especially in portable devices. Proton exchange membrane (PEM) fuel cells are suitable for low and medium temperature applications. They have several desirable properties, such as ability to start-up at a room temperatures and low weight. The key component in such fuel cell is a proton exchange membrane, which must obtain the following criteria, such as good proton conductivity, mechanical and thermal durability, and low cost. For long time, the Nafion has been possessed as the industry's polymer membrane of choice in PEM fuel cells, due to it's good conductive properties and overall stability. However, this conductivity is retained only up to about 80 °C, since it's proton conductivity depends on water content in membrane [1,2]. For this reason, alternatives are sought after. One approach was to modify the Nafion itself, for example, with SiO_2 particles [3], TiO_2 nanoparticles [4] and ionic liquids (ILs) [5]. Other approach is to search for alternative materials for PEM. As possible candidates worth mentioning are polyethersulphone, polyetheretherketone (PEEK), and polybenzimidazoles [6]. Due to it's properties and commercial availability PEEK seems to be among most promising. For PEEK to possess the proton conductivity, it needs to be sulfonated. Degree of sulfonation (DS) then directly correlates with conductivity. Unfortunately, with higher DS some key properties of SPEEK severely detoriate [7]. Again, various attempts to modify SPEEK have been made, e.g. the impregnation with ILs, yielding composite membranes with a notable conductivity even in temperatures above 100 °C, making them potentially suitable for medium temperature PEM fuel cells [8–11]. Mechanical strength is another important factor for successful operation of PEMs in FC [12]. Typical gas pressure in FCs is about 0,1–0,3 MPa [13].

In this work, the influence of IL modification degree on SPEEK composite mechanical properties was investigated in a temperature range from 20 to 200 °C.

2. Experimental

2.1. Materials

PEEK in granular form was purchased from Sigma-Aldrich. SPEEK was then synthesized by stirring a mixture of PEEK and concentrated sulphuric acid for 3 hours at 50 °C. Solution was then poured into large excess of ice cold deionized water. Resulting SPEEK precipitate was then washed until neutral pH, filtered, and dried in oven for 24 h at 80 °C. Sulphuric acid used was analytical grade (Sigma-Aldrich). DS of SPEEK was determined to be 45 % using method described in [14]. ILs were synthesized in reaction of amine and trimethylphosphate as described elsewhere [15]. Chemical structures of synthesized ILs are shown in figure 1.

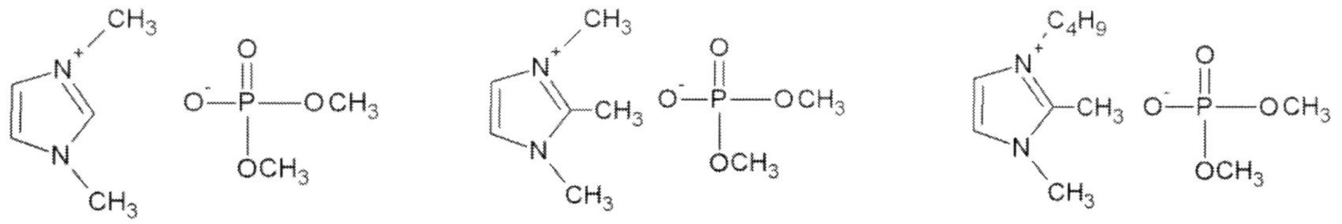

[MMIM][Me_2PO_4] [MMMIM][Me_2PO_4] [BMMIM][Me_2PO_4]

Figure 1. Chemical structures of ILs used

2.2. Impregnation with ionic liquids

0.5 g SPEEK and appropriate amount of IL (see table 1) was dissolved in dimethylformamide by stirring the solution for 2 h. Solution was poured in Petri dish which then was left in oven for 24 h at 70 °C to evapotate the solvent.

Table 1. Composition of prepared membranes

Sample	IL	m_{SPEEK}, g	m_{IL}, g	W_{IL}, %
A0	-	0.4901	0	0
B15	[BMMIM][Me_2PO_4]	0.4750	0.0833	14.9
B25	[BMMIM][Me_2PO_4]	0.4927	0.1677	25.4
B40	[BMMIM][Me_2PO_4]	0.5155	0.3680	41.7
C15	[MMMIM][Me_2PO_4]	0.4892	0.0879	15.2
C25	[MMMIM][Me_2PO_4]	0.5059	0.1678	24.9
C40	[MMMIM][Me_2PO_4]	0.5026	0.3345	40.0
D15	[MMIM][Me_2PO_4]	0.5110	0.0920	15.3
D25	[MMIM][Me_2PO_4]	0.5067	0.1697	25.1
D40	[MMIM][Me_2PO_4]	0.5254	0.3481	39.9

2.3. Equipment

For measuring the stress-strain curves at a room temperature, the Tinius Olsen H1K-S universal testing machine was used. For measurements, the membranes were cut in bone shaped pieces with width of about 5 mm in narrowest point. Measurements then were taken with deformation speed 1 mm/min (for determining Young's modulus) and 10 mm/min for determination of tensile characteristics up to the sample rupture (until sample breaks).

Testing at elevated temperatures was done using in-house equipment consisting of tension testing

IOP Conf. Series: Materials Science and Engineering **77** (2015) 012043 doi:10.1088/1757-899X/77/1/012043

stand with controllable motor and heating camera, autotransformer for maintaining heating mode, Pico ADC-24 data logger, and termopair interface for controlling temperature. Deformation speed was set to 50 mm/min and measured at 100 °C, 160 °C, and 200 °C.

3. Results and discussion

For membrane usage in FCs, the tensile strength at yield (σ_t) is one of the most important mechanical properties, describing the maximal acceptable load the material can withstand before it starts to deform irreversibly, therefore it will be a central point in this discussion.

Figure 2 shows the impact of structure and content of IL on strain–stress curves at room temperature, considering the changes of elastic and viscoelastic transitions up to the yield point, followed by the neck formation. It is clearly seen for composites containing all three IL modifiers that the values of elastic modulus are decreasing with an increase of IL content from 0 to 40 wt.% in composites – the slope length decrease, comparing to entire SPEEK matrix (see the curves at deformation <1% in Figure 2). As IL content increases, the yield strength decrease, while the deformation at the yield increase due to the change of composite deformational behavior from semi crystalline (SPEEK) to elastomer type. The data indicate an effective plasticizer effect of IL in SPEEK composite. The impact of chemical structure of IL isn't so straightforward due to complexity of structure. For that reason, three similar structures are compared varying the length of alkyl radical in IL. The mechanical strength (Fig. 3) decreases by increasing the IL concentration in composites. At lower IL concentrations the mechanical strength of membranes decreases by increasing the length of alkyl radical in IL. Other researchers have comparable results. In one case, the tensile strength of the IL-polymer composite (with polymer structure being different but similar to SPEEK, and IL content of 40 %) was reported to be about 9 to 16 MPa [14]. Others reported significantly lower figures – about 2 to 3 MPa, even with lower IL content [11].

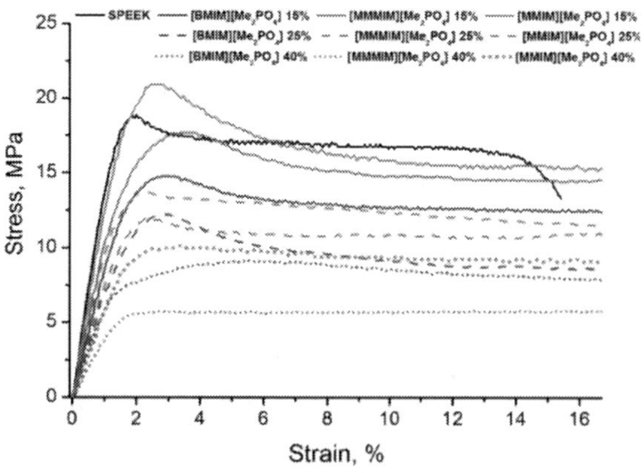

Figure 2. Strain-stress curves at room temperature (expanded the yielding region of stress-strain curve)

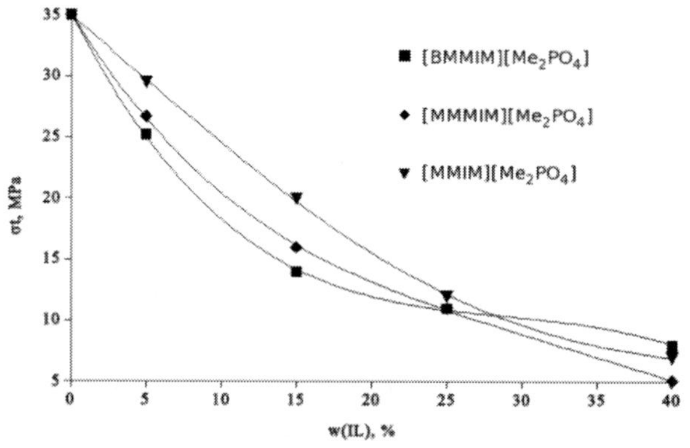

Figure 3. Tensile strength depending on IL content at room temperature

Since PEMs must operate at elevated temperatures, care was taken to do measurements under such circumstances. Results for [MMIM][Me_2PO_4]-SPEEK composite membrane can be seen in Figure 4.

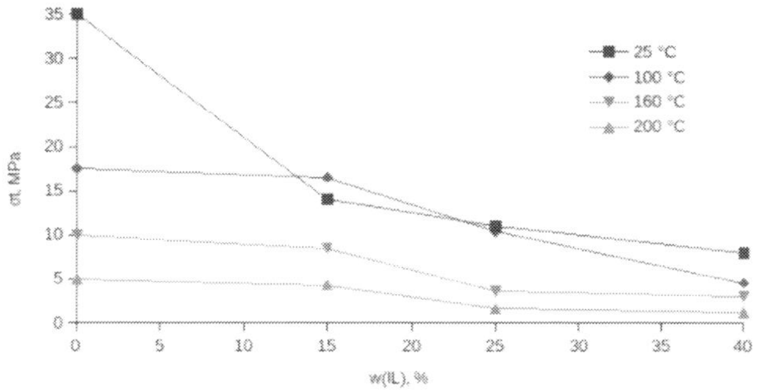

Figure 4. Tensile strength depending on IL content for [MMIM][Me_2PO_4]-SPEEK composite

The mechanical strength of composite was rather stable during heating to 100 °C. At higher temperatures the membrane strength rapidly deteriorated.

4. Conclusion

The mechanical strength of SPEEK composite membrane with IL is decreasing by increasing the content of IL, by increasing the temperature and by increasing the length of alkylradical in IL. At highest temperatures use of these membranes in PEM FCs might be questionable, especially for longer periods of time, but they are promising for applications in a temperature range 100 – 150 °C, where typically the proton conducting membranes are loosing water.

5. Acknowledgment

All authors acknowledge Latvian Council of Science Cooperation Project COPOCO No 666/2014 for financial support.

References

1. Bose, S.; Kuila, T.; Nguyen, T. X. H.; Kim, N. H.; Lau, K.; Lee, J. H. *Prog. Polym. Sci.* **2011**, *36*, 813–843.
2. Peighambardoust, S. J.; Rowshanzamir, S.; Amjadi, M. *Int. J. Hydrog. Energy.* **2010**, *35*, 9349–9384.
3. Amjadi, M.; Rowshanzamir, S.; Peighambardoust, S. J.; Sedghi, S. *J. Power Sources.* **2012**, *210*, 350–357.
4. Amjadi, M.; Rowshanzamir, S.; Peighambardoust, S. J.; Hosseini, M. G.; Eikani, M. H. *Int. J. Hydrog. Energy.* **2010**, *35*, 9252–9260.
5. Garaev, V.; Pavlovica, S.; Reinholds, I.; Vaivars, G. *IOP Conf. Ser. Mater. Sci. Eng.* **2013**, *49*, 012058.
6. Namazi, H.; Ahmadi, H. *J. Power Sources.* **2011**, *196*, 2573–2583.
7. Wu, X.; Wang, X.; He, G.; Benziger, J. *J. Polym. Sci. Part B Polym. Phys.* **2011**, *49*, 1437–1445.
8. Che, Q.; He, R.; Yang, J.; Feng, L.; Savinell, R. F. *Electrochem. Commun.* **2010**, *12*, 647–649.
9. Sasikala, S.; Meenakshi, S.; Bhat, S. D.; Sahu, A. K. *Electrochimica Acta.* **2014**, *135*, 232–241.
10. Che, Q.; Sun, B.; He, R. *Electrochimica Acta.* **2008**, *53*, 4428–4434.
11. Chen, J.; Guo, Q.; Li, D.; Tong, J.; Li, X. *Prog. Nat. Sci. Mater. Int.* **2012**, *22*, 26–30.
12. Sgreccia, E.; Chailan, J.-F.; Khadhraoui, M.; Di Vona, M. L.; Knauth, P. *J. Power Sources.* **2010**, *195*, 7770–7775.
13. Balasubramanian, B., Barbir, F., Neutzler, J. Optimal Operating Temperature and Pressure of PEM

Fuel Cell Systems In Automotive Applications. Proc. 218th ACS National Meeting, Vol. 44, No. 4, American Chemical Society. **1999**, pp. 977-981.

14. Chen, B.-K.; Wong, J.-M.; Wu, T.-Y.; Chen, L.-C.; Shih, I.; others. *Polymers.* **2014**, *6*, 2720–2736.
15. Kuhlmann, E.; Himmler, S.; Giebelhaus, H.; Wasserscheid, P. *Green Chem.* **2007**, *9*, 233–242.

IOP Conf. Series: Materials Science and Engineering 77 (2015) 012044 doi:10.1088/1757-899X/77/1/012044

Effect of bismuth oxide dispersivity on the dielectric properties of zinc oxide ceramics

O.V. Malyshkina[1], A.I. Ivanova[1], R. Luzin[1], I. Makarenkov[1], S.I. Pugachev[2], E. Rytov[2]

[1]Tver State University, Sadovij per.35, 170002 Tver, Russia
[2]State Marine Technical University of St. Petersburg, St.- Peterburg, Russia

Olga.Malyshkina@mail.ru

The dielectrics properties and microstructure of fractured surfaces of ZnO electroceramics samples with different Bi_2O_3 dispersivity was studied. It is shown that the change of dispersion leads to an insignificant change in the grain structure, but has no effect on character of the dispersion of the permittivity ZnO ceramic.

1. Introduction

Ceramics on the base of zinc oxide (ZnO) is of great practical use in protective facilities for power engineering [1, 2]. There exist a number of theories explaining the mechanism of conductivity in ZnO varistors [1]. At the present time it is recognized that nonlinearity in ZnO varistors is conditioned by intergrain boundaries characterized by a barrier in the depleted layers of adjacent grains for maul charge carriers (electrons). It is believed that the most feasible model of the intergrain boundary is that of double Schottky barier. As this takes place the negative surface charge at the grain boundary (electron trap) compensates the positive charge in the depleted layer at both sides of the boundary surface. Thereby the properties of zinc-oxide ceramics are substantially affected by the quality of raw materials, composition and component ratio as well as preparation technology. According to literature [3, 4], the concentration of additives are introduced into the ZnO ceramics could effect on the grain growth of zinc oxide.

We have studied ceramics produced by JSC "Plant energozaschitnyh devices". In addition to ZnO in the burden included the following oxides: Bi_2O_3, Sb_2O_3, Cr_2O_3, MnO_2, Co_3O_4, SiO_2. Bismuth oxide (Bi_2O_3) is an important additive necessary for the production of ZnO varistors. Another is the Sb_2O_3, which affects the structural features and the grain size of ZnO ceramics [5, 6]. Hereupon it is of interest to compare the properties of zinc-oxide ceramics prepared using bismuth oxide of different dispersivity in the charge.

In the present work we researched the effect of Bi_2O_3 dispersivity on the structure and dielectric properties of zinc oxide ceramics. A study was made of the samples prepared with the aid of Bi_2O_3 of usual dispersivity (particle size 3…10) and highly disperse one (particle size < 1 μm). The structure and element composition were studied by the methods of scanning electron microscopy (SEM) (JEOL 6510LV) by inspection of the topography and energy-dispersive analysis (EDA) of the sample cross sections.

2. Results and discussion

The distinctions due to different Bi_2O_3 dispersivity are clearly revealed by the images of lateral cleavages of ZnO ceramics obtained in the regime of back scattered electrons (BEC) (Fig. 1), thus indicative of the variations in the element composition. In samples with high-disperse Bi_2O_3 the interlayer enters the ceramics by large portions (Fig. 1b), while in samples with usual additive the interlayer distributes in the material in a more uniform way and contains inclusions of main zinc oxide (Fig. 1a).

Table 1 and 2 presents the element composition of both types of ceramics. The results are given in weight percent. The spectra were measured for the regions marked in Fig. 1. It is seen that the main grains contain only zinc oxide, while all other oxides of the charge are concentrated in the interlayer between grains.

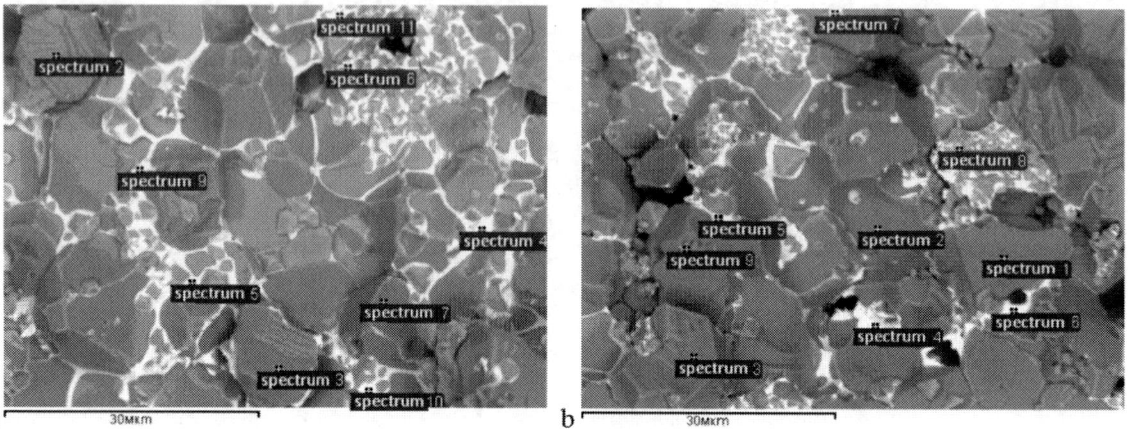

a b

Figure 1. Lateral cleavage images ZnO ceramics for ordinary (a) and finely dispersed Bi_2O_3 (b). Scale mark length 30 μm

Table 1. The element composition (in weight %) of ordinary dispersed Bi_2O_3.

Spectrum	O	Si	Cr	Mn	Co	Zn	Sb	Bi	Total
2	21.66					78.34			100.00
3	21.95					78.05			100.00
4	11.10	0.71				13.68	4.22	70.30	100.00
5	21.14	0.77	0.68			20.84	5.68	50.89	100.00
6	21.21		1.20			20.24	9.47	47.87	100.00
7	21.35					78.65			100.00
9	24.62					71.67		3.71	100.00
10	25.53	0.27	1.47	0.65	2.94	55.82	13.33		100.00
11	13.83		5.17	5.71	1.12	28.98	19.57	25.61	100.00

Spectra: 2, 3, 7 – grains, 4, 5, 6, 9, 10, 11 – layer between the grains.

Table 2. The element composition (in weight %) of finely dispersed Bi_2O_3.

Spectrum	O	Si	Cr	Co	Zn	Sb	Bi	Total
1	20.32			1.25	78.43			100.00
2	20.75				79.25			100.00
3	22.45				77.55			100.00
4	12.93	0.50			3.60		82.97	100.00
5	13.14	0.64			41.05		45.17	100.00
6	15.71		1.02		7.04		76.23	100.00
7	19.51				66.37		14.12	100.00
8	21.95	0.37	1.64	2.78	29.56	14.79	28.92	100.00
9	16.25	0.43			66.87	4.96	11.49	100.00

Spectra: 1, 2, 3 – grains, 4, 5, 6, 7, 8, 9 – layer between the grains.

IOP Conf. Series: Materials Science and Engineering **77** (2015) 012044 doi:10.1088/1757-899X/77/1/012044

In spite of the decrease of both dielectric permittivity and dielectric loss tangent values which are observed when finely dispersed Bi_2O_3 is used for the preparation of the samples instead of standard ingredient, the shape of the corresponding frequency dependences remains unchanged (Fig. 2). Two regions having different behaviour of the complex dielectric permittivity may be distinguished at the dispersion diagrams ε″(ε′) (Fig. 3). It means that there are two different relaxation processes: one dominate in low-frequency part of the spectra and one – in high frequency part of the spectra [7, 8]. The frequency of ~ 8 kHz corresponding to the change of the dispersion type is independent of the Bi_2O_3 dispersivity. In the low frequency region a linear dependence of the imaginary component on the real part of dielectric permittivity is observed, while in the high frequency region this dependence may be approximated by a circular arc. On the graph of the dielectric loss frequency dependence this region corresponds to the smeared maximum (Fig. 2). This type of dielectric response is described by the Cole-Cole theory [7]. It means that there is a strongly pronounced relaxation process with set of relaxation times. It is usually caused by dipole relaxation [8]. The most probable relaxation time is defined by the condition $\tau_o = 1\ \omega_{max}$, where $\omega_{max} = 2\pi f_{max}$. f_{max} is shown on Fig.3 (it equal to 100 kHz). It is independent of the Bi_2O_3 dispersivity. Therefore the most probable relaxation time is $1.6 \cdot 10^{-6}$ s for ceramics of both dispersivity.

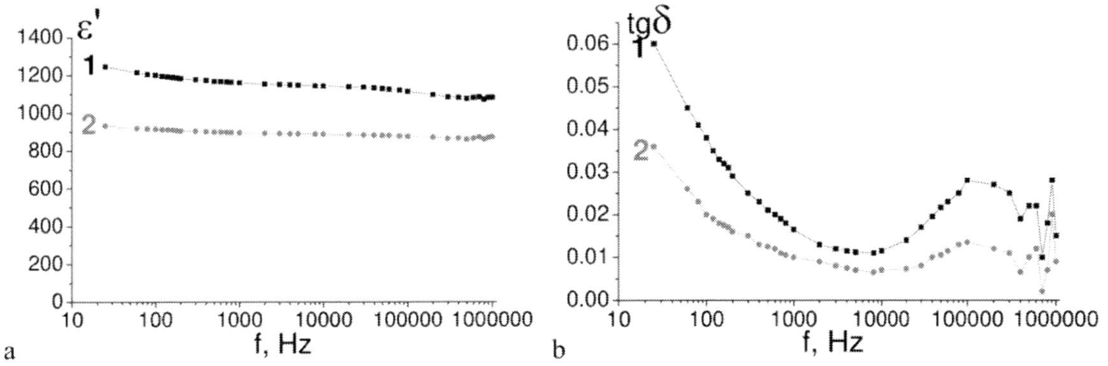

Figure 2. Permittivity (a) and dielectric loss tangent (b) of ZnO ceramics.
Ordinary (curve 1) and finely dispersed (curve 2) Bi_2O_3.

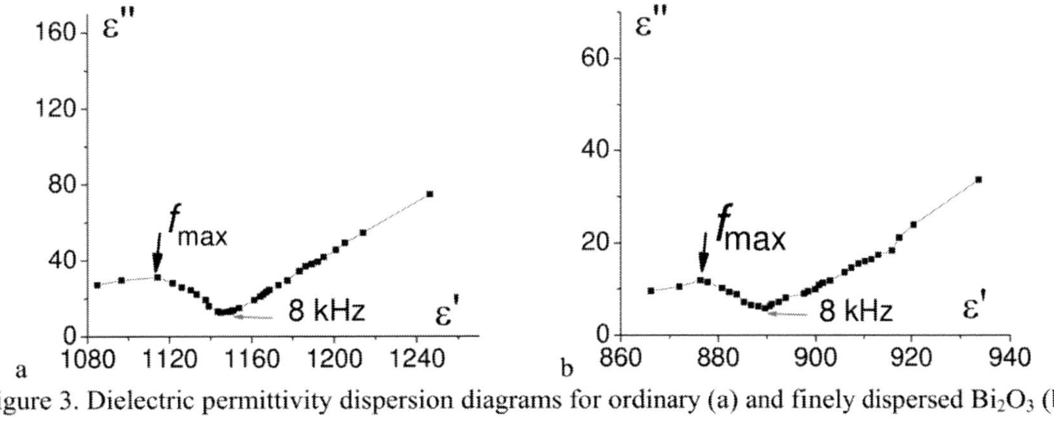

Figure 3. Dielectric permittivity dispersion diagrams for ordinary (a) and finely dispersed Bi_2O_3 (b)

Conclusion

The results of the present study show that the grains of zinc-oxide ceramics are composed of zinc oxide while other oxides from the charge are concentrated in the grain interlayer. The existing structural differences have an influence on the dielectric permittivity magnitudes and do not change the character of the dispersivity. The observed behaviour of the dielectric response is indicative of two mechanisms of dispersion. With the increase of the measurement frequency one mechanism is replaced by another.

According to Jonscher [7, 8] the experimental dielectric responses can be distinguished into two

IOP Conf. Series: Materials Science and Engineering **77** (2015) 012044 doi:10.1088/1757-899X/77/1/012044

cases. The first one is characterized by a maximum on the $\varepsilon''(f)$ (or tgδ(f)) dependences, while it is absent in the second case. According to our experiments both types of dielectric response behaviour are realized in the studied ZnO ceramics. Former occurs for the high frequency region, which relates to dipole relaxation. Another takes place for the low frequency regions. The linear dependence $\varepsilon''(\varepsilon')$, occurring for ZnO ceramics at low frequencies, usually associated with pass-through conductivity [7]. We did not find the relation between dielectric response and Bi_2O_3 dispersivity.

References

[1] Gupta TK 1990 *J. of the American Ceramic Society* **73** 1817–1840
[2] Clarke DR 1999 *J. Am. Ceram. Soc.* **82** 485–502
[3] Yangin H, Meidong L 2001 *J. Chin. Ceram. Soc.* **29 (6)** 580–583
[4] Senda T, Bradt R 1990 *J. Am. Ceram. Soc.* **73 (1)** 106–114
[5] Achard N, Counioux J 2000 *J. Therm. Anal. and Calorim* **59 (3)** 785–788.
[6] Ni Yu-Hua, Ma Li-Dun, Sheu-Xiao-Liang, Shi Guo-Shun 2000 *Chem. J. Chin. Univ.* **21 (5)** 785–788.
[7] Jonscher AK 1996 *Universal relaxation law* (London: Chelsea Dielectrics Press Ltd)
[8] Jonscher AK 1983 *Dielectric relaxation in solids* (London: Chelsea Dielectrics Press Ltd)

IOP Conf. Series: Materials Science and Engineering **77** (2015) 012045 doi:10.1088/1757-899X/77/1/012045

Surface relief and domain structure of ferromagnetic shape memory alloys

R M Grechishkin[1,2], **O V Gasanov**[1], **E T Kalimullina**[3], **S E Ilyashenko**[1,4], **O M Korpusov**[5] **and A B Zalyotov**[1,5]
[1]Tver State University, 170100 Tver, Russia
[2]National University of Science and Technology "MISIS", 119049 Moscow, Russia
[3]Kotel'nikov Institute of Radio Engineering and Electronics of RAS, 125009 Moscow, Russia
[4]Tver State Technical University, 170023 Tver, Russia
[5]Tver State Medical Academy, 170100 Tver, Russia

e-mail: rostislav.grechishkin@tversu.ru

Abstract. A study is made of the surface corrugation during thermal cycling of ferromagnetic shape memory alloys (FSMA). This specific feature is a property of FSMA alloys and is a consequence of martensitic phase transformations not necessarily connected with surface defects of the material. The surface relief structure was studied together with martensite and magnetic domain structure changes during thermal cycling of the samples with the aid of differential polarized light microscopy. The analysis was facilitated making use of auxiliary reference grids applied to the surface of the samples.

1. Introduction

In recent studies [1-2] it was shown that the structural, thermal, magnetomechanical and mechanical properties of ferromagnetic shape memory (FSMA) alloys may be substantially affected by defects at or near the surface introduced by different surface treatments such as abrasive grinding, spark eroding, wire cutting. In the present work we examine another type of surface effects manifesting themselves in surface corrugation during thermal cycling of initially planar FSMA samples [3]. This specific feature is inherent to FSMA alloys and is basically a consequence of martensitic phase transformations not necessarily connected with surface defects of the material. The key experiments were performed by examination of the surface relief structure and magnetic domain structure during thermal cycling with the aid of differential polarized light microscopy.

2. Experimental

2.1. Corrugation of planarized surfaces of bulk samples

The experiments have revealed a very specific behaviour of FSMA samples during the metallographic preparation of planar polished surfaces. Normally the planarization is performed at room temperature when the samples are in martensite (M) state (figure 1(*a*)) (faint images of M twins are due to optical contrast in polarized light). Heating the planarized samples to temperatures above the phase transition point when they transform into the cubic austenite phase (A) causes a corrugation of the planar surface (figure 1 (*b*)). The corrugation remains upon cooling back to martensite (figure 1(*c*)); now the contrast of the twin structure becomes much stronger due to the effect of relief. Two additional heating/cooling cycles (figure 2) have shown that the recovery of the martensite structure is not full so that each time the structural details undergo changes.

(*a*) martensite 1, RT (*b*) austenite, 90°C (*c*) martensite 2, RT

Figure 1. Microstructure of the polycrystalline $Ni_{2.19}Mn_{0.81}Ga$ sample in initial planarized state (martensite 1), heated to 90°C (austenite) and cooled back to RT (martensite 2)

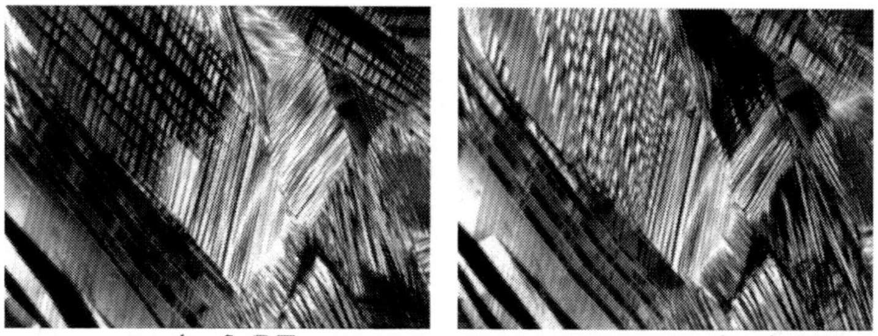

martensite 3, RT martensite 4, RT

Figure 2. Residual martensite structures after two additional heatings to 95°C

During further experiments we studied the surface relief of the samples for which the standard metallographic procedures (lapping, grinding, polishing) were applied at a temperature above the phase transition point, i.e. when the samples acquire the A state. It was found that although corrugation occurs upon cooling of these samples to M state the repeated heating allows to restore the planarized surface.

Figure 3 shows the motion of the A/M phase boundary from right to left toward the cold side of the sample (see the arrow) under the action of adjustable temperature gradient produced by a differential pair of Peltier elements. It is seen that on passing across the sample the initially corrugated A/M boundary leaves the uncorrugated surface behind. This process is quite reproducible and may be repeated many times.

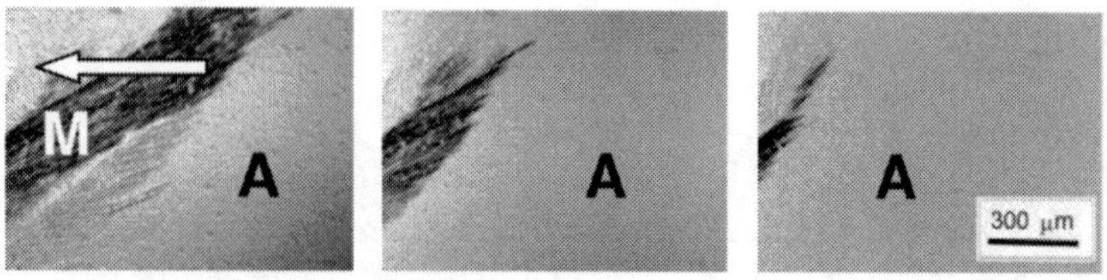

Figure 3. Video sequence showing the phase boundary motion at the surface of $Ni_{2.12}Mn_{0.88}Ga$ sample under the action of temperature gradient. Sample was planarized at 80 °C

2.2. Surface relief of thin film

As distinct from bulk samples, in film specimens the repeatibility of the structural details is much better. Figure 4 illustrates the direct and converse transformation process in a $Co_{48}Ni_{22}Ga_{30}$ thin film prepared by RF magnetron sputtering (thickness ~10 μm) starting from martensite up to austenite and back. Comparison shows that the final martensite structure coincides in detail with the starting one. Presumably this may be explained by the restriction of the degrees of freedom imposed by the large (>10) ratio of grain size/film thickness.

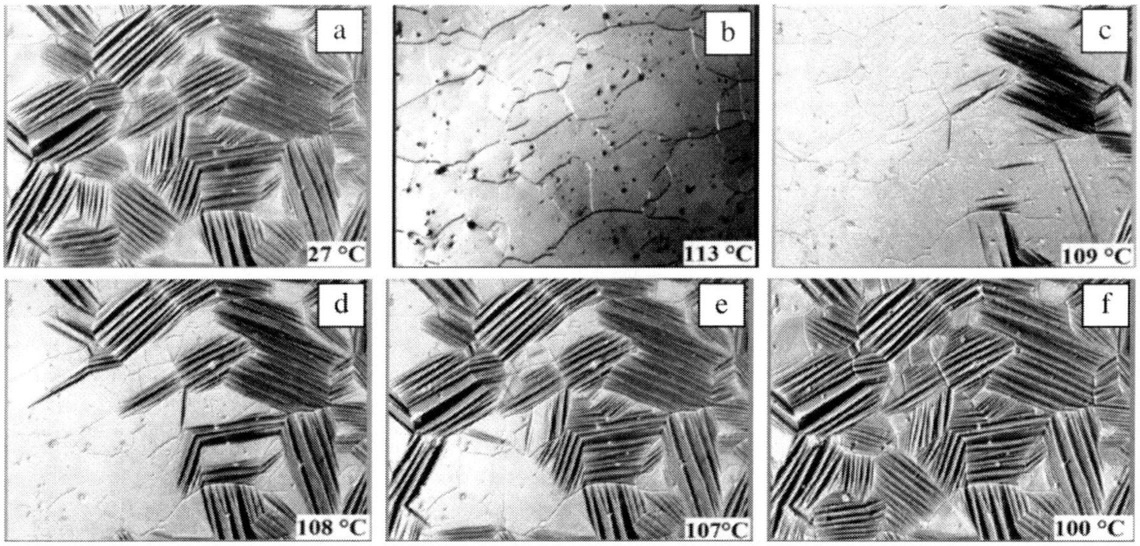

Figure 4. Dynamics of the martensite transformations of a $Co_{48}Ni_{22}Ga_{30}$ thin film. (*a*) starting point at RT (100% martensite); (*b*) 100% austenite state at 113 °C; (*c*), (*d*) and (*e*) – intermediate states; (*f*) – final return to 100% martensite

2.3. Observation of local deformation

Optical observations provide a possibility to perform quantitative measurements of local deformations caused by martensitic transformation. To this end the sample is supplied by some kind of reference marks. In the present work we used a diamond indentor of a microhardness testing instrument to apply a rectangular grid of light scratches on the sample's surface. Figure 5(*a*) shows these scratches applied at RT when the sample is in martensite state.

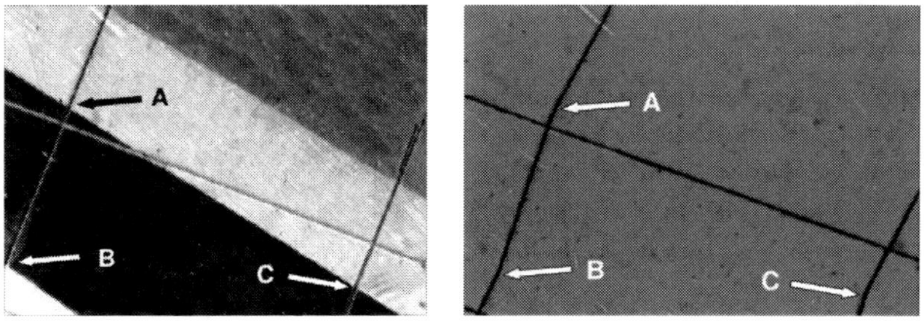

(*a*) (*b*)

Figure 5. Microstructure of $Ni_{2.16}Mn_{0.84}Ga$ FSMA alloy observed at RT (martensite state) (*a*) and after transformation into the austenite state at T = 370 K (*b*). The arrows show the points of martensite twin boundary intersections with a square grid of scratches in (*a*) and the points of the grid inflection in (*b*)

By using polarized light we were also able to observe the martensite twins. On passing to the austenite state by heating the polarized contrast of the martensite disappears because the cubic austenite phase is optically isotropic. The deformation is characterized by the changes in the geometry of the grid by the kink of scratch lines in characteristic points of intersection with the martensite boundary labeled as A, B, and C in figure 5. Comparison of the two images provides quantitative information on the localized deformation values of the material during martensitic transformation.

2.4. Absorption and desorption of liquids at the surface of FSMA

Another unusual feature of the FSMA was discovered during examination of metallographically processed samples. It was found that heating the sample polished at RT to a temperature above A/M transition (~80 °C) is accompanied by an appearance (desorption) of liquid drops with a diameter of 10 to 20 μm (figure 6). Surprisingly the effect appeared to be reproducible during at least 50 cycles of cooling and heating resembling the work of some kind of micropump.. The origin of the oil is definitely from the diamond polishing pastes which are prepared on the base of oil. The same behavior was observed with other polycrystalline FSMA compositions, in particular, $Ni_{2.19}Mn_{0.81}Ga$.

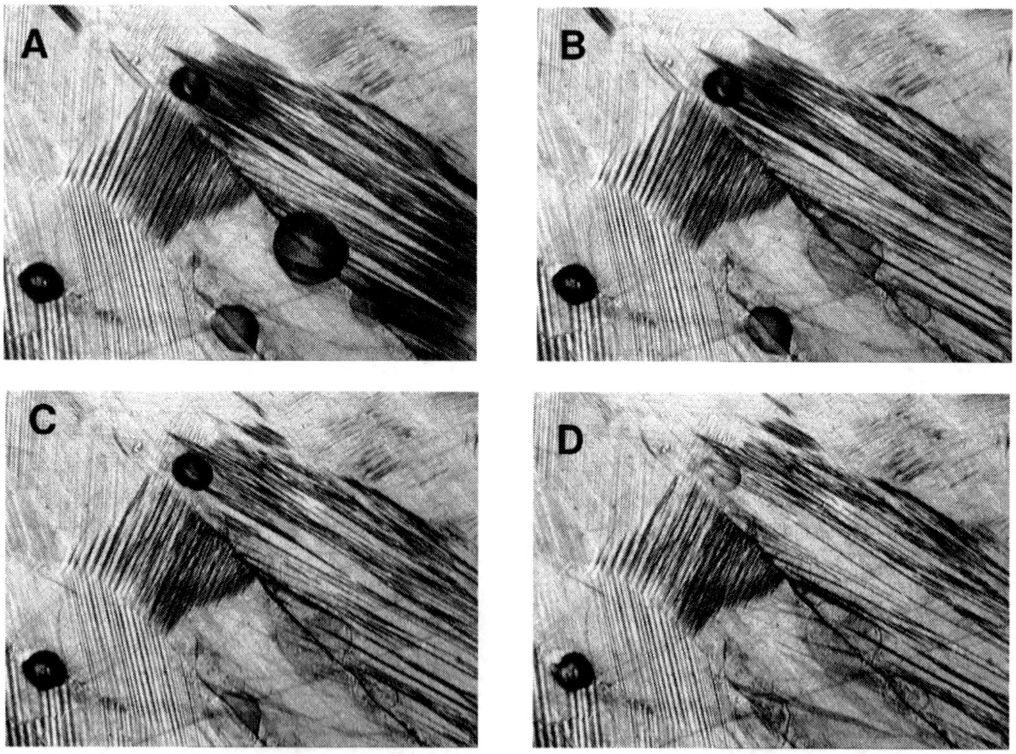

Figure 6. Desorption of four drops of oil at the grain boundaries of a heated polycrystalline $Ni_{2.16}Mn_{0.84}Ga$ sample heated to ~80 °C (A) and their one-by-one absorption during further cooling to 60 (B), 40 (C) and 22 °C (D). On next heating the drops arrive again one-by-one at the same places. The whole process may be repeated many times (up to 50 cycles were made in the experiment)

4. Conclusion

An experimental study is made of the effect of sample preparation on the surface relief of shape memory alloys of the Heusler family and its behaviour in the course of direct and converse martensite-austenite transformations. The transition from tetragonal twinned martensite to a cubic austenite causes a corrugation in the planar surfaces. Performing the planarization of the samples at elevated temperatures in the austenite state enables to conserve the planarity at this state, but reverse is not true for the samples planarized in the martensite state. In addition to the experiments on planarizing and ridging of the samples an unusual effect of reversible absorption and desorption of liquids by the surface of the specimen was found to exist. This effect may be useful for understanding the mechanical behaviour of shape memory alloys.

Acknowledgement

The work was carried out with financial support in part from the Ministry of Education and Science of the Russian Federation in the framework Program of NUST "MISIS" (No. K-2014-008) and RFBR Research Project No. 14-07-00888 A.

References

[1] Chmielus M, Rolfs K, Wimpory R, Reimers W, Mullner P and Schneider R 2010 Effects of surface roughness and training on the twinning stress of Ni–Mn–Ga single crystals *Acta Materialia* **58** 3952–3962
[2] Chmielus M, Witherspoon C, Ullakko K, Mullner P and Schneider R 2011 Effects of surface damage on twinning stress and the stability of twin microstructures of magnetic shape memory alloys *Acta Materialia* **59** 2948–2956

IOP Conf. Series: Materials Science and Engineering **77** (2015) 012046 doi:10.1088/1757-899X/77/1/012046

Synthesis of Nickel and Cobalt Ferrite Nanopowders and Selected Properties of Polycarbonate Composites with Nickel Ferrite

I Bockovs[1], I Zalite[2], M Kodols[2], G Heidemane[2], J Grabis[2], J Zicans[1], R Merijs-Meri[1], A K Bledzki[3]

[1]Institute of Polymer Materials, Riga Technical University, Riga, Latvia
[2]Institute of Inorganic Chemistry, Riga Technical University, Salaspils, Latvia
[3]Institute of Materials Science and Engineering, West Pomeranian University of Technology Szczecin, Poland

e-mail: ilmars@nki.lv

Abstract. $NiFe_2O_4$ and $CoFe_2O_4$ have been synthesized by sol-gel self-combustion method and high frequency plasma chemical synthesis. Structure and magnetic properties of the synthesized products have been characterized. All the synthesized ferrites are nanocrystalline single phase materials with specific surface area of 30–40 m^2/g and calculated particle size of 30–40 nm. Ni and Co ferrites synthesized by high-temperature plasma route are characterized by saturation magnetization M_S of 44,2 emu/g and 75,4 emu/g, remanent magnetization M_r of 10,0 and 32,0 emu/g and coercivity H_c of 74 Oe and 780 Oe, respectively. $NiFe_2O_4$ (2 - 10 wt. %) has been introduced in a polycarbonate (PC) matrix by melt compounding. Magnetic as well as elastic properties of PC matrix nanocomposites are increased along with growing $NiFe_2O_4$ content.

1. Introduction

Ferrites, are widely applied in life sciences, biochemical processes (magnetic liquids, hyperthermia etc.) and special coatings (antistatics, electromagnetic interference shielding). The effectiveness of the spinel ferrites in most of these applications can be increased, if their dimensions are in nanolevel, i.e., below single domain sizes [1]. Properties of ferrites, can be tailored also depending on it morphology and nanostructure. Consequently, it is important to analyse the effects of manufacturing methods/conditions on structural features of ferrite nanoparticles. Several liquid phase and gas phase synthesis have been developed to synthesize ferrite nanoparticles – hydrolysis, hydrothermal synthesis [2], pyrolysis, microwave synthesis [3], co-precipitation method [4], sol-gel method [5], combustion [6], and plasma synthesis [7]. In this research, synthesis of nickel and cobalt ferrite nanopowders have been performed by using sol-gel self-combustion method (liquid phase reaction) and high frequency plasma synthesis (gas phase reaction).

Application potential of synthesized ferrite nanostructures can be increased by their introduction in suitable matrix materials. Consequently the aim of the current research is to find out the effects of the nanofiller type and concentration on certain structural, mechanical and thermal characteristics of the polycarbonate based composites.

2. Experimental

Two distinct ferrite types ($NiFe_2O_4$ and $CoFe_2O_4$) are obtained by using high frequency plasma chemical synthesis and sol-gel self-combustion method. Plasma synthesized ferrites are synthesized by using the technological equipment, developed in RTU Institute of Inorganic Chemistry [7]. The synthesis of aforementioned ferrite nanostructures has been realized by evaporating commercial metal and metal oxide (Ni, Co, NiO, CoO and FeO) powders in high frequency plasma. Sol-gel self-combustion method synthesized ferrites have been obtained by using reagent grade chemicals: $Co(NO_3)_2{\cdot}6H_2O$, $Ni(NO_3){\cdot}6H_2O$, $Fe(NO_3)_3{\cdot}9H_2O$, glycine, nitric acid.

All samples have been analysed by using X-ray diffractometer Advance 8 (Bruker AXS). Crystallite sizes have been determined by using Scherer's equation. Magnetic properties of the synthesized ferrites and nanocomposites have been analysed by vibrating sample magnetometry (VSM Lake Shore Cryotronics, Inc., model 7404 VSM). Specific surface area (SSA) has been measured with the BET single point method. Measurements were made at the boiling point of liquid nitrogen (–196 °C), the adsorbing gas was Ar (~7 % Ar gas mixture in He gas).

Plasma synthesized $NiFe_2O_4$ nanoparticles in the amount of 2 to 10 wt. % have been introduced in a polycarbonate (PC) matrix (Trirex 3022IR) by using two-roll mill technique. Specimens for tensile tests have been manufactured by injection moulding. Specimens for dynamic mechanical thermal analysis, density characterization and magnetic property determination have been manufactured by compression moulding. Density has been determined at 23 °C by hydrostatic weighting. Room temperature (23 °C) tensile properties have been determined by using Zwick/Roell BDO020 universal testing machine. Dynamic mechanical thermal analysis have been performed by using TA Instruments TA Q800 apparatus.

3. Results and discussion

The characteristics for synthesized ferrite products are given in Table 1 and in Figures 1-3. It is found that all the synthesized ferrites are nanocrystalline single phase materials with specific surface area of 30–40 m^2/g and calculated particle size of 30–40 nm.

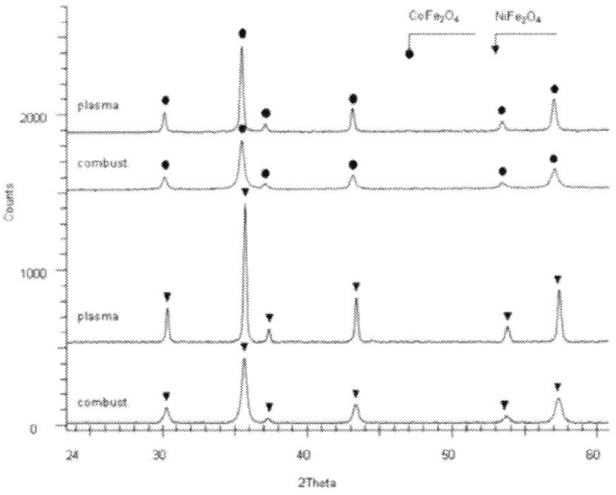

Figure 1. XRD patterns of ferrite nanopowders.

Figure 2. Microstructure of plasma $CoFe_2O_4$.

Analyzing XRD patterns there are slight differences between relative intensities and width of reflexes comparing ferrite samples depending on synthesis method which indicates to the differences of crystallite size. The sol-gel self-combustion method gives nanopowders with smaller crystallite size (Table 1).). One should however consider rather broad particle size distribution of the synthesized ferrites (in range 10 - 100 nm with some particles of 200 nm) as shown in Figure 2 for plasma synthesized $CoFe_2O_4$.

IOP Conf. Series: Materials Science and Engineering **77** (2015) 012046 doi:10.1088/1757-899X/77/1/012046

The magnetic properties (saturation magnetization M_S, remanent magnetization M_r, emu/g and coercivity H_c, Oe) (Table 1, Figure 3) of nanopowders obtained by using sol-gel self-combustion method differ from those, obtained by plasma synthesis. Probably, it is attributed to the differences in particle sizes of the synthesized nanopowders. It is interesting to note that although all of the nanoparticles, synthesized within the framework of this research, have particle sizes below the critical single-domain limit (ca. 70 nm [8, 9]), quasi-supermagnetic behavior is observed only in the case of plasma synthesized $NiFe_2O_4$ nanoparticles.

Table 1. Properties of synthesized ferrite nanopowders

Sample	SSA, m^2/g	d_{50}, nm*	Crystallite size, nm	Phase content	Magnetic properties		
					M_S, emu/g	M_r, emu/g	H_c, Oe
$CoFe_2O_4$ (plasma)	29	39	40	100% $CoFe_2O_4$	75,4	32,0	780
$CoFe_2O_4$ (combust.)	37	31	20	100% $CoFe_2O_4$	53,4	20,3	1170
$NiFe_2O_4$ (plasma)	29	38	40	100% $NiFe_2O_4$	44,2	10,0	74
$NiFe_2O_4$ (combust.)	43	26	10	100% $NiFe_2O_4$	29,0	6,0	140

*calculated from SSA

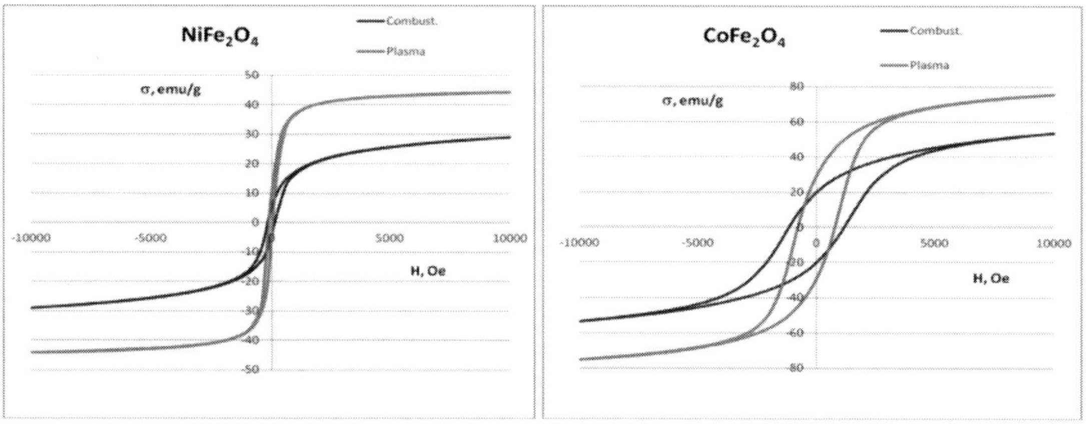

Figure 3. Magnetic properties of the ferrites obtained by sol-gel self-combustion method and high-frequency synthesis.

The effects of plasma synthesized $NiFe_2O_4$ nanoparticles on the dynamic elastic characteristics, stress-strain properties as well as density are shown in Figures 4-7. As shown in Figure 4, certain magnetic properties are assigned to otherwise diamagnetic PC matrix already at the lowest investigated nanofiller content. It is important to mention that saturation magnetization M_S of PC composition containing 10 wt. % of $NiFe_2O_4$ (1.6 emu/g) is more than one tenth from that of as-syntesized nanofiller. Remanent magnetization M_r of the same PC composite, in its turn, is close to zero, which is not surprising by considering low M_r value of neat $NiFe_2O_4$. Coercivity H_c of the PC composite with 10 wt. % of $NiFe_2O_4$ (190 Oe) is almost 2.6 times higher than that of neat $NiFe_2O_4$ (74 Oe). Such behaviour could be connected with certain change in particle dimensions after their introduction in PC matrix.

In addition to magnetic properties elastic behaviour of PC nanocomposites is studied in a broad temperature interval by using dynamic mechanical thermal analysis in a single cantilever mode (Figure 5). It is an effective tool to study not only elastic properties of the materials but also their structural relaxation behaviour.

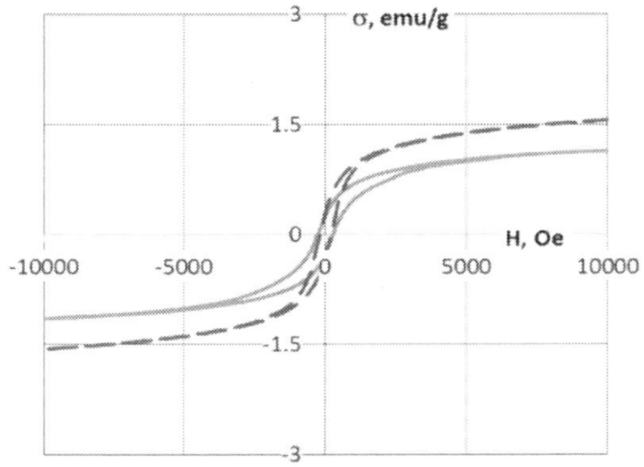

Figure 4. Magnetization hysteresis of $PC/2\%NiFe_2O_4$ (red solid curve) and $PC/10\%NiFe_2O_4$ (blue dashed curve) nanocomposites.

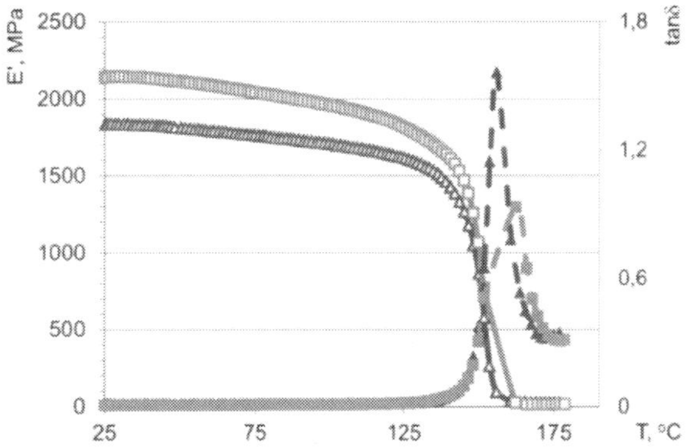

Figure 5. Storage modulus *E'* (□,Δ) and *tanδ* (▲,■) of PC (Δ,▲) and $PC/2wt.\%NiFe_2O_4$ (□,■) nanocomposite.

As one can see, there is considerable increase in storage modulus E' of PC nanocomposite with $NiFe_2O_4$ wt. content of 2%, especially at lower temperatures. By increasing temperature, elastic modules of neat PC and the nanocomposite with 2wt. % of $NiFe_2O_4$ approach each other and close to glass transition of the polymer matrix have practically the same values. However, it should be mentioned that upon addition of the nanostructured filler, glass transition of the composite is increased, the behavior characteristic for nanocomposite materials and generally explained with stiffening of polymer macromolecules in the proximity of nanostructured filler [10].

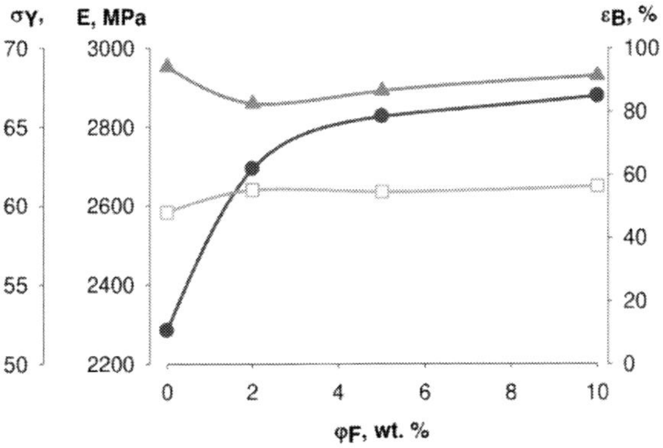

Figure 6. Tensile modulus E (○), stress at yield σ_Y (□) and strain at break ε_B (Δ) of $PC/NiFe_2O_4$ nanocomposite.

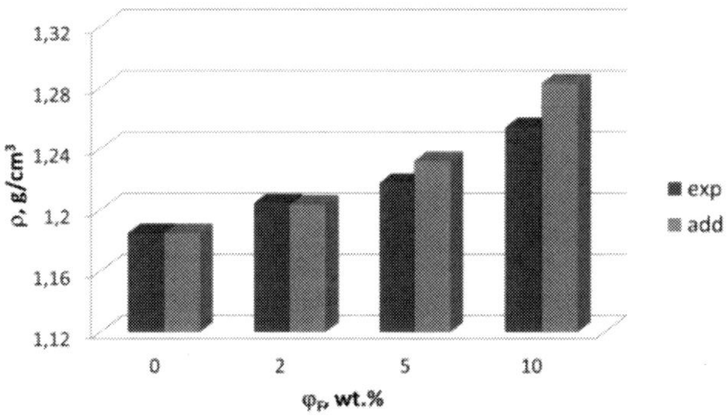

Figure 7. Experimental (exp) and additive (add) densities of $PC/NiFe_2O_4$ nanocomposite.

Elastic as well as stress-strain properties of $PC/NiFe_2O_4$ nanocomposites more in detail are investigated at room temperatures. As it is demonstrated in Figure 6, tensile modulus E and yield strength σ_Y of $PC/NiFe_2O_4$ nanocomposites are increased along with increasing nanofiller content up to 10 wt.%. Increase of stress-strain characteristics is the steepest up to the nanofiller content of 2 wt.%. At higher nanofiller contents rise of the aforementioned stress-strain characteristics is somewhat hindered, which most probably could be explained with greater possibility of the formation of $NiFe_2O_4$ agglomerates in the PC matrix. At the same time strain at break ε_B of the nanocomposite decreases in comparison to neat PC matrix. It is, however, worth mentioning that character of deformation remains unchanged for all of the investigated nanocomposites, i.e., nanocomposites deform through yielding, accompanied with expressed necking, and followed by unexpressed strain hardening until the rupture of the test specimens at specified ultimate deformations.

In addition to magnetic, elastic and stress-strain behaviour, PC/$NiFe_2O_4$ nanocomposites are investigated also in respect to their densities, determined experimentally and calculated according to the rule of mixtures (Figure 7). In the calculations it is assumed that densities of polycarbonate matrix and the ferrite nanofiller are 1.18 g/cm^3 and 5 g/cm^3, respectively. As expected, experimental density of the investigated PC nanocomposite increase along with growing $NiFe_2O_4$ content and is lower than that calculated according to the rule of mixtures. Differences between the experimentally detected and arithmetically calculated densities increase by rising the nanofiller content in the polymer matrix (up to 2.2 % for the nanocomposite with 10wt.% of $NiFe_2O_4$). This most probably is attributed to the void content in the nanocomposite.

4. Summary

In the current research $NiFe_2O_4$ and $CoFe_2O_4$, synthesized either by sol-gel self-combustion method or by high frequency plasma chemical synthesis, are characterized in respect to their structure and magnetic properties. In addition, selected structural, magnetic, elastic as well stress-strain properties of the melt compounded PC matrix based nanocomposites, modified with various amounts (0-10 wt. %) of plasma synthesized $NiFe_2O_4$ nanoparticles, are characterized.

In general it is shown that (a) high purity nanostructured ferrite fillers with average crystallite sizes between 10 – 40 nm are synthesized, (b) certain magnetic traits are assigned to the investigated polycarbonate nanocomposites already at $NiFe_2O_4$ content of 2 wt.%, (c) considerable increase in stiffness of $NiFe_2O_4$ modified PC nanocomposites is observed already at the ferrite content of 2 wt. %, (d) gain in the elastic modulus E and yield strength σ_Y of the nanocomposites, however, decreases, above the ferrite content of 2 wt. %.

References

[1] Magnetic Nanoparticles. Ed. by S. Gubin. Wiley VCH Verlag GmbH&Co KgaA, Weinheim, 2009

[2] N. Millot, S. Le Gallet, D. Aymes, F. Bernard, Y. Grin: J. Europ. Ceram. Soc. Vol. 27 (2007), pp. 921-926

[3] B. Xue, R. Liu, Z. De Xu, Y. Fan Zheng: Chemistry Letters, Vol. 37, No.10 (2008), pp. 1058-1059

[4] T.A.S. Ferreira, J.C. Waerenborgh, M.H.R.M. Mendonça, M.R. Nunes, F.M. Costa: Solid State Sciences, Vol. 5 (2003), pp. 383–392

[5] D. Q. Tang, D. J. Zhang, H. Ai: Chemistry Letters, Vol.35, No.11 (2006), pp. 1238-1239.

[6] Zhongpo Zhou, Yue Zhang, Ziyu Wang, Wei Wei, Wufeng Tang, Jing Shi, Rui Xiong: Applied Surface Science, Vol. 254 (2008), pp. 6972–6975

[7] J. Grabis, I. Zalite: Mater. Sci. Forum, Vol. 555 (2007), pp. 267-272

[8] R.M. Cornell and U. Schwertmann, in: The Iron Oxides. Structure, Properties, Reactions, Occurrences and Uses, Wiley-VCH, Weinheim (2003)

[9] Q. Song, Z.J. Zhang: J. Phys. Chem. B110 (2006), pp. 11205–11209

[10] L.S. Schadler, S.K. Kumar, B.C. Benicewicz, S.L. Lewis, and S.E. Harton: Designed Interfaces in Polymer Nanocomposites: A Fundamental Viewpoint. MRS Bulletin, Vol. 32, April (2007), pp. 335-340

IOP Conf. Series: Materials Science and Engineering **77** (2015) 012047 doi:10.1088/1757-899X/77/1/012047

Optical anisotropy and domain structure of multiferroic Ni-Mn-Ga and Co-Ni-Ga Heusler-type alloys

A I Ivanova[1], O V Gasanov,[1] E I Kaplunova[1], E T Kalimullina[2], A B Zalyotov[3] and R M Grechishkin[1,4]
[1]Tver State University, 170100 Tver, Russia
[2]Institute of Radio Engineering and Electronics of RAS, 125009 Moscow, Russia
[3]Tver State Medical Academy, 170001 Tver, Russia
[4]National University of Science and Technology "MISIS", 119049 Moscow, Russia

e-mail: alex.ivanova33@yandex.ru

Abstract. A study is made of the reflectance anisotropy of martensitic and magnetic domains in ferromagnetic shape memory alloys (FSMA) Ni-Mn-Ga and Co-Ni-Ga. The reflectance of metallographic sections of these alloys was measured in the visible with the aid of standard inverted polarized light microscope with a 360° rotatable specimen stage. Calculations are presented for the estimation of image contrast values between neighboring martensite twins. Qualitative and quantitative observations and angular measurements in reflected polarized light proved to be useful for the analysis of specific features of the martensite microstructure of multiferroic materials.

1. Introduction

Ferromagnetic shape memory Heusler-type alloys Ni-Mn-Ga and Co-Ni-Ga are a relatively new class of ferroic materials which may change their shape and magnetization by a magnetically induced martensite transformation or by a magnetically induced reorientation of martensite twin structure [1-3]. The behaviour of these materials is determined by the magnetoelastic interaction of correlated ensembles of structural (martensite) and ferromagnetic domains. The details of this interaction in real materials depend on a large number of microstructural parameters and crystal defects [4-6]. In this view it is important to gain detailed experimental information on both static configurations of coexisting martensite and magnetic domain structure (DS) and their dynamic behavior under the action of external magnetic, mechanical and thermal influences.

The studies of the martensite and magnetic domain structure (DS) were attempted by a number of techniques including Bitter patterns [7,8], interference optical microscopy [9], magnetooptic indicator films (MOIF) [10-13], scanning electron microscopy [14], magnetic force microscopy [15], Lorentz transmission electron microscopy and electron holography [16], and some others.

In spite of rather large number of publications the problem of DS observation appeared to be more difficult than expected, so the necessary experimental data are still lacking. This is explained partly by the fact that in addition to the known limitations of existing DS techniques [17] the FSMA put forward a number of additional problems. First of all one faces the situation of revealing in a single experiment the coexisting DS's of different physical origin – martensitic and magnetic. Another problem is the proper preparation of defect-free metallographic sections [4-6]. Even if the FSMA

IOP Conf. Series: Materials Science and Engineering **77** (2015) 012047 doi:10.1088/1757-899X/77/1/012047

surface is successfully planarized by thorough polishing it readily becomes corrugated during martensite variant rearrangement. Among other peculiarities is the "ghost" martensite-like residual relief observed by optical contrast in a sample heated well above the martensite-austenite phase transition point where no martensite should exist (S J Murray [18]). One more unexpected feature mentioned by Y.Ge et al. is the formation of a relief delineating the magnetic DS and making it observable in ordinary optical microscope even without the use of polarized light illumination [19].

Thinking about the further development of martensite and magnetic DS observation techniques for FSMA our attention was attracted by the polarized light studies of twin structure of optically anisotropic YBaCuO high temperature superconductors [20]. In the present work we examine the possibility of exploiting this technique to the case of MSMA.

2. Experimental

Polycrystalline samples of the Ni-Mn-Ga and Co-Ni-Ga alloys examined earlier by the method of MOIF [12] were prepared by arc melting in argon followed by vacuum annealing. The composition of the alloys was determined by X-ray energy dispersive microanalysis (Oxford Instruments). Various abrasive, chemico-mechanical and electrolytic polishing were employed to prepare the metallographic sections of the samples. The martensite and austenite start and finish temperatures were estimated by the temperature dependence of AC initial susceptibility. The angular variations of the reflectivity data were measured in the visible with the aid of a silicon photodiode or a CCD video camera (integral and local light intensity studies, respectively) in conjunction with a light polarizing microscope with a 360° rotatable specimen stage.

3. Results and discussion

Figure 1 shows the low magnification grain structure image of a longitudinal cross section of polycrystalline $Ni_{2.12}Mn_{0.88}Ga$ button prepared by ordinary metallographic abrasive polishing routine. The angular orientation of the sample on the specimen stage is arbitrary while analyzer and polarizer of the microscope are set at nearly crossed position. The existence of uniaxial crystallographic texture is clearly evidenced by the elongation of the grains in a direction perpendicular to the copper hearth. Rotation of the sample results in a complicated interplay of the image contrast between the individual grains and (at higher magnifications) martensite twins. At the same time the polarized light observations generally lead to a great enhancement of the images of noninformative artifacts (scratches, etch pits, pores) which may remain unseen under bright field illumination.

Figure 1. Grain structure of the longitudinal cross section of polycrystalline $Ni_{2.12}Mn_{0.88}Ga$ button prepared by arc melting on a copper hearth observed in polarized light

To understand the mechanism of optical polarization contrast formation it is convenient to consider a plane parallel plate splitted into two martensite domains with different optical indicatrices having ellipse projections on the plane of observation with the same semiaxes n' and n'' but different orientation. For such an optical system there is a known solution given by Born and Wolf [21]:

$$I_{1,2} = I_0\left(\cos^2\theta - \sin(2\varphi \pm \chi)\sin(2\varphi \pm \chi - 2\theta)\sin^2\frac{\delta}{2}\right), \tag{1}$$

where θ is the orientation of the transmission axis of the analyzer with respect to the plane of the incident light, and δ is the phase difference between the output rays, χ is the angle between the large semiaxes of the ellipses. For a plate of thickness h the phase difference is $\delta = 2\pi\, h\,(n'' - n')/\lambda$, where λ is the light wavelength. From this the contrast C is written as

$$C \equiv \frac{I_1 - I_2}{I_1 + I_2} = \frac{\sin(4\varphi - 2\theta)\sin 2\chi \sin^2\frac{\delta}{2}}{(\cos 2\theta - \cos(4\varphi - 2\theta)\cos 2\chi)\sin^2\frac{\delta}{2} - 2\cos^2\theta} \quad (2)$$

It is seen that C is zero for the angles φ and θ related as $\theta = 2\varphi - (\pi n/2)$, thus resulting in fourfold change of the contrast with the rotation of the analyzer by 360^0 and eight-fold changes for the sample rotation by the same angle:

$$C \equiv \frac{I_1 - I_2}{I_1 + I_2} = \frac{\sin(4\varphi - 2\theta)\sin 4\omega \sin^2\frac{\delta}{2}}{(\cos 2\theta - \cos(4\varphi - 2\theta)\cos 4\omega)\sin^2\frac{\delta}{2} - 2\cos^2\theta}\,. \quad (3)$$

To check the predictions of the calculations detailed angular measurements of the light intensities were performed experimentally for a pair of adjacent martensitic variants (twins). Figure 2 shows a local area of a flat sample surface containing the twin made visible by polarized light observation.

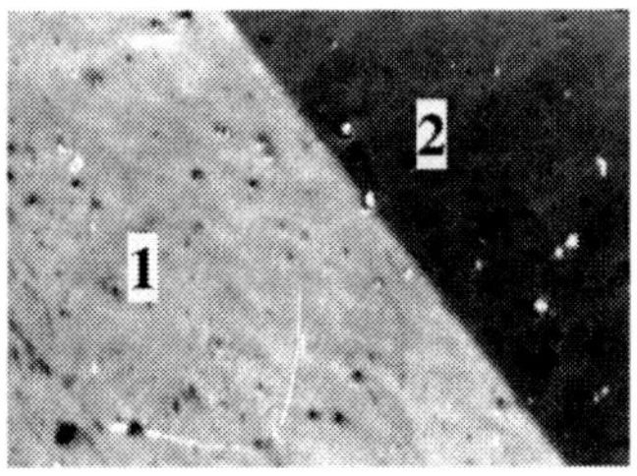

Figure 2. The optical polariztion contrast during observation of martensite twin in polarized light

The experiments have shown that the contrast between the martensite variants labeled above as 1 and 2 may varies in a regular way. The detailed data of the measurements made in a localized field of view separately for the two adjacent martensite variants are presented by the graphs of figure 3.

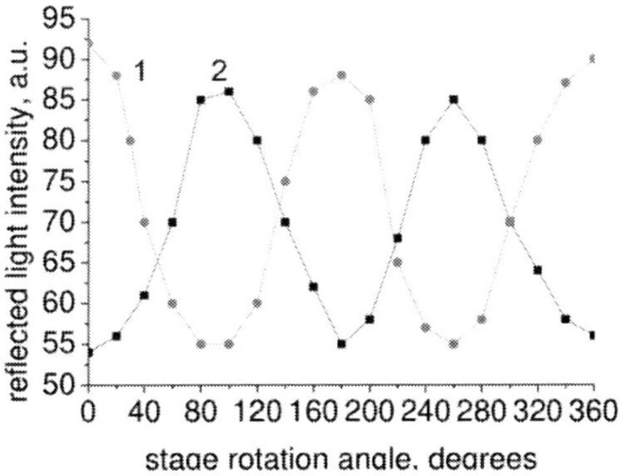

Figure 3. Dependence of the intensity of the polarized light reflected from two adjacent martensite variants of a $Ni_{2.16}Mn_{0.84}Ga$ sample (curves 1 and 2). The points of curves intersection correspond to the condition of zero contrast between the variants.

It is seen from figure 3 that the optical contrast $C = (I_1 - I_2)/(I_1 + I_2)$ may be set to zero, be maximized or inverted simply by rotation of the sample on the microscope stage. It is also evident that the symmetry of this optical effect differs from that of the magnetooptic polar Kerr or Faraday effect which are invariant to sample rotation. This feature opens the way of enhancing the conditions for both martensite and magnetic domain structure observations by the digital methods of differential polarized microscopy [17, 22]. The differential procedure of [17, 22] starts with a digitized image of the magnetically saturated state. This domain-free image serves as a background which is subtracted from a state containing domain information, so that in the difference image a clear micrograph of the domain pattern is obtained. This procedure is not applicable to FSMA because of great difficulties in obtaining the saturation. Alternatively, for FSMA our data show that it is possible to produce two images of the same field of view and invert the contrast of one of them through rotation of the sample by an angle multiple to 45°. In this way it becomes possible to form a differential pair of images. The resulting effect of such background subtraction is shown figure 4.

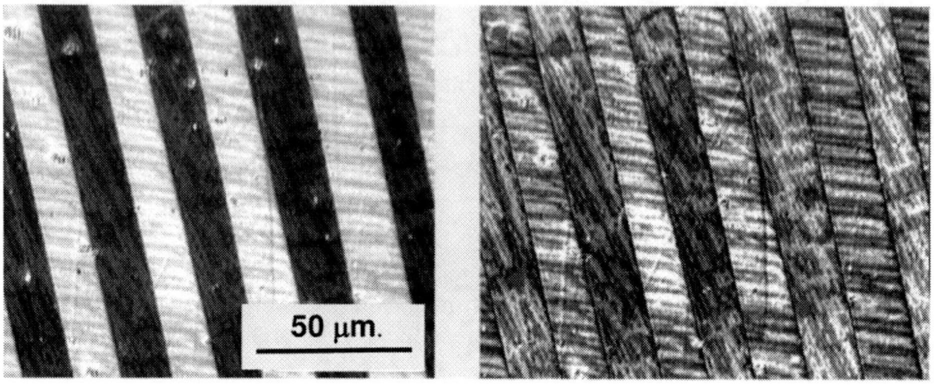

Figure 4. Polarized optical image of martensitic and magnetic domain structure on a flat surface of $Co_2Ni_{0.85}Ga_{1.15}$ sample. (*a*) original image, (*b*) after background compensation

The left part of figure 4 demonstrates the original image of $Co_2Ni_{0.85}Ga_{1.15}$ sample with band structure of martensite variants and hardly discernible internal structure. Background subtraction (at the right) demonstrates many-fold increase in the contrast of this internal structure bringing into evidence its relation to the magnetic DS of 180° degree type which is invariant to sample rotation.

4. Conclusion

Polarized light microscopy was applied to study the martensite twin domain structure of ferromagnetic shape memory alloys of the Heusler type. It is shown that due to anisotropic character of the reflection of light in these alloys the polarization optical contrast depende on a number of factors among which is the angular orientation of the sample on the microscope stage. By proper choice of these factors the optical contrast of martensite twins may be either maximized or set to zero or intermediate values. This behaviour is in contrast to that of the magnetooptic polar Kerr effect which is invariant to the rotation of the sample about the normal to the plane of observation. This peculiarity opens ways for manipulating the contrast of polarized light images by their segmentation and using the principles of differential polarized microscopy.

References

[1] Heczko O, Scheerbaum N and Gutfleisch O 2009 Magnetic shape memory phenomena. In *Nanoscale magnetic materials and applications* (New York: Springer US) pp 399-439

[2] Thomas M, Heczko O, Buschbeck J, Lai Y W, McCord J, Kaufmann S and Fähler S 2009 Stray-Field-Induced Actuation of Free-Standing Magnetic Shape-Memory Films *Advanced Materials* **21** 3708-3711

[3] Vasil'ev A N, Buchelnikov V D, Takagi K, Khovailo V V and Estrin E I 2003 Shape memory ferromagnets *Physics-Uspekhi* **46** 559–588.

[4] Chmielus M, Rolfs K, Wimpory R, Reimers W, Mullner P and Schneider R 2010 Effects of surface roughness and training on the twinning stress of Ni–Mn–Ga single crystals *Acta Materialia* **58** 3952–3962.
[5] Aaltio I,X.W. Liu X W, Valden M,K. Lahtonen K, Söderberg O,Y. Ge Y and S-P. Hannula S P 2013 Nanoscale surface properties of a Ni–Mn–Ga 10M magnetic shape memory alloy *Journal Alloys and Compounds* 577S S367-371
[6] Chmielus M, Witherspoon C, Ullakko K, Mullner P and Schneider R 2011 Effects of surface damage on twinning stress and the stability of twin microstructures of magnetic shape memory alloys *Acta Materialia* **59** 2948–2956
[7] Korpusov O M, Grechishkin R M, Koledov V V, KhovailoV V, Takagi T and Shavrov V G 2004 Simultaneous magnetooptic observation and thermomagnetic analysis of phase transitions in shape-memory Ni–Mn–Ga alloys *Journ. Magn. Magn. Mater.* **272** 2035-2037
[8] Heczko O, Jurek K and Ullakko K 2001 Magnetic properties and domain structure of magnetic shape memory Ni–Mn–Ga alloy *Journ. Magn. Magn. Mater.* **226** 996-998
[9] Chopra H D, Ji C and Kokorin V V 2000 Magnetic-field-induced twin boundary motion in magnetic shape-memory alloys *Physical Review B*, **61**(22), R14913
[10] Grechishkin R M, Goosev M Yu, Ilyashenko S E and Neustroev N S 1996 High resolution sensitive magnetooptic ferrite garnet films with planar anisotropy *Journ. Magn. Magn. Mater.* **157-158** 305-306
[11] Lai Y W, Schäfer R, Schultz L and McCord J 2010 Volume magnetic domain mirroring in magnetic shape memory crystals *Applied Physics Letters*, **96**(2) 022507
[12] Breczko T M, Grechishkin R M, Ilyashenko S E, Nelaev V V, Dovzhik K N and Korpusov O M 2010 Martensitic and magnetic domain structures of ferroics of the Ni-Mn-Ga and Co-Ni-Ga families *Physics of the Solid State* **52**(1) 101-104
[13] Ge Y, Heczko O, Söderberg O and Hannula S P 2008 Comparison of different methods for studying magnetic domains in Ni–Mn–Ga martensites *Materials Science and Engineering: A*, **481**, 302-305
[14] Ge Y, Heczko O, Söderberg O and Hannula S P and Lindroos V K 2005 Investigation of magnetic domains in Ni-Mn-Ga alloys with a scanning electron microscope Smart. Mater. Struct. **14** S211-S215
[15] Pan Q and James R D 2000 Micromagnetic study of Ni_2MnGa under applied field *Journal of Applied Physics* **87**(9) 4702-4706
[16] Park H S, Murakami Y, Shindo D, Chernenko V A and Kanomata T 2003 Behavior of magnetic domains during structural transformations in Ni_2MnGa ferromagnetic shape memory alloy *Applied Physics Letters* **83**(18) 3752-3754
[17] Hubert A and Schafer R 2008 *Magnetic Domains* (Springer, Berlin)
[18] Murray S J 2000 Magneto-mechanical properties and applications of Ni-Mn-Ga FSMA (Theses, MIT)
[19] Ge Y, Heczko O, Söderberg O and Hannula S P 2006 Direct optical observation of magnetic domains in Ni–Mn–Ga martensite *Applied Physics Letters* **89**(8) 082502
[20] Dorosinsky L, Farber B, Indenbom M, Nikitenko V, Polyanskii A and Vlasko-Vlasov V 1990 Polarized-optics study of twins in $YBa_2Cu_3O_{7-\delta}$ *Ferroelectrics* **111** 321-331
[21] Born M and Wolf E 1968 *Principles of Optics* (Cambridge: England)
[22] McCord J, Hubert A 1999 Normalized differential Kerr microscopy an advanced method for magnetic imaging *phys. stat. sol.* (a) **171** 555-562

Acknowledgements

The work was carried out with financial support in part from the Ministry of Education and Science of the Russian Federation in the framework Program of NUST “MISIS” (No. K-2014-008) and RFBR Research Project No. 14-07-00888 A.

IOP Conf. Series: Materials Science and Engineering **77** (2015) 012048 doi:10.1088/1757-899X/77/1/012048

Heat losses and thermal imaging of ferroic components

S E Ilyashenko[1], **A I Ivanova**[2], **O V Gasanov**[2], **R M Grechishkin**[2,3], **S A Tretiakov**[2,3], **K B Yushkov**[3] and **B B J Linde**[4]
[1]Tver State Technical University, 170026 Tver, Russia
[2]Tver State University, 170100 Tver, Russia
[3]National University of Science and Technology "MISIS", 119049 Moscow, Russia
[4]Institute of Experimental Physics, University of Gdansk, 80-952 Gdansk, Poland

e-mail: SvIlyashenko@yandex.ru

Abstract. A study is made of spatial and temporal temperature variations in working devices based on ferroic functional materials. The measurement of the sample's temperature is complemented with direct observation of its distribution over the sample surface. For the latter purpose a thermovision infrared videocamera technique was employed. Specific features of the temperature distribution and its evolution during heating and cooling of a number of piezoelectric, acoustooptic and shape memory components are revealed. Examples of hot spot observations indicative of structural defects in the samples under study are given thus suggesting the use of thermal vision for nondestructive testing. A proposal is made to combine the thermovision method with that of thermomagnetic analysis for the study of ferromagnetic shape memory alloys.

1. Introduction

The magnitude of the electrical and mechanical losses in components made of functional ferroic materials which are currently widely used in electronic circuits often determines the application of these materials [1-3]. Temperature and thermal characterization of ferroic materials, components and modules represents an important part in quantifying and minimizing temperature stresses for efficient thermal management. Diverse array of methods for local temperature measurements and thermography were developed to satisfy the requirements of materials science researchers and designers of electromechanical, optoelectronic, acoustic, acoustooptic, magnetooptic, magnetic and magnetoelectric devices, etc. Among the different thermal imaging technologies are imaging crystal thermography, fluorescent microthermography, scanning thermal microscopy, infrared radiation thermometry, thermoreflectance, acoustic thermography, photothermal deflection spectroscopy and many others [4].

Nowadays special interest is paid to the methods of infrared radiation thermometry. In addition to a number of great benefits inherent to these methods the researchers are inspired by rapid progress in the development of commercially available high-sensitive IR videocameras which are more and more employed in R&D activities.

In the present work we report on the experimental results of applying the IR thermal imaging technique for the examination of several types of ferroic-based components including ordinary piezoceramic actuators, piezoelectric macro fiber composites (NASA MFC), acoustooptic components on the base of piezoelectric single crystals, and ferromagnetic shape memory alloys [5].

2. Experimental

2.1. LZT piezoceramics

Thermal imaging and quantification of standard piezoceramic $PbZr/TiO_3$ (LZT) disk shaped samples excited at their frequency of resonance ~ 80 kHz. was performed with the aid of FLIR ResearchIR camera (IR resolution 160×120 pixels, spatial resolution 2.72 mrad, thermal sensitivity <0.07 °C, spectral range 7.5…13 μm) with proper calibration procedure taking into account the coefficient of reflection of the materials under study. Figure 1 shows the thermal images of the surface of freely suspended LZT disk (diameter ~25 mm) together with the temperature line profiles starting from the moment of power switching on to the moment of establishing a steady-state regime. The presented data clearly show that the heating of the sample starts from its centre while the temperature distribution is much more uniform during unforced cooling under ambient conditions. This feature may have an influence on the choice of working conditions for components exploited in a pulsed mode.

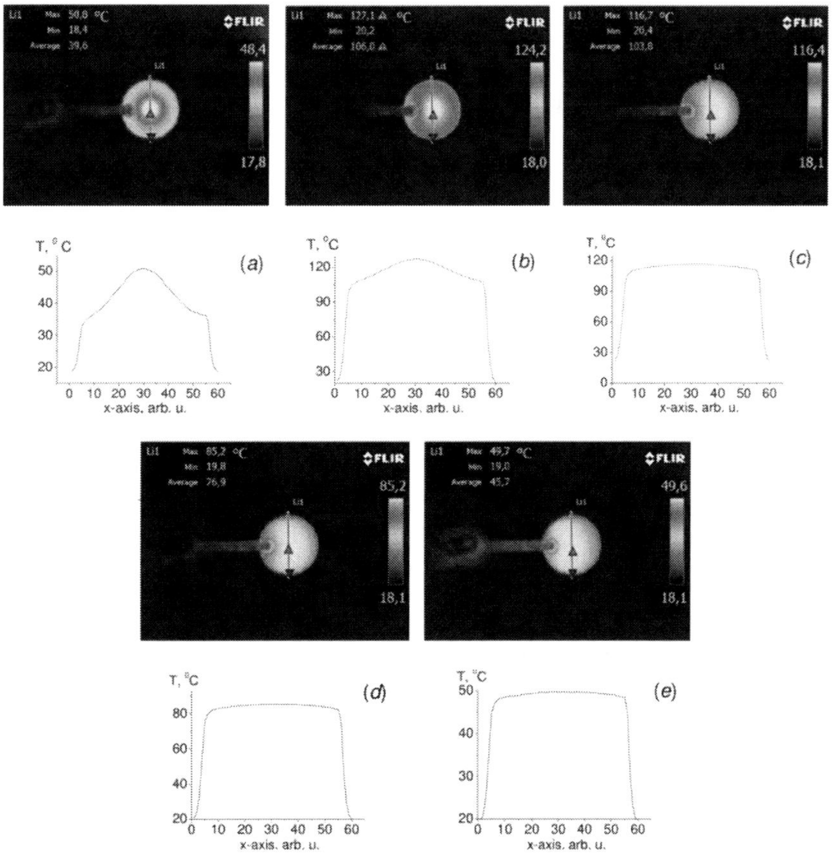

Figure 1. Thermal images and temperature line profiles for a disk-shaped piezoceramic sample during heating (*a, b*) and cooling (*c, d, e*). Note the change of the *T* (*x*) curve shape from conoidal (a) to a hat-like one (*c – e*)

2.2. Piezoelectric Macro Fiber Composite (MFC)

MFC's belong to a relatively new class of smart structures utilizing the piezoelectric effect for sensing, actuation, energy harvesting and damping [5], so the information about their temperature behavior is important both from the academic and engineering point of view.

Figure 2 presents the images and temperature line profiles of an MFC flat sample with a size of 20 by 30 mm. It is seen that at the early stage of heating after activation the temperature distribution is rather asymmetrical (Figure 2, *a, b*). This asymmetry is much less pronounced during the cooling process. The microstructural control of this sample by means of optical and scanning electron microscopy enabled to reveal a subsurface defect (delaminated flat spot ~5 mm in diameter) in this particular sample just corresponding to the region of thermal anomaly. This result demonstrates the usefulness of thermal imaging method for nondestructive testing of piezoelectric composites.

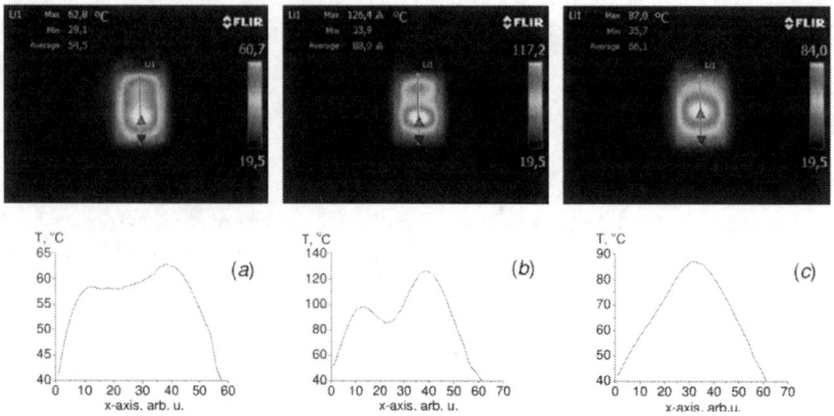

Figure 2. Temperature line profile of flexible MFC sample (size 20×30 mm) at T_{max} 62.8 (*a*), 126.4 (*b*) and 87 °C (*c*). State (*c*) was achieved by cooling from (*b*)

2.3. Acoustooptic devices on the base of piezoelectric single crystal $LiNbO_3$ and α-TeO_2

Heat release in working acousto-optical (AO) device is inevitably accompanied by a change of its acousto-optical parameters. In the visible and medium-infrared the most called-for and effective materials for acoustooptic devices are paratellurite (α-TeO_2) and lithium niobate ($LiNbO_3$) single crystals.

In the present work we focus on the specific features of the temperature distribution in large-size (36×25×22 mm) acousto-optic line utilized as a part of tunable acoustooptic filter (Figure 3). The filter was intended for astrophysical spectral studies described in detail elsewhere [6]. Large size of acousto-optic line necessary for high spectral resolution imposes strict requirements on the device thermostat control and temperature drift compensation of filter parameters exploited in terrestrial telescopes, all the more for spacecraft applications.

(*a*) (*b*) (*c*)

Figure 3. Acoustooptic tunable filter on the base of paratellurite single crystal with lithium niobate actuator. (*a*) schematic, (*b*) general view (cover removed), (*c*) thermal image. Cross marks correspond to measurement points Sp1...Sp5 allocated in the center and at the periphery of the image to obtain information on the temperature distribution over the output face

The main results of the thermal testing are presented in figure 4, illustrating the time and frequency dependence of the temperature at reference points Sp1...Sp5 of the AO filter crystal. The data were registered at a speed of 25 frames/s with 10 MHz increments from 50 to 250 MHz (*a*, *b*...u) each 300 s and constant power. In addition, the temperature changes at the same points of observation

Sp1…Sp5 are plotted for different input power values changed stepwise by 0.5 W increments starting from 0.5 W (region (a), 1.0 W (b), 1.5 W (c), … etc.

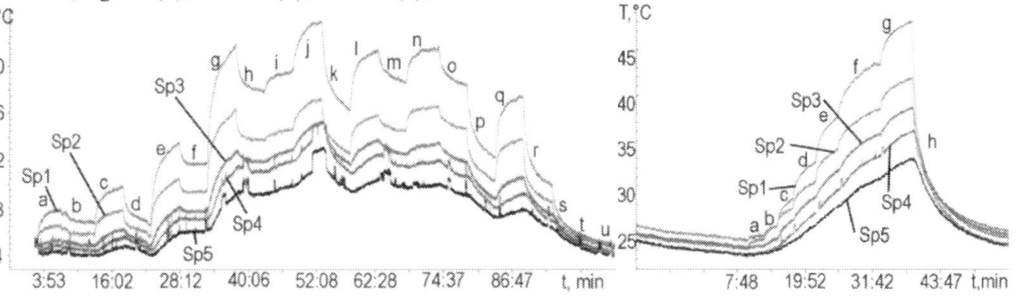

(*a*) (*b*)

Figure 4. (a) Time and frequency dependence of the temperature at reference points Sp1...Sp5 of the AO filter crystal registered a 10 MHz increments from 50 to 250 MHz (*a*, *b*…u) each 300 s and constant power; (b) temperature changes at Sp1…Sp5 observation points for different power values changed stepwise by 0.5 W increments starting from 0.5 W (region (*a*), 1.0 W (*b*), 1.5 W (*c*), … etc. Generator frequency 140 MHz.

The performed tests show that relatively large temperature gradients of 4…5 K/cm are arising during the exploitation of the AO devices on the base of paratellurite. Evidently this fact is explained by the low heat conductivity of the latter. This unfavourable circumstance should be taken into account in the design of AO devices because temperature-induced distortion of the acoustic and light wave fronts results in additional distortion of the filter transmission function within the limits of designated spectral range.

2.4. *Temperature effects in ferromagnetic shape memory alloys Ni-Mn-Ga*

Ferromagnetic shape memory alloys on the base of Heusler alloys of the $NiMn_2Ga$ family possess a unique set of magnetic, magnetocaloric, electric and mechanical properties,. making them perspective candidates for a large number of various applications (see, e.g., [5]).

Among the characteristic features of these alloys is a martensite-type structural transformation from highly symmetric high-temperature cubic phase to a phase to a phase with lower symmetry. This transformation is accompanied by a number of anomalies of physical properties, in particular, thermal ones.

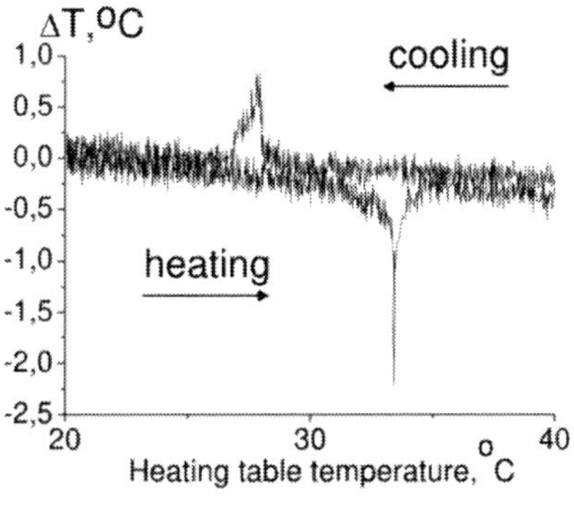

Figure 5. Temperature difference ΔT between the heating table and the $Ni_{2.16}Mn_{0.84}Ga$ sample during heating and cooling

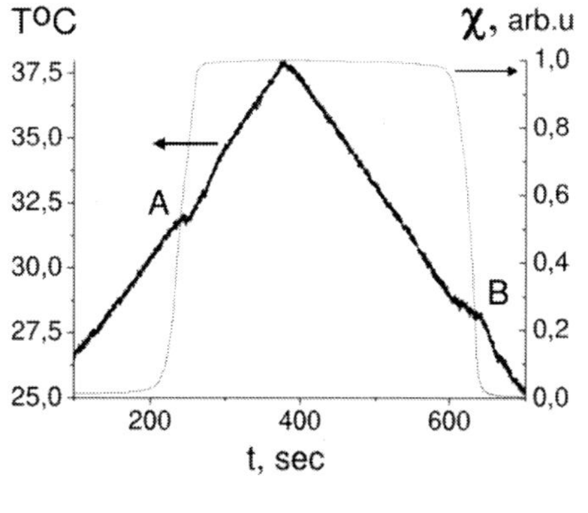

Figure 6. Superimposed time dependence of the $Ni_{2.16}Mn_{0.84}Ga$ sample temperature and initial ac magnetic susceptibility

Figure 5 shows the variation of the temperature of a $Ni_{2.16}Mn_{0.84}Ga$ sample in the neighborhood of the phase transformation region measured with the aid of thermovision camera. Two characteristic peaks are clearly seen at the temperatures corresponding to those of martensite start and finish of this alloy. Moreover, due to remote character of thermal testings it became possible to combine this measurement with the differential thermal analysis (TMA) by the temperature dependence of the ac initial magnetic susceptibility χ (figure 6). The presented measurements were performed in the following way: starting from RT the sample was smoothly heated up to 37.5 °C and then allowed to cool to RT in a time of ~600 s. Both sample temperature T and susceptibility χ were recorded as a function of time (t). Characteristic kinks are seen at the points A and B of the T(t) curve (figure 6) indicative of magnetocaloric effect due to direct and reverse phase transitions at these temperatures. It is worthwhile to mention that this information is not available from the $\chi(t)$ data taken separately.

3. Conclusions

Based on the thermal vision infrared camera technique we demonstrated the spatial and temporal temperature variations in working components based on piezoceramic, piezoelectric macro fiber composite, acoustooptic single crystals and ferromagnetic shape memory alloys. Specific features of the temperature distribution and its evolution with time after switching on the exciting voltage (heating) followed by cooling are shown. Hot spot detection indicative of defects in the structure of materials are observed thus suggesting the use of thermal vision for nondestructive testing. A proposal is made to combine the direct thermal observations of the surface of ferromagnetic shape memory alloys with their simultaneous thermomagnetic analysis by the temperature dependence of initial ac magnetic susceptibility. This combination adds a new dimension in the characterization of structural phase transformations.

Acknowledgement

The work was carried out with financial support in part from the Ministry of Education and Science of the Russian Federation in the framework Program of NUST«MISIS»(№K1-2014-008) and RFBR research project No. 14-07-00888 A.

References

[1] Uchino K 2001 Loss mechanisms in piezoelectrics: how to measure different losses separately *IEEE Trans. Ultrasonics, Ferroelectrics and Freq. Control.* **48** 307-

[2] VanGordon J A, Kovaleski S D, Norgard P and Gall BB 2014 Measurement of the internal stress and electric field in a resonating piezoelectric transformer for high-voltage applications using the electro-optic and photoelastic effects *Rev. Sci. Instr.* **85** 02310

[3] Hardtl K H 1982 Electrical and mechanical losses in ferroelectric ceramics *Ceram. Intern.* **8** 121

[4] Jorge Aldave I 2013 Review of thermal imaging systems in composite defect detection *Infrared Phys. Techn.* **61** 167

[5] Otsuka K and Ren X 1999 Recent developments in the research of shape memory alloys *Intermetallics* **7**, 511-515

[6] Molchanov V Ya, Lyuty V M, Esipov V F, Anikin S P, Makarov OYu and Solodovnikov N P 2002 An acousto-optical imaging spaectrophotometer for astrophysical observations *Astronomy Letters* **28** (10) 813